Advances in Intelligent Systems and Computing

Volume 265

Series editor

Janusz Kacprzyk, Polish Academy of Sciences, Warsaw, Poland
e-mail: kacprzyk@ibspan.waw.pl

For further volumes:
http://www.springer.com/series/11156

About this Series

The series "Advances in Intelligent Systems and Computing" contains publications on theory, applications, and design methods of Intelligent Systems and Intelligent Computing. Virtually all disciplines such as engineering, natural sciences, computer and information science, ICT, economics, business, e-commerce, environment, healthcare, life science are covered. The list of topics spans all the areas of modern intelligent systems and computing.

The publications within "Advances in Intelligent Systems and Computing" are primarily textbooks and proceedings of important conferences, symposia and congresses. They cover significant recent developments in the field, both of a foundational and applicable character. An important characteristic feature of the series is the short publication time and world-wide distribution. This permits a rapid and broad dissemination of research results.

Sirapat Boonkrong · Herwig Unger
Phayung Meesad
Editors

Recent Advances in Information and Communication Technology

Proceedings of the 10th International Conference on Computing and Information Technology (IC^{2}IT2014)

Editors
Sirapat Boonkrong
Faculty of Information Technology
King Mongkut's University of Technology North Bangkok
Bangkok
Thailand

Phayung Meesad
Faculty of Information Technology
King Mongkut's University of Technology North Bangkok
Bangkok
Thailand

Herwig Unger
Communication Networks
University of Hagen
Hagen
Germany

ISSN 2194-5357 ISSN 2194-5365 (electronic)
ISBN 978-3-319-06537-3 ISBN 978-3-319-06538-0 (eBook)
DOI 10.1007/978-3-319-06538-0
Springer Cham Heidelberg New York Dordrecht London

Library of Congress Control Number: 2014936765

Printed on acid-free paper

Springer is part of Springer Science+Business Media (www.springer.com)

Preface

This volume contains the papers of the 10th International Conference on Computing and Information Technology (IC^{2}IT2014). IC^{2}IT is an annual conference, which is held in conjunction with the National Conference on Computing and Information Technology (NCCIT), one of the leading Thai national events in the area of Computer Science and Engineering. IC^{2}IT provides a venue for the presentation and discussion of current research in the field of computing and information technology.

IC^{2}IT2014 took place between 8th and 9th May at Angsana Laguna, Phuket, Thailand. This is the first time that the conference is located in the South of Thailand. Following the interests of our participants of the last events, IC^{2}IT2014 proceedings has been structured into five main tracks: Data Mining Algorithms and Methods, Application of Data Mining, Infrastructures and Performance, Text Analysis and Search, and Security. Also, we are delighted to announce that the conference got a large financial support by the Thai government with the aim of encouraging and improving research in the ASEAN countries.

Although the support for the development of ASEAN and AEC (ASEAN Economic Community) are in the focus of the conference, the committee received 96 submissions from authors of 25 countries at 5 continents. The stable number of contributions over the last years is an indicator that our event is well established in the scientific community; with respect to the exploding number of conferences we are proud on that in particular. As usual, each submission was reviewed by at least 2-4 members of the program committee to avoid contradictory results. On these suggestions, the committee decided to accept 32 papers for oral presentation and inclusion in the conference proceedings.

Again, Springer agreed to publish our proceedings in its well-established and worldwide distributed series on Advances in Intelligent Systems and Computing. Last but not least, two internationally well-known scientists, from Germany and Japan, have been invited and accepted to give keynote talks to our participants.

A special thank is given to KMUTNB and its President, Professor Dr. Teeravuti Boonyasopon for his support of our conference from the first year on, and for providing us with a lot of resources from KMUTNB. We hope that IC^{2}IT again provides great opportunities for academic staff, students and researchers to present their work. IC^{2}IT is also a platform for exchange of knowledge in the field of computer and information technology and shall inspire researchers to generate new ideas and findings and meet partners for future collaboration. We also hope that our participants use this occasion to learn more about Phuket and its beautiful scenery, people, culture and visit its famous beaches before or after the conference.

We would also like to thank all authors for their submissions and the members of the program committee for their great work and valuable time. The staff members of the Faculty of Information Technology at King Mongkut's University of Technology North Bangkok have done a lot of technical and organisational works. A very special and warm thank you is given to our web masters: Mr. Jeerasak Numpradit, and Mr. Armornsak Armornthananun. Without the meticulous work of Ms. Watchareewan Jitsakul the proceedings could not have been completed in the needed form at the right time.

After so much preparation, all of the organisers of course hope and wish that IC^{2}IT2014 will again be a successful event and will be remembered by the participants for a long time.

February 27, 2014
Bangkok

Sirapat Boonkrong
Herwig Unger
Phayung Meesad

Organization

In Cooperation with

King Mongkut's University of Technology North Bangkok (KMUTNB)
FernUniversität in Hagen, Germany (FernUni)
Chemnitz University, Germany (CUT)
Oklahoma State University, USA (OSU)
Edith Cowan University, Western Australia (ECU)
Hanoi National University of Education, Vietnam (HNUE)
Gesellschaft für Informatik (GI)
Mahasarakham University (MSU)
Ubon Ratchathani University (UBU)
Kanchanaburi Rajabhat University (KRU)
Nakhon Pathom Rajabhat University (NPRU)
Mahasarakham Rajabhat University (RMU)
Rajamangala University of Technology Lanna (RMUTL)
Rajamangala University of Technology Krungthep (RMUTK)
Rajamangala University of Technology Thanyaburi (RMUTT)
Prince of Songkla University, Phuket Campus (PSU)
National Institute of Development Administration (NIDA)
Council of IT Deans of Thailand (CITT)

IC^{2}IT 2014 Organizing Committee

Conference Chair	Phayung Meesad, KMUTNB, Thailand
Technical Program Committee Chair	Herwig Unger, FernUni, Germany
Secretary and Publicity Chair	Sirapat Boonkrong, KMUTNB, Thailand

Program Committee

M. Aiello	RUG, The Netherlands
T. Anwar	UTM, Malaysia
T. Bernard	Syscom CReSTIC, France
W. Bodhisuwan	KU, Thailand
A. Bui	Uni Paris 8, France
T. Böhme	TU Ilmenau, Germany
M. Caspar	CUT, Germany
P. Chavan	MMMP, India
T. Chintakovid	KMUTNB, Thailand
H. K. Dai	OSU, USA
D. Delen	OSU, USA
N. Ditcharoen	URU, Thailand
T. Eggendorfer	HdP Hamburg, Germany
R. Gumzej	Uni Maribor, Slovenia
H. C. Ha	HNUE, Vietnam
M. Hagan	OSU, USA
P. Hannay	ECU, Australia
W. Hardt	Chemnitz, Germany
C. Haruechaiyasak	NECTEC, Thaialnd
S. Hengpraprohm	NPRU, Thailand
K. Hengproprohm	NPRU, Thailand
U. Inyaem	RMUTT, Thailand
M. Johnstone	ECU, Australia
T. Joochim	URU, Thailand
J. Kacprzyk	PAS, Poland

A. Kongthon	NECTEC, Thailand
S. Krootjohn	KMUTNB, Thailand
P. Kropf	Uni Neuchatel, Switzerland
M. Kubek	FernUni, Germany
S. Kukanok	MRU, Thailand
G. Kulkarni	MMMP, India
K. Kyamakya	Klagenfurt, Austria
J. Laokietkul	CRU, Thailand
U. Lechner	UniBw, Germany
M. Lohakan	KMUTNB, Thailand
J. Lu	UTS, Australia
A. Mikler	UNT, USA
A. Mingkhwan	KMUTNB, Thailand
C. Namman	URU, Thailand
P. P. NaSakolnakorn	MRU, Thailand
C. Netramai	KMUTNB, Thailand
K. Nimkerdphol	RMUTT, Thailand
S. Nitsuwat	KMUTNB, Thailand
S. Nuanmeesri	RSU, Thailand
N. Porrawatpreyakorn	KMUTNB, Thailand
P. Prathombutr	NECTEC, Thailand
A. Preechayasomboon	NECTEC, Thailand
P. Saengsiri	TISTR, Thailand
P. Sanguansat	PIM, Thailand
R. Shelke	MMMP, India
S. Smanchat	KMUTNB, Thailand
M. Sodanil	KMUTNB, Thailand
S. Sodsee	KMUTNB, Thailand
B. Soiraya	NPRU, Thailand
T. Srikhacha	TOT, Thailand
W. Sriurai	URU, Thailand
P. Sukjit	FernUni, Germany
D.H. Tran	HNUE, Vietnam
K. Treeprapin	UBU, Thailand
D. Tutsch	Uni Wuppertal, Germany
N. Utakrit	KMUTNB, Thailand
C. Valli	ECU, Australia
M. Weiser	OSU, USA
N. Wisitpongphan	KMUTNB, Thailand
A. Woodward	ECU, Australia
K. Woraratpanya	KMITL, Thailand
P. Wuttidittachotti	KMUTNB, Thailand

Contents

Invited Paper

Session 1: Data Mining Algorithms and Methods

Session 2: Application of Data Mining

Session 3: Infrastructures and Performance

Wireless Mesh Networks and Cloud Computing for Real Time Environmental Simulations

Peter Kropf[1], Eryk Schiller[1], Philip Brunner[2], Oliver Schilling[2], Daniel Hunkeler[2], and Andrei Lapin[1]

[1] Université de Neuchâtel, Computer Science department (IIUN), CH-2000 Neuchâtel, Switzerland
{peter.kropf,eryk.schiller,andrei.lapin}@unine.ch

[2] Université de Neuchâtel, Centre for Hydrogeology and Geothermics (CHYN), CH-2000 Neuchâtel, Switzerland
{philip.brunner,oliver.schilling,daniel.hunkeler}@unine.ch

Abstract. Predicting the influence of drinking water pumping on stream and groundwater levels is essential for sustainable water management. Given the highly dynamic nature of such systems any quantitative analysis must be based on robust and reliable modeling and simulation approaches. The paper presents a wireless mesh-network framework for environmental real time monitoring integrated with a cloud computing environment to execute the hydrogeological simulation model. The simulation results can then be used to sustainably control the pumping stations. The use case of the Emmental catchment and pumping location illustrates the feasibility and effectiveness of our approach even in harsh environmental conditions.

Keywords: wireless mesh network, cloud computing, data assimilation, environmental measurements, hydrogeological modelling and simulation, ground water abstraction.

1 Introduction

Climatic or hydrological systems are driven by highly dynamic forcing functions. Quantitative numerical frameworks such as simulation models are powerful tools to understand how these functions control the systems' response. Models are, however, always imperfect descriptions of reality and therefore model calculations increasingly deviate from the "real" physical conditions of the environmental system simulated. We can alleviate these biases by a real time integration of field data into the modeling framework (data assimilation). To accomplish this goal, we have to constantly monitor the environment through dense networks of sensors deployed over the geographical area concerned. The technology should provide high performance even in the case of harsh meteorological conditions (snow, low temperatures, fog, strong winds, etc.) and other location and infrastructure related limitations like high altitude, lack of access to the power grid, and limited accessibility (resulting in long access delays inducing significant installation/maintenance costs).

S. Boonkrong et al. (eds.), *Recent Advances in Information and Communication Technology,* Advances in Intelligent Systems and Computing 265,
DOI: 10.1007/978-3-319-06538-0_1, © Springer International Publishing Switzerland 2014

1.1 Wireless Infrastructure

In principle, communication networks can be wired or wireless. However, building a vast and complex wired infrastructure is costly and can be technically impossible in remote locations. An alternative are radio-based technologies, which do not require expensive cabled infrastructures. Moreover, this technological choice is extremely portable, because one can easily transfer equipment from one location to another when necessary. The first choice transport technology would be GSM/UMTS, however, this solution suffers from significant shortcomings. On the one hand, it is infeasible to equip every station with a GSM/UMTS connection in the case of vast measuring networks because of the associated cost of this operation, while the provider may charge for every additional SIM card. On the other hand, there exist important locations from an environmental perspective that have poor or non-existent coverage (e.g., highly elevated regions in Swiss Alps). These drawbacks force us to search for another scalable transport technology, which may grow to reach large proportions and provide us with good coverage over remote locations. Because of recent progress in the domain of low power wireless devices we may operate Wireless Mesh Networks that allow us to significantly cut operational expenses.

Wireless Mesh Networking is an interesting communication scheme which can provide cheap Internet connectivity delivered to end users at the last mile, an easily deployable multi-hop wireless bridge between distant bases in no-direct line of sight scenarios, or a wireless back-haul connecting sensors of different purposes such as environmental monitoring or smart-home applications. To properly deploy a wireless network, there are numerous hardware and software challenges. The hardware has to be properly selected to operate under a specific power consumption regime [1], e.g., when a node is solar powered, it has to harvest and store enough energy during the day-light operation to work uninterruptedly at night. Wireless cards and antennas have to provide an acceptable signal strength to allow for high throughput, while the node setup has to provide satisfactory performance such as computational power for ciphering and packet forwarding or other network adapters able to accommodate traffic coming from wireless interfaces. The experience obtained from pilot projects installed in remote and mountainous regions for environmental monitoring [2,3] and backup backbones in urban areas illustrates that mesh networks perfectly integrate into the existing AAA (Authentication, Authorization, Accounting) [4,5], monitoring and cloud infrastructure schemes of Swiss universities. For the purpose of this work, we use Wi-Fi based backhauls to transport information from environmental sensors to Internet storage facilities in real time.

1.2 Data Storage and Monitoring

In addition to provisioning the transmission infrastructure, facilities for data storage and processing have to be developed. Our studies reveal several similarities between environmental monitoring in the wireless mesh setup and network monitoring provided by typical monitoring agents such as Nagios, Zabbix, and

SNMP. In all these cases, current peripherals' status is reported to central storage for future analysis and visual presentation, while the information retrieval is triggered on a time basis. Our experience shows that we can re-use network monitoring for environmental purposes by configuring monitoring agents to constantly read out values provided by a sensor, e.g., every 15 mins. However, this methodology requires specific counter measures against Wi-Fi backhaul failures as we cannot afford the loss of environmental data if the network is inaccessible at a given time.

1.3 Going to the Clouds

The data collected can be processed in numerical models. In our case, we are simulating the use of groundwater water resources in a highly dynamic river-aquifer system—the Emmental[1] in Switzerland. The purpose of the simulation approach is to provide a quantitative basis for sustainable water resource management. Groundwater in the aquifer of the Emmental is pumped to supply the city of Bern with drinking water. However, the abstraction of groundwater[2] causes the water table in the aquifer to drop and can increase the infiltration from the river with adverse impacts on the stream ecosystems, but to what extent and how fast groundwater abstraction influences the flow in the river depends on the system state[3] (i.e., how much water is stored in the aquifer). To optimize the amount of water pumped (pumping scheme), predictions on how groundwater abstraction will affect the system are required. This paper describes recent developments in data acquisition and transmission infrastructures, integrated in the data assimilation system with the goal of generating a real-time pumping scheme for the Emmental. As discussed in the next section, predictions are generated using computationally expensive models that simulate the interactions and feedback mechanisms between the river, the aquifer, the pump, and climatic forcing functions such as precipitation. The models are continually updated with acquired field data. The data assimilation approach implemented for this task requires that multiple models are run in parallel. This allows us to assess the reliability of the proposed pumping rates in a stochastic way. However, computational costs for such an approach are significant. Running a few models in parallel is challenging since even a single one may require a few days of computations on an ordinary desktop machine. We decided to use a recently developed cloud computing paradigm for our computations to speed up the whole process, while the cloud allows us to run a few parallel computing workers. Cloud computing is a growing business with many established players such as Salesforce, Amazon, Akamai, and Google. We integrated our work with the ongoing SwissACC project [6], which aims to establish a Swiss nationwide cloud computing infrastructure. Cloud providers offer several service models,

---

[1] In German, the Emmental means the valley of the Emme river.

[2] i.e., pumping groundwater from the aquifer.

[3] This system state should be understood in the physical sense, i.e., similar to states in thermodynamics and definitely not as stored information in automata theory.

however, we concentrate on Infrastructure as a Service (IaaS), which provides us with the required number of powerful Virtual Machines (VMs) on-demand and remote control through the Internet. The VMs are extremely useful, because they can accommodate any generic type of computations, while they do not actually require any physical maintenance from the user side. SwissACC builds upon the open-source infrastructure, OpenStack[4]. OpenStack comes along with the Nova controller, which automates pool managing of worker resources. For storing data in the cloud, SwissACC integrates the S3 (Simple Storage Service) driver provided by Amazon[5] with the necessary API.

This work is organized in the following way. Section 2 provides a detailed problem description. In Section 3, we specify our proposed solution and we provide the most important implementation details. The results are gathered in Section 4. Finally, we conclude in Section 5.

2 Use Case: Groundwater Pumping

Drinking water supply in Switzerland is largely based on groundwater (about 80%). Numerous water supply systems abstract groundwater close to rivers. Surface water and groundwater systems interact in highly dynamic and complex ways [7,8], and therefore abstracting groundwater in the vicinity of rivers can substantially influence these dynamics [9]. Environmental laws demanding minimum water levels are in place in Switzerland. This gives rise to a challenging optimization problem. The critical parameters are the discharge in a river[6] (which is minimal due to strict environmental laws), the amount of water stored in the aquifer and the drinking water requirements. Balancing these target functions by adjusting the pumping rate thus requires a solid and quantitative understanding of the dynamics and the interactions of the river-aquifer system.

The Emmental is a perfect example that illustrates the tradeoff between the need for drinking water supply and minimal discharge in the river. The Emmental is a pre-alpine river catchment (about 200 km^2) in central Switzerland (Fig. 1). The catchment features steep hydraulic gradients with rapid groundwater flow rates (up to 100 m/d). The Emme River itself is highly dynamic (discharge between 0 and 300 (m^3/s). The aquifer pumped close to Emme provides roughly 45% of the drinking water for Bern, the Swiss capital. Groundwater abstraction in the Emmental increases the infiltration from the river to the aquifer. In fact, during low flow periods, groundwater abstraction often causes the river to dry up. The stream water levels in the the upper Emmental are strongly affected by seasonality and are highly sensitive to dry periods. In 2003 and 2011 large stretches of the river ran completely dry, as illustrated in Fig. 2. This pronounced seasonality adds an additional level of complexity to the system.

The efficiency and sustainability of water resources management in the Emmental is directly linked to the amount of water pumped from the aquifer. A

[4] There obviously exist other open-source infrastructures such as OpenNebula.

[5] `http://aws.amazon.com/s3/`

[6] The discharge in a river is the volumetric flow rate.

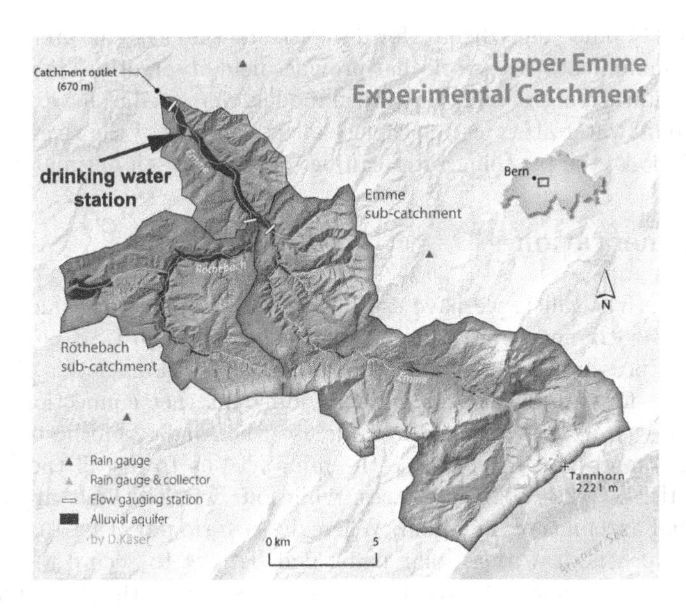

Fig. 1. The upper Emmental catchment is located close to the Swiss capital Bern (shown on the map in the top right corner). The Emme and the Roethebach rivers flow downhill from southeast and continue through the main upper Emmental after their confluence. Figure provided by D. Käser.

Fig. 2. A comparison between a high and a low water level situation in the upper Emmental, right below the pumping station. Photos provided by D. Käser.

quantitative approach to simulate the system is required to optimize pumping rates in this dynamic environment. Numerical models are therefore necessary for this task. By continuously incorporating field observations in the model simulation, any potential model biases can be identified, quantified and corrected. This process is called data assimilation [10]. While data-assimilation approaches are widely used in climatological models, they are rarely applied to hydrogeological simulations [11]. The “ingredients” for data assimilation systems are a measurement and communication network that provides field observations in real time; a data storage infrastructure; and numerical models that predict for example how different groundwater abstraction schemes affect the flow of the river. Based on these simulations, the pumping rates can be regulated in an optimal way.

3 Implementation

In the Emme river valley, we have established a wireless setup which contains a few stations with environmental sensors attached through USB (such as temperature and pressure meters) and other necessary stations acting as wireless backhaul thus forwarding packets and providing Internet connection (Fig. 3). From the hardware perspective, every node uses the Alix3d2 motherboards with two on-board mini-pci slots. We use the mini-pci bus to install the Winstron DNMA-92 IEEE 802.11abgn interfaces, while our wireless links are provided by directional antennas of high-gain. When the electric power grid is not available, we equip a node with a solar panel and battery to secure a continuous 24 hours operation (normally, the battery charges during the day-light operation). Our nodes are placed in a special-purpose enclosure which protects them against outdoor conditions, e.g., humidity. When a high number of Wi-Fi interfaces are required, we gather a few mother boards together in a single box. The Linux based ADAM system[7] serves as the operating system platform. Due to the installed OLSR and IEEE 802.11s, the network is easily expandable, i.e., the installation of a new node requires little attention from the administration perspective.

3.1 Environmental Monitoring

Zabbix[8] provides a client-server infrastructure which allows us to monitor and control remote machines. There are a large number of predefined parameters, while Zabbix also provides an opportunity to launch user-defined commands to support user-specific peripherals. Due to this feature, we are able to equip Zabbix agents with drivers, i.e., special purpose applications which read out environmental parameters from the sensors through USB and provide the agent with the received data. We deployed one running instance of the Zabbix agent on every node in the mesh and one instance of the Zabbix server at the central storage (online database). To control the Zabbix server (e.g., including another sensor), we are provided with an advanced back-end web interface and rich logging system.

The online database allows the access of the measured environmental system state in real-time, providing the basis for a real-time forecasting system to control the groundwater abstraction rates. Periodically, our software asks sensors about current values of the measurements. Then, again periodically, the Zabbix server

[7] Developed by the University of Bern: `http://cds.unibe.ch`
[8] `http://www.zabbix.org`

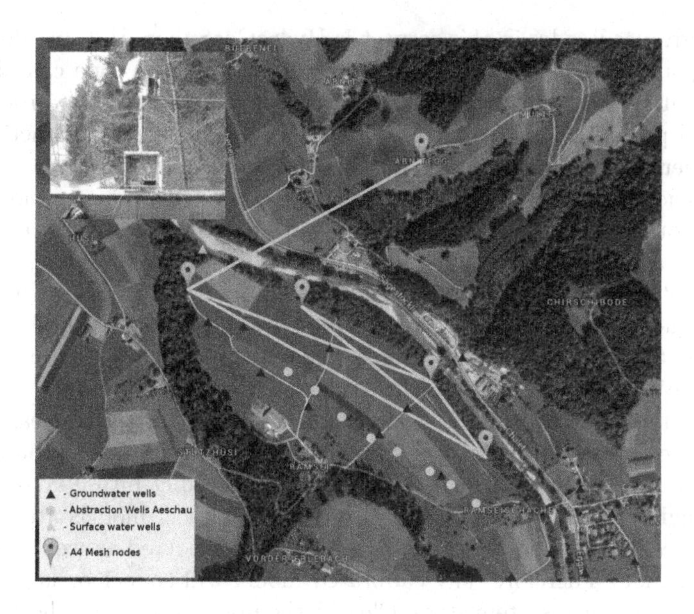

Fig. 3. Environmental setup of the network

executes a remote driver on the nodes and obtains values of the measurements. This data is transferred to the database on the server. Finally, the data can be accessed through the web interface (Fig. 4). The Zabbix infrastructure provides us with wide variety of tools for drawing plots and applying simple formulas to the data; it fits well to the requirements of our application context.

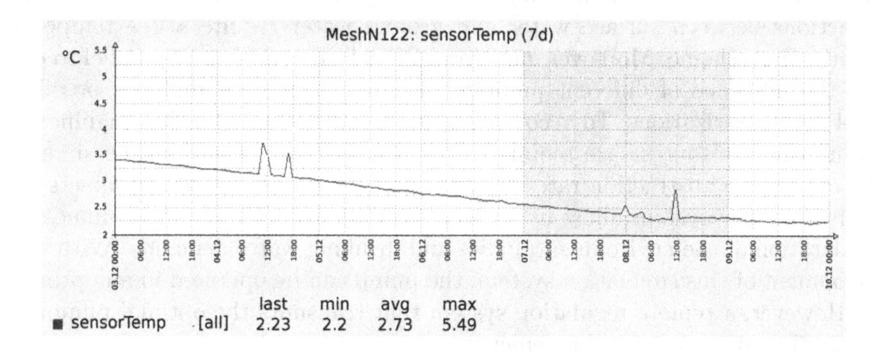

Fig. 4. Online environmental data output (the larger peaks appear to be artefacts)

3.2 Real-Time Modeling

The latest generation of numerical models is now able to simulate the interactions between surface water and groundwater in a fully coupled way [12]. One of the most advanced codes in this respect is HydroGeoSphere [13]. In addition to simulating surface water and groundwater interactions, the code can also simulate vegetation dynamics as well as the recharge[9] and discharge processes in response to precipitation, evapotranspiration or groundwater abstraction. HydroGeoSphere is therefore used to simulate the Emmental system.

The geometric setup is based on a high resolution digital terrain model. The numerical coupling between the surface and subsurface domain is conceptualized through a dual node approach, as described in [13]. The model requires a very large amount of parameters, such as hydraulic properties of the streambed, the soil or the aquifer. These parameters cannot be measured in the field at the required spatial resolution and therefore have to be estimated. Numerous approaches are available in this regard. A "classic" way is to adjust parameters in order to minimize the mismatch between the available historical measurement data and the corresponding model simulations. Once a model reproduces historic measurement data satisfactorily, it is used to predict future system states under changing forcing functions. However, all numerical models are a simplification of reality, both in terms of the processes considered as well as in their parameterization. Therefore, any calibrated model will sooner or later deviate from the real, physical system state. Clearly, the model state (i.e., the simulated water levels or the actual discharge in the river) has to be as close as possible to the real system in order to provide reliable predictions on how a planned pumping scheme will affect the system in the near future. Therefore, the "classical" calibration approach is not well suited for this application. By using a data assimilation approach, the model is continuously updated in terms of its state and parameters. We implement a data assimilation approach similar to the work of [11]. Currently, the assimilated measurements consist of stream level and water tables along the river.

HydroGeoSphere is a numerically demanding code, and the highly dynamic interactions between surface water and ground water require a fine temporal discretization scheme. Moreover, numerous models are running in parallel to explore the influence of different pumping schemes, as well as different possible model parameterizations. To accomplish this computational burdon, significant computational resources are required (see next section). The multiple simulations of possible abstraction rates with the corresponding predicted impacts on the river-aquifer system allow us to identify the optimal pumping volumes in consideration of the environmental laws and drinking water demands. With the development of this simulation system, the pump can be operated in an optimal way. However, a remote regulation system that transmits the optimal pumping rates to the pump must be implemented.

3.3 Swiss Academic Compute Cloud

Our Cloud based solution allows researchers to perform resource consuming computations with minimal efforts. Firstly, the data collected with the environmental

[9] Water infiltrating the soil reaching the underground water table.

sensors is stored on the pilot-project Swiss Academic Compute Cloud (SACC), a unified cloud service providing storage and computation resources for Swiss academic institutes, which makes use of a specialized S3-based cloud repository—the Object Storage (OS). The user front-end is developed in Django[10], which is a free open source Python based web application framework that provides us with the model-view-controller architectural paradigm. Our currently implemented front-end allows a researcher to visit the web-page, choose required input data, initiate required tasks, and download the results of completed computations. Behind the user front-end, we integrate a Python engine—GC3Pie[11], which is developed by the GC3 group at the University of Zürich[12] and enables all cloud related operations such as starting and stopping new workers. We configured our framework to run several instances of HydroGeoSphere on allocated working VMs. Due to the infrastructure configuration, every instance of the HydroGeoSphere is provided with the input files from the OS, which in turn also acts as a storage facility for models returned by completed instances. This OS-based data organization scheme is important, because it provides portability as there are many different cloud providers supporting this storage manner.

4 Results

Firstly, we deployed a measuring and transporting mesh network in the Emmental which proves its high performance and reliability in harvesting environmental data. Secondly, the first numerical HydroGeoSphere model that is capable of simulating the interactions and feedback mechanisms between the river, the aquifer and the pumps has been set up. It includes the integration of the HydroGeoSphere binary with the cloud computing workers, implementing the web-interface for running tasks, and integrating the OS for maintaining both the input and output of the HydroGeoSphere. All parts of the so far implemented infrastructure fully correspond to our requirements. Due to the integration with cloud infrastructures, simultaneous running of different models showed us significant profit in comparison with the usual sequential running. Also, the web interface for controlling the computations, greatly simplified the whole process of launching models.

The current infrastructure is under ongoing developments. In the future, we plan to strongly integrate all the technological pieces to allow for fully automated model computations thus providing valuable pumping scheme predictions in real time. We also plan to develop our web interface to allow for any generic computational use-case. One of the identified improvements relates to precise definition of input and output to support many different applications (e.g., by employing XML to define program options, input/output files, etc.).

[10] `http://www.djangoproject.com`
[11] `http://code.google.com/p/gc3pie/`
[12] `www.gc3.uzh.ch`

5 Conclusions

The integration of advanced Information Technologies in environmental simulation systems allows for a new dimension of natural resource management. Our proposed solution is especially interesting for remote locations with harsh environmental conditions in which wireless mesh network prove to provide a reliable network infrastructure. When the transporting infrastructure is developed, one can employ cloud computing to solve any computationally expensive problem, while the network monitoring application (e.g., Zabbix) can transport information in different use-cases such as environmental monitoring or smart-home applications.

Acknowledgements. This work is partially funded by the Swiss State Secretariat for Education and Research through SWITCH and the Swiss National Science Foundation through NRP 61 on Sustainable Water Management. We particularly thank Torsten Braun (the leader of the A4Mesh project) and his team as well as Sergio Mafioletti (the leader of the SwissACC project) for their contributions to this work.

References

1. Badawy, G., Sayegh, A., Todd, T.: Solar powered wlan mesh network provisioning for temporary deployments. In: Wireless Communications and Networking Conference, WCNC 2008, pp. 2271–2276. IEEE (March 2008)
2. Wu, D., Mohapatra, P.: Qurinet: A wide-area wireless mesh testbed for research and experimental evaluations. In: 2010 Second International Conference on Communication Systems and Networks (COMSNETS), pp. 1–10 (January 2010)
3. Jamakovic, A., Dimitrova, D.C., Anwander, M., Macicas, T., Braun, T., Schwanbeck, J., Staub, T., Nyffenegger, B.: Real-world energy measurements of a wireless mesh network. In: Pierson, J.-M., Da Costa, G., Dittmann, L. (eds.) EE-LSDS 2013. LNCS, vol. 8046, pp. 218–232. Springer, Heidelberg (2013)
4. Anwander, M., Braun, T., Jamakovic, A., Staub, T.: Authentication and authorisation mechanisms in support of secure access to wmn resources. In: 2012 IEEE International Symposium on a World of Wireless, Mobile and Multimedia Networks (WoWMoM), pp. 1–6 (June 2012)
5. Schiller, E., Monakhov, A., Kropf, P.: Shibboleth based authentication, authorization, accounting and auditing in wireless mesh networks. In: LCN, pp. 918–926 (2011)
6. Kunszt, P., Maffioletti, S., Flanders, D., Eurich, M., Schiller, E., Bohnert, T., Edmonds, A., Stockinger, H., Jamakovic-Kapic, A., Haug, S., Flury, P., Leinen, S.: Towards a swiss national research infrastructure. In: Proceedings of the 1st International Workshop on Federative and Interoperable Cloud Infrastructures 2013, FedICI 2013 Organized in Conjunction with Euro-par (August 2013)
7. Partington, D., Brunner, P., Frei, S., Simmons, C.T., Werner, A.D., Therrien, R., Maier, H.R., Dandy, G.C., Fleckenstein, J.H.: Interpreting streamflow generation mechanisms from integrated surface-subsurface flow models of a riparian wetland and catchment. Water Resources Research 49(9), 5501–5519 (2013)

8. Brunner, P., Cook, P.G., Simmons, C.T.: Disconnected surface water and groundwater: from theory to practice. Ground Water 49(4), 460–467 (2011)
9. Winter, T.C., Harvey, J.W., Franke, O.L.: Alley, W.M.: Ground Water and Surface Water A Single Resource. USGS, Circular 1139, Denver, Colorado (1998)
10. Evensen, G.: Data assimilation: the ensemble Kalman filter, 2nd edn. Springer, New York (2009)
11. Hendricks Franssen, H.J., Kinzelbach, W.: Ensemble kalman filtering versus sequential self-calibration for inverse modelling of dynamic groundwater flow systems. Journal of Hydrology 365(3-4), 261–274 (2009)
12. Brunner, P., Simmons, C.T.: Hydrogeosphere: A fully integrated, physically based hydrological model. Ground Water 50(2), 170–176 (2012)
13. Therrien, R., McLaren, R., Sudicky, E., Panday, S.: HydroGeoSphere. Groundwater Simulations Group (2013)

Attribute Reduction Based on Rough Sets and the Discrete Firefly Algorithm

Nguyen Cong Long[1], Phayung Meesad[1], and Herwig Unger[2]

[1] Faculty of Information Technology, King Mongkut's University of Technology North Bangkok, Thailand
nclong.c52@moet.edu.vn, pym@kmutnb.ac.th
[2] Faculty of Mathematics and Computer Science, FernUniversitat in Hagen, Germany
herwig.unger@fernuni-hagen.de

Abstract. Attribute reduction is used to allow elimination of redundant attributes while remaining full meaning of the original dataset. Rough sets have been used as attribute reduction techniques with much success. However, rough set applies to attribute reduction are inadequate at finding optimal reductions. This paper proposes an optimal attribute reduction strategy relying on rough sets and discrete firefly algorithm. To demonstrate the applicability and superiority of the proposed model, comparison between the proposed models with existing well-known methods is also investigated. The experiment results illustrate that performances of the proposed model when compared to other attribute reduction can provide comparative solutions efficiently.

Keywords: Attribute Reduction, Reduction, Feature selection, Rough sets, Core, Firefly Algorithm.

1 Introduction

Attribute reduction is a very important issue in many fields such as data mining, machine learning, pattern recognition and signal processing [1-4]. That is a process of choosing a subset of significant attributes (features) and elimination of the irrelevant attributes in a given dataset in order to build a good learning model. The subset is a retaining high accurate representation of original features and sufficient to describe target dataset.

Quite often, abundance of noisy, irrelevant or misleading features is usually presented in real-world problems. The ability to deal with imprecise and inconsistent information has become one of the most important requirements for attribute reduction. Rough sets theory can be utilized as a tool to discover data dependencies and reduce the number of attribute in inconsistent dataset [1]. Rough sets are applied to attribute reduction to remove redundant attributes and select sets of significant attributes which lead to better prediction accuracy and speed than systems using original sets of attributes.

Due to the NP-hard problem, designing an efficient algorithm for minimum attribute reduction is a challenging task [5]. There are many rough sets algorithms that

S. Boonkrong et al. (eds.), *Recent Advances in Information and Communication Technology*, Advances in Intelligent Systems and Computing 265,
DOI: 10.1007/978-3-319-06538-0_2, © Springer International Publishing Switzerland 2014

have been proposed for attribute reduction in past literature. Generally, there are two categories of rough set methods for attribute reduction, greedy heuristics and meta-heuristic algorithms. The greedy heuristics approaches usually select the significant attributes or eliminate redundant attributes as heuristic knowledge. Algorithms in this category are fast. However, they usually find a reduction than a minimum reduction [6-9]. On the other hand, meta-heuristic algorithms have been applied to find minimal reductions [5][10-12]. In many systems that require minimal subset of attributes, the meta-heuristic algorithms are necessary to use.

There are many meta-heuristic algorithms that have been proposed to find minimal reductions based on rough sets. Wang, X. *et al* [11] proposed feature selection based on rough sets and particle swarm optimization (PSO). In those techniques, PSO is applied to find optimal feature selections. However, the fitness function applied in this algorithm may not correspond to a minimum reduction [5]. Inbaria, H. H. *et al* [12] proposed a hybrid model to combine the strength of rough sets and PSO to find a reduction. In this model, relying on two existing algorithms, quick reduct and relative reduct algorithm, PSO is applied to find optimal reductions. Nevertheless, the fitness function was only considered by correctness of attribute reduction without minimal reduction. As a result, reductions may not be minimal reductions that were found in this research. Ke, L. *et al* [10] investigated a model that combined rough sets and ant colony optimization (ACO) to find minimal attribute reduction. The experiment results shown that ACO applied to rough sets can provide competitive solutions efficiently. Ye, D. *et al* [5] proposed a novel fitness function for meta-heuristic algorithms based on rough sets. Genetic algorithm (GA) and PSO were applied to find the minimal attribute reduction using various fitness functions. The experiment results illustrated that PSO outperformed GA in terms of finding minimal attribute reduction. In addition, the novel fitness function was considered by correctness and minimal attribute reduction.

This paper proposes a new attribute reduction mechanism, which combine rough sets and the firefly algorithm to find minimal attribute reduction. Firefly algorithm (FA) is a new meta-heuristic algorithm that relies on flashing behavior of fireflies in nature to find global optimal solution in search space for special problems [13]. FA has been successfully applied to a large number of difficult combinational optimization problem [14-19]. Preliminary studies suggest that the FA outperforms GAs and PSO [13][19] in terms of accuracy and running time.

The remainder of this paper is organized as follows: Section 2 reviews rough sets preliminaries. The original firefly algorithm is summarized in section 3. In section 4 the attribute reduction using rough sets and the firefly algorithm is presented. Experiment results and comparison of differential models are discussed in section 5. Finally, conclusions are summarized in section 6.

2 Rough Sets Preliminaries

This section reviews some basic notions in the rough sets theory [1][20] which are necessary for this research.

2.1 Decision Table

Let a decision table $DT = (U, A = C \cup D)$, where $U = \{x_1, x_2, \dots, x_n\}$ is non-empty finite set of objects called the universe of discourse, C is a non-empty finite set of condition attributes, D is a decision attribute.

2.2 Indiscerbility Relation

$\forall a \in A$ determine a function $f_a = U \to V_a$, where V_a is the set of values of a. if $P \subseteq A$, the *P-indiscernibility relation* is denoted by $IND(P)$, is defined as:

$$IND(P) = \{(x, y) \in U | \forall a \in P, f_a(x) = f_a(y)\} \quad (1)$$

The partition of U generated by *IND*(*P*) is denoted by *U*/*P*. If $(x, y) \in IND(P)$, x and y are said to be indiscernibility with respect to P. The equivalence classes of the *P-indiscernibility relation* are denoted by $[x]_P$. The indiscernibility relation is the mathematical basic notion of rough sets theory.

$$\text{Let } U/C = \{Y_1, Y_2, \dots, Y_N\}, N \le n \quad (2)$$

where the equivalence classes Y_i are numbered such that $|Y_1| \le |Y_2| \le \dots \le |Y_N|$

2.3 Lower and Upper Approximation

Let $X \subseteq U$ and $P \subseteq A$, X could be approximated by the lower and upper approximation. P-lower and P-upper approximation of set X, is denoted by $\underline{P}$~ and $\overline{P}X$, respectively, is defined as:

$$\underline{P}X = \{x \in U: [x]_P \subseteq X\} \quad (3)$$

$$\overline{P}X = \{x \in U: [x]_P \cap X \neq \emptyset\} \quad (4)$$

2.4 Positive, Negative and Boundary Region

Let $P, Q \subseteq A$, be equivalence relations over U, then the positive, negative and boundary regions, denoted $POS_P(Q)$, $NEG_P(Q)$, $BN_P(Q)$, respectively, can be defined as:

$$POS_P(Q) = \cup_{X \in U/Q} \underline{P}X \quad (5)$$

$$NEG_P(Q) = U - \cup_{X \in U/Q} \overline{P}X \quad (6)$$

$$BN_P(X) = \cup_{X \in U/Q} \overline{P}X - \cup_{X \in U/Q} \underline{P}X \quad (7)$$

A set is said to be rough (imprecise) if its boundary region is non-empty, otherwise the set is crisp.

2.5 Dependency of Attributes

Let $P \subseteq C$, D depends on P in a degree k $(0 \leq k \leq 1)$ denoted by $P \Rightarrow_k D$, is determined by

$$k = \gamma_P(D) = \frac{|POS_P(D)|}{|U|} \quad (8)$$

where $|.|$ denotes the cardinality of a set, $\gamma_P(D)$ is quality of classification. If k=1, D depends totally on P; if $0 < k < 1$, D depends partially on P, if k=0, D is not depends on P. Decision table DT is consistent if $\gamma_C(D) = 1$, otherwise DT is inconsistent.

2.6 Attribute Reduction and Core

Generally, there are often existing redundant condition attributes. So, these redundant attributes can be eliminated without losing essential classificatory information [20]. The goal of attribute reduction is to eliminate redundant attributes leading to the reduced set that provides the same quality of classification as the original.

A given decision table may have many attribute reductions, the set of all reductions are defined as

$$Red(C) = \{R \subseteq C | \gamma_R(D) = \gamma_C(D), \forall \mathrm{B} \subset \mathrm{R}, \gamma_{\mathrm{B}}(D) \neq \gamma_{\mathrm{R}}(D)\} \quad (9)$$

A set of minimal reductions is defined as

$$Red(C)_{min} = \{R \in Red(\mathrm{C}) | \forall R' \in Red(C), |R| \leq |R'|\} \quad (10)$$

Core of condition attributes is an intersection of all reductions, defined as

$$Core(C) = \bigcap Red(æ) \quad (11)$$

3 Original Firefly Algorithm

The Firefly algorithm is a kind of stochastic, meta-heuristic algorithm to find the global optimal solution in search space for special problems. This is inspired by the flashing behavior of fireflies in nature and is originally proposed by Yang, X.S. in 2008 and relies on three key ideas [13]

- All fireflies are unisex and there may be an attractive in any two fireflies
- Their attractiveness is proportional to their light intensity. A firefly with lower light intensity will move toward the fireflies with higher light intensity. If there is not firefly with higher light intensity, the firefly will randomly move in search space.
- The light intensity of a firefly is related to fitness function in genetic algorithms.

FA starts with randomly generated positions of fireflies (population). In each time step t, positions of fireflies with lower light intensity move toward fireflies with higher light intensity by Eqs. (12) and (13). That mean, for any two fireflies, if a firefly has lower light intensity, it will move toward the other firefly in the search space.

$$X_i(t+1) = X_i(t) + \beta(X\ \ (t) - X_j(t)) + \alpha\left(rand - \tfrac{1}{2}\right) \tag{12}$$

$$\beta = \beta_0 e^{-\gamma r_{ij}^2} \tag{13}$$

where $X_i(t)$ and $X_j\ (t)$ are positions of firefly with lower light intensity and firefly with higher intensity at time t respectively, α is a random parameter which determines randomly behavior of movement, rand is a random number generator uniformly distributed in [0, 1], γ is a light absorption coefficient, β_0 is the attractiveness at $r = 0$, and r_{ij} is Euclidean distance between any two fireflies i and j at X_i and X_j, respectively.

After movements, all fireflies move toward the firefly with the highest light intensity and their light intensity improves. After stopping criteria are satisfied, the firefly with the highest light intensity will be considered as the best solution.

More details of firefly algorithm can see in [13].

4 Firefly Algorithm for Attribute Reduction

The idea of firefly algorithm is used to find minimal attribute reduction problem. The process of firefly algorithm is to find a minimal attribute reduction, is illustrated in Fig. 1.

4.1 Encoding Method

The position of each firefly represents the possible solution of the problem as binary strings of length m ($m = |C|$). Every bit represents an attribute, the value '1' shows that the corresponding attribute is selected while the value '0' illustrates that the corresponding attribute is not selected.

For example, suppose that $C = \{a_1, a_2, \ldots, a_{10}\}$ and a firefly $X = 1010001101$, then an attribute subset is $\{a_1, a_3, a_7, a_8, a_{10}\}$.

4.2 Fitness Function

There are many definitions of fitness function for this problem in past literature [5][11]. There are some drawbacks with these fitness functions. These may considered by correctness of attribute reduction without minimal reductions [5]. The fitness function used in this research is defined as Eq. (14) [5]

$$Fitness(X) = \frac{m-|X|}{m} + \frac{n|R|\,\gamma_X(\mathrm{D})}{m\Gamma} \tag{14}$$

where $m = |C|, n = |U|$, $\gamma_X(\mathrm{D})$ is quality of classification. R is a reduct of condition attribute C, R is computed by an efficient attribute reduction algorithm in [21].

$\Gamma = |Y_1| + |Y_2|$ if the decision table *DT* is consistent and $\Gamma = |Y_1|$ if not where Y_1 and Y_2 are defined in Eq. (2) .

This fitness function not only considers quality of classification but also considers minimal reductions [5].

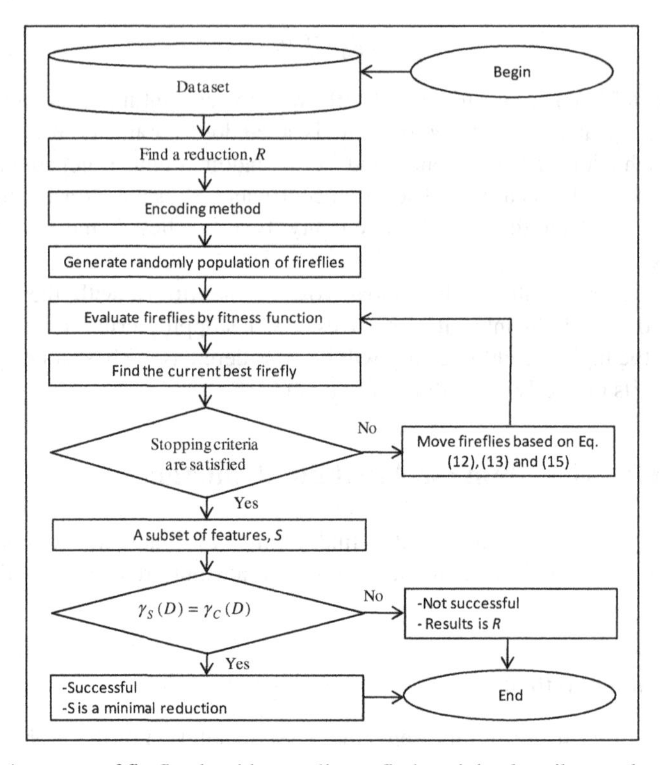

Fig. 1. A process of firefly algorithm applies to find a minimal attribute reduction

4.3 Position Update Strategies

In the original firefly algorithm, position of fireflies with lower light intensity will move towards fireflies with higher light intensity. Those fireflies will change from binary number to a real number when they move in the search spaces. Therefore, this real number must be replaced by a binary number. For this purpose, a sigmoid function can be applied [22]. However, in order to improve the binary firefly algorithm, a tang function is used in this research [23]. The tang function is defined in Eq. (15) as:

$$f(X_i^k) = \frac{\exp(2X_i^k)-1}{\exp(2X_i^k)+1}, i = 1, \dots, N; k = 1, \dots, d \quad (15)$$

This function scale the X_i value in the [0, 1] range. The final value of each part of fireflies after movement is determined by: If $f(X_i^k) \geq rand$ then $X_i^k = 1$ otherwise $X_i^k = 0$. $rand$ is a random number generator uniformly distributed in [0, 1]

5 Simulations

To evaluate the proposed model, attribute reduction based on rough sets and discrete firefly algorithm (ARRSFA), various datasets with different numbers of condition attributes and objects are used to test the model. In addition, in order to demonstrate the superiority of proposed model, the comparison between the model and genetic algorithm for rough set attribute reduction (GenRSAR) [5] and particle swarm optimization for rough set attribute reduction (PSOAR) [5] is also investigated.

5.1 Data Sets

In this paper, 6 well-known datasets are collected from UCI Repository Machine Learning Database those are used to test the models. Most of these datasets are used for evaluating attribute reduction algorithms in the past literature [5][10-11]. Basic information about datasets is shown in Table 1.

5.2 Parameters Setting

The proposed model and other attribute reduction models are implemented in MATLAB. In the experiments, the parameters, except when indicated differently, were set to the following values: Initially, 25 fireflies are randomly generated in a population, $\alpha = 0.2, \gamma = 1, \beta_0 = 0.2$, number of generations is equal to 100. The FA MATLAB code for each dataset ran 100 times with different initial solutions, same as [5].

Table 1. Basic information about datasets

No.	Datasets name	Number of objects (n)	Number of condition attributes (m)	Number of classifications
1	Audiology	200	69	10
2	Bupa	345	6	2
3	Corral	32	6	2
4	Lymphography	148	18	4
5	Soybean-small	47	35	4
6	Vote	300	16	2

5.3 Results and Discussion

A number of results from the experiments are recorded, consisting of minimal (Min) and average (AVG) length of output attribute reduction during 100 runs of the algorithms. The fitness function uses in this research not only considers quality of classification but also considers minimal reductions [5]. Therefore, results of minimal and average length of attribute reduction during 100 runs are adequate for simulation. The experiment results are illustrated in Table 2 and Fig. 2.

Table 2. Length of minimal and average attribute reduction of differential models

Datasets name	GenRSAR		PSOAR		ARRSFA	
	Min	AVG	Min	AVG	Min	AVG
Audiology	12	14.35	12	14.32	11	11.5
Bupa	3	3	3	3	3	3
Corral	4	4.04	4	4.02	4	4.02
Lymphography	5	5.66	5	5.6	6	6.6
Soybean-small	2	2.9	2	2.24	2	2.1
Vote	8	8.28	8	8.2	5	5.63

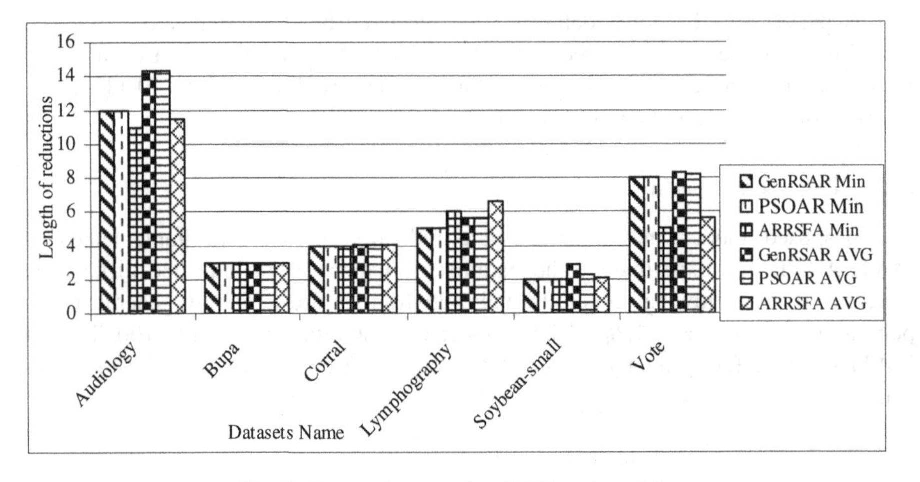

Fig. 2. Comparison results of different models

Looking at the Table 2 and Fig 2, it can be seen that all of the algorithms could find minimal reductions of the Bupa dataset with the same length. ARRSFA and PSOAR had the same results in terms of min and average of attribute reduction length and they outperform GenRSAR in terms of average of attribute reduction length of the Corral dataset. In addition, ARRSFA could find the best minimal reductions for the all of other tested datasets except for Lymphography. ARRSFA outperforms both GenRSAR and PSOAR in 3 datasets namely Audiology, Soybean-small and Vote. However, ARRSFA sometimes could not obtain the best solutions same as the other models. It is not better when compared to the other methods for the Lymphography dataset. There is no single model that always find the best solution for all data sets, but ARRSFA outperforms in terms of obtaining better solutions for such datasets as Vote and Audiology.

6 Conclusion and Future Work

This paper proposed a model to find minimal reductions based on rough sets and the discrete firefly algorithm. In this model, rough sets are used to build fitness function of the firefly algorithm as well as verifying correctness of reductions. The discrete firefly algorithm is applied to find minimal reductions. To demonstrate the superiority of the proposed model, numerical experiments have been conducted on 6 well-known datasets. Comparisons of performance with the proposed model and two meta-heuristic algorithms have revealed that the proposed model has a superior performance.

The proposed model is only tested on 6 differential datasets. Further investigation will concentrate on two other aspects, namely running time and classification accuracy. Furthermore, there are existing several extended types of attribute reduction in the concepts of rough sets such as entrope-based reducts [24], distribution reducts [25]. These extension may assist to improve the performance of the proposed model. In conclusion, all the future work will contribute to further improve the proposed model, making it a more robust technique for attribute reduction.

References

1. Pawlak, Z.: Rough Sets: Theoretical Aspects of Reasoning About Data. Springer (1991)
2. Cheng, C.H., Chen, T.L., Wei, L.Y.: A hybrid model based on rough sets theory and genetic algorithms for stock price forecasting. Inf. Sci. 180(9), 1610–1629 (2010)
3. Chen, H.L., Yang, B., Liu, J., Liu, D.Y.: A support vector machine classifier with rough set-based feature selection for breast cancer diagnosis. Expert Syst. Appl. 38(7), 9014–9022 (2011)
4. Zhao, S., Tsang, E.C.C., Chen, D., Wang, X.: Building a Rule-Based Classifier, A Fuzzy-Rough Set Approach. IEEE Trans. Knowl. Data Eng. 22(5), 624–638 (2010)
5. Ye, D., Chen, Z., Ma, S.: A novel and better fitness evaluation for rough set based minimum attribute reduction problem. Inf. Sci. 222, 413–423 (2013)
6. Hoa, N.S.: Some Efficient Algorithms For Rough Set Methods. In: Proceedings IPMU 1996 Granada, Spain, pp. 1541–1457 (1996)
7. Degang, C., Changzhong, W., Qinghua, H.: A new approach to attribute reduction of consistent and inconsistent covering decision systems with covering rough sets. Inf. Sci. 177(17), 3500–3518 (2007)
8. Wang, C., He, Q., Chen, D., Hu, Q.: A novel method for attribute reduction of covering decision systems. Inf. Sci. 254, 181–196 (2014)
9. Meng, Z., Shi, Z.: A fast approach to attribute reduction in incomplete decision systems with tolerance relation-based rough sets. Inf. Sci. 179(16), 2774–2793 (2009)
10. Ke, L., Feng, Z., Ren, Z.: An efficient ant colony optimization approach to attribute reduction in rough set theory. Pattern Recognit. Lett. 29(9), 1351–1357 (2008)
11. Wang, X., Yang, J., Teng, X., Xia, W., Jensen, R.: Feature selection based on rough sets and particle swarm optimization. Pattern Recognit. Lett. 28(4), 459–471 (2007)
12. Inbarani, H.H., Azar, A.T., Jothi, G.: Supervised hybrid feature selection based on PSO and rough sets for medical diagnosis. Comput. Methods Programs Biomed. 113(1), 175–185 (2014)

13. Yang, X.-S.: Firefly algorithms for multimodal optimization. In: Watanabe, O., Zeugmann, T. (eds.) SAGA 2009. LNCS, vol. 5792, pp. 169–178. Springer, Heidelberg (2009)
14. Fister, I., Fister Jr., I., Yang, X.S., Brest, J.: A comprehensive review of firefly algorithms. Swarm Evol. Comput., 34–46 (2013)
15. Luthra, J., Pal, S.K.: A hybrid Firefly Algorithm using genetic operators for the cryptanalysis of a monoalphabetic substitution cipher. In: 2011 World Congress on Information and Communication Technologies (WICT), pp. 202–206 (2011)
16. Mohammadi, S., Mozafari, B., Solimani, S., Niknam, T.: An Adaptive Modified Firefly Optimisation Algorithm based on Hong's Point Estimate Method to optimal operation management in a microgrid with consideration of uncertainties. Energy 51, 339–348 (2013)
17. Dos, L., Coelho, S., Mariani, V.C.: Firefly algorithm approach based on chaotic Tinkerbell map applied to multivariable PID controller tuning. Comput. Math. Appl. 64(8), 2371–2382 (2012)
18. Kazem, A., Sharifi, E., Hussain, F.K., Saberi, M., Hussain, O.K.: Support vector regression with chaos-based firefly algorithm for stock market price forecasting. Appl. Soft Comput. 13(2), 947–958 (2013)
19. Long, N.C., Meesad, P.: Meta-heuristic algorithms applied to the optimization of type-1 and type 2 TSK fuzzy logic systems for sea water level prediction. In: 2013 IEEE Sixth International Workshop on Computational Intelligence Applications (IWCIA), Hiroshima, Japan, pp. 69–74 (2013)
20. Pawlak, Z.: Rough set approach to knowledge-based decision support. Eur. J. Oper. Res. 99(1), 48–57 (1997)
21. Shi, Z., Liu, S., Zheng, Z.: Efficient Attribute Reduction Algorithm. In: Bramer, M., Devedzic, V. (eds.) Artificial Intelligence Applications and Innovations. IFIP AICT, vol. 154, pp. 211–222. Springer, Heidelberg (2004)
22. Sayadi, M.K., Hafezalkotob, A., Naini, S.G.J.: Firefly-inspired algorithm for discrete optimization problems: An application to manufacturing cell formation. J. Manuf. Syst. 32(1), 78–84 (2013)
23. Chandrasekaran, K., Simon, S.P., Padhy, N.P.: Binary real coded firefly algorithm for solving unit commitment problem. Inf. Sci. 249, 67–84 (2013)
24. Kryszkiewicz, M.: Comparative studies of alternative type of knowledge reduction in inconsistent systems. Int. J. Intell. Syst. 16, 105–120 (2001)
25. Wang, G.: Rough reduction in algebra view and information view. Int. J. Intell. Syst. 18(6), 679–688 (2003)

A New Clustering Algorithm Based on Chameleon Army Strategy

Nadjet Kamel[1,2] and Rafik Boucheta[2]

[1] Univ-Setif, Fac-Sciences, Depart. Computer Science, Setif, Algeria
[2] USTHB, LRIA, Algiers, Algeria
nkamel@usthb.dz, rafik911@yahoo.fr

Abstract. In this paper we present a new clustering algorithm based on a new heuristic we call Chameleon Army. This heuristic simulates a Army stratagem and Chameleon behavior. The proposed algorithm is implemented and tested on well known dataset. The obtained results are compared to those of the algorithms K-means, PSO, and PSO-kmeans. The results show that the proposed algorithm gives better clusters.

Keywords: Clustering algorithm, K-means, PSO, metaheuristic.

1 Introduction

Clustering is grouping objects such that similar objects are within a same group, and dissimilar objects are in different groups. The main problem of clustering is to obtain optimal grouping. This issue arises in many scientific applications, such as biology, education, genetics, criminology, etc… Several approaches [1], [2], [3], [4], [5], [6], [7] have been developed in this regard.

Many clustering methods have been proposed, and they are classified into major algorithms classes: hierarchical clustering, partitioning clustering, density based clustering and graph based clustering.

In this paper we propose a new clustering algorithm based on a new heuristic we call Chameleon Army. This heuristic simulates an Army stratagem and some Chameleon behavior.

The efficiency of the proposed algorithm is tested on different datasets issued from literature [8]. The obtained results are compared with those of the algorithms kmeans [9], PSO [10], and PSO-kmeans.

The remaining of the paper is organized as follows: the next section presents the related works. The section 3 presents the new algorithm and its implementation. The results of our algorithm are presented in section 4. Finally, section 5 presents our conclusion and future works.

2 Related Works

Many clustering algorithms are defined in the literature. In general, they are classed into two classes: partitioned algorithms, and hierarchical algorithms. The kmeans

S. Boonkrong et al. (eds.), *Recent Advances in Information and Communication Technology*, Advances in Intelligent Systems and Computing 265,
DOI: 10.1007/978-3-319-06538-0_3,

algorithm [9] is the most popular clustering algorithm for its simplicity and efficiency (O(n)). Its main problem is its sensitivity to the initial cluster centroids. This influences seriously its optimal solution which may be local rather than global. Several solutions were proposed to overcome these shortcomings. In [11] the authors proposed the Global kmeans algorithm which is an incremental approach that dynamically adds one centroid at a time, followed by the processing of the kmeans algorithm until the convergence. In this algorithm, the centroids are chosen one by one in the following way: the first centroid is chosen to be the centroid of all the data. Other centroids are chosen in the data set where every data is a candidate to become a centroid. The latter centroid will be tested with the rest of the data set. The best candidate is the one who minimizes the objective function. In [12] the authors proposed to use the hierarchical clustering to choose the initial centroids. This approach proposes to use a hierarchical clustering at a first step, then we compute the centroids of each resulting cluster, and at the end we use the kmeans algorithm with the obtained results. In [13] the authors propose to use bootstrap to determine the initial centroids. It consists of dividing the data set into a set of samples and then we use the kmeans algorithm on each sample. Each cluster of the different groups produces a set of candidate centroids to be used for the initialization of the kmeans algorithm. Another approach proposed in [14], uses MaxMin method for the initialization. Many Hybridizations approaches are also used to face the problem of the initialization of kmeans. These approaches use metaheuristics such genetic algorithms [15], Particle Swarm Optimization (PSO) [10, 16], and Ant Colony Organization (ACO) [17, 18], to choose the initial centroids. Recently, nature inspired approaches have received increased attention from researchers dealing with data clustering problems [7].

3 Chameleon Army

In this section we present a new approach that we use to propose a new clustering algorithm. This approach is inspired from a strategy used by the militaries, and a tactic used by chameleons changing their color according to their states.

The implementation of the strategy consists in defining and in leading coherent actions occurring in sequence to reach one or several goals. At the operational level, it declines in action plans by domains and by periods, including possibly alternative plans usable in case of events changing strongly the situation.

The military strategy appears under two forms: the organizational level and the mode of conduct.

1. As mode of conduct, the strategy is a way of behaving in the uncertainty by including the uncertainty in the action conduct. So, it is close to guiles and stratagems.
2. At the organizational level, there is a political level which chooses between the war or the peace and attributes them necessary resources. This level directs and bounds the possible peace or war strategies with the resources attributed by the political level.

Chameleons form a family of reptiles. They are known for their varied colors and especially their capacity to quickly change it. This capacity is due to the presence of chromatophores, cells of skin were endowed with colored pigments. There are four types: black, blue, red and yellow. These various pigments are activated thanks to hormones. It is an instinctive mechanism of communication. So Chameleons indicate their state to their congeners: stress, gestation, disease, temperature, etc.

The combination of these two approaches leads to a group of soldiers which executes their strategy to reach one or several goals, and which can put their zone in red, blue or black, by meaning their state (good, average, bad), according to their position with regard to the goal and to the performance of the group which is evaluated after every movement. So the soldiers can communicate between them and help to reach the goal thanks to this characteristic.

3.1 The Chameleon Army Approach

The strategy is defined through two steps. At the first step, a first group A of soldiers is sent for a global evaluation of the ground. Once on the ground, the soldiers can estimate their situations, and specify the best points. These points will serve as references for the second step.

At the second step a second group *B* is sent to reference zones (neighborhood of the previous soldiers). Once there, they can ask the soldiers of their closest group. In this way each is going to try to follow the red zone (the most good), by taking account the blue zone (the good) and avoid the black zone (the closest). This process is repeated after every movement of the group with a new evaluation of every soldier. The evaluation of the performance of every soldier is performed according to the global performance of the group and to the looked goal (or the best solution found until now).

3.2 The Chameleon Army Algorithm

The approach described above is translated to be applied to the clustering algorithm. We propose a clustering algorithm based on the concepts of this approach. For this, we consider D the set on n objects, d the dimension of the space, pop the population of soldiers.

$$\mathrm{D} = [instance_1, \ldots, instance_n] \text{ where } instance_i = [attribut_1, \ldots, attribut_d]$$

$$pop = \begin{bmatrix} s_1 \\ \vdots \\ s_m \end{bmatrix} \quad m \text{ soldiers}$$

$$s_j = [centroid_1, \ldots\ldots\ldots, centroid_k] \text{ where } centeroid_l = [c_1, \ldots\ldots, c_d]$$

$$Best = [best_1, \ldots\ldots\ldots, best_k]$$

To evaluate the resulting clusters, different metrics have been defined in the literature. In this paper we use the following metrics:

1. *Intra class* given by the following formula is the variance of the points in the same cluster. It must be minimized.

$$J_w = \sum_g \sum_{i \in C_g} d^2(x_i, \mu_g) \tag{1}$$

where $d(x_i, \mu_g)$ is the distance between the objects x_i and the centroid μ_g.

2. Inter class is the variance of the cluster centroids. We maximize the following value:

$$J_b = \sum_c N_g d^2(u_g, \bar{x}) \tag{2}$$

The Davies and Bouldin [19] measures the within scatter and the separation of the clusters using the average of the similarity between clusters. It is defined as follows:

$$DB(C) = \frac{1}{K} \sum_{i=1}^{K} \max_{j = 1, \dots, K, i \neq j} \left(\frac{S(C_i) + S(C_j)}{d(\mu_i, \mu_j)} \right) \tag{3}$$

Where $d(\mu_i, \mu_j)$ represents the distance between the centroids of C_i and C_j, and $S(C_i)$ the average distance between each objet in C_i and its centroïd μ_i.

The soldier state is determined and evaluated after each moving of the group. We check if the soldier follows the group and contributes to reach the goal. If maxF and minF are respectively the bests and the worst performance in the group, we normalize the fitness of the group by the following formula:

$$NF_{i1} = \frac{fitness - minF}{maxF - minF}$$

Even the soldier performance in the group can be good, we aim at approaching the most possible the goal (best solution) BestF. By consequence, we evaluate its performance in the group according to the best solution. This last is normalized as follows:

$$NF_{i2} = \frac{fitness - minF}{BestF - minF}$$

The values NF_{i1} and NF_{i2} are in [0, 1]. The Table 1 illustrates an example of the color choice according to NF_{i1} and NF_{i2}.

Table 1. An example of color choice (state) for a soldier in the CA algorithm

Color		NF_{i1}		
		1/3	2/3	1
NF_{i2}	2/3	Black	Blue	Red
	3/4	Blue	Blue	Red
	1	Red	Red	Red

The soldiers in the black area are replaced to improve the search. Once the colors are updated, each soldier tries to follow, to approach or to avoid its neighbors. In this case the soldier will check the area around him to choice its orientation according to its neighbors. We can define this orientation by the following formula:

$$p' = Omega * p + Alpha * (Red - p) + Beta * (Blue - p) - Gamma * (Black - p) \quad (4)$$

Where p' and p are respectively, the new and the old position of the soldier.

Red and Blue are respectively the best positions of the best red and blue soldiers.

Black is the closest soldier.

These are parameters to be defined in order to have some flexibility:

Omega: [0.8, 1.2], *Alpha*: [0.8, 2], *Beta* : [0, 0.8], and *Gamma*: [0, 0.2]

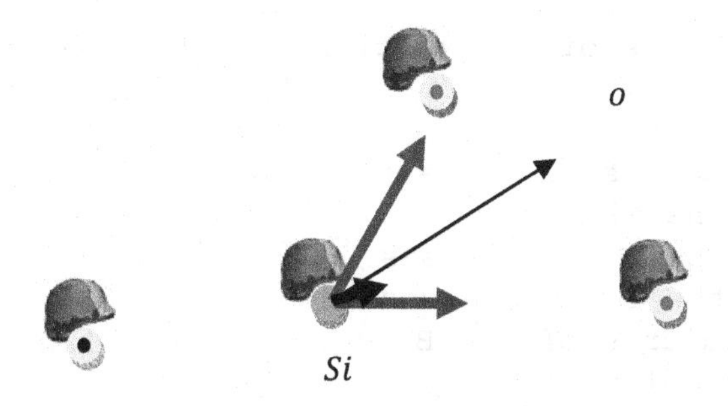

Fig. 1. Movement of soldier *Si* using the Neighbors information

To adapt this strategy to the clustering problem, we used the values of intra-classes and inter-classes given by the formulae (1) and (2). We have normalized the two values for each soldier according to the group. So, we have:

$$Nintra_i = \frac{Intra_i - minIntra}{maxIntra - minIntra}$$

$$Ninter_i = \frac{Inter_i - minInter}{maxInter - minInter}$$

The following table presents an example how to define the color of the soldier according to the $Nintra_i$ and $Ninter_i$.

Table 2. An example how to define the color of a soldier in the clustering problem

Color		$Nintra_i$		
		1/3	2/3	1
$Ninter_i$	1/3	Blue	Black	Black
	2/3	Red	Blue	Black
	1	Red	Blue	Black

Pseudo-code of the Chameleon Army Algorithm:

- **Step 1:**

```
Assign randomly the soldiers of the section A to th
search space
  For each soldier s of the section A do:
    Calculate the fitness          {using the formula (3)}
    Calculate the intra-classes    {using the formula (1)}
    Calculate the inter-classes    {using the formula (2)}
  For each soldier s of the section A do:
    Define its color
Best = s (best Fitness)
```

- **Step 2:**

```
Assign the soldiers of the section B to the areas red and
blue.
Repeat:
  For each soldier s of the section B
    Calculate the fitness
    Calculate the intra-classes
    Calculate the inter-classes
  For each soldier s of the B
    Define its color
    Define its new position using the formula (4)
  Update Best
Until  (condition satisfaction)
Return Best {the best found solution}
```

3.3 Implementation

We start by adjusting the parameters: Alpha, Beta, Gamma and Omega. To do this, we tested 108 combinations. For each one, we performed 10 executions. The combined values are: *Alpha*= [0.8, 1.0, 1.2, 1.5], *Beta* = [0.4, .6, .8, 1.0], *Gamma* = [0.1, .2], *Omega* = [0.9, 1, 1.1]

The figures 2 and 3 show an example of a simulation of the CA algorithm on the win dataset. We can see how the curve of fitness varies, and then decreases little by little. The point 14 represents the best solution where the intra classes is rather low and the inter classes is enough high. At this point the value of the Davies Bouldin fitness is the lowest. This means that the clusters are the most compact.

We choose the values: *Alpha* = 1, *Beta* = 0.4, *Gamma* = 0.2 and *Omega* = 1.

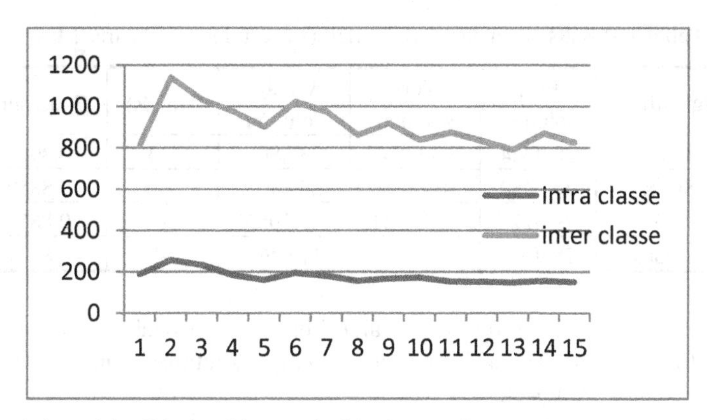

Fig. 2. Simulation of the CA algorithm on the iris dataset. Curves of the intra and inter classes.

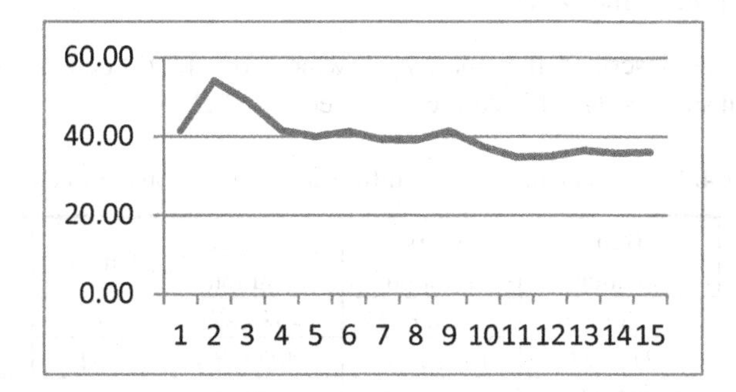

Fig. 3. Simulation of the CA algorithm on the iris dataset: curve of the Davies Bouldin fitness

4 Results Analysis

To analyze the results, we implemented our proposed CA algorithm, and the algorithms: K-means, PSO, and PSO-Kmeans. The tests are made on benchmarks of quantitative data (Iris, Wine, and Contraceptive Method Choice) [8]. We compared the results of the algorithms on these benchmarks.

The comparison of the results for every dataset is based on the solutions found by 100 different executions for every algorithm, and the processing time of convergence to reach the best solution. The quality of the respective clusters will be also compared using the F-measure. The results are presented in the Tables 3, 4, and 5.

4.1 Results on the Iris Dataset

The Iris dataset contains 150 objects with four attributes. They are unscrewed into 3 classes of 50 instances, where each class represents a type of iris plant.

Table 3. Results obtained for 100 different executions on the iris dataset

Algorithm	Best solution	Worst solution	Average solution	Time(s)	F-measure
CA	28.0264	43.6994	35.86	~ 4	0.8905
PSO-K	78.9450	100.3366	89.6408	~14	0.8852
PSO	79.6770	115.7341	97.7055	~7	0.8852
K-means	78.9408	149.4080	114.408	~1	0.8917

The best results are given by the CA algorithm at a reasonable time. It produces the lowest fitness values (Best, Worst, and Average solution), which means that the clusters are the most compact.

4.2 Tests on the Wine Dataset

The wine dataset describes the quality of wine from physicochemical properties. There are 178 instances with 13 features grouped into 3 classes.

Table 4. Results obtained for 100 different executions on the win dataset

Algorithm	Best solution	Worst solution	Average solution	Time(s)	F-measure
CA	3969.92	6009.823	4989.8717	~7	0.7209
PSO-K	13318.4813	14826.245	14072.764	~31	0.7147
PSO	13332.2429	14860.606	14096.424	~14	0.6811
K-means	13318.4813	14795.254	14056.867	~2	0.7147

We notice here also the CA algorithm can give very good results with a reasonable time (best time after K-means). It produces the lowest fitness values (Best, Worst, Average solution) which means that the clusters are the most compact.

4.3 Tests on the Contraceptive Method Choice (CMC) Dataset

The Contraceptive Method Choice dataset contains 1473 instances with 10 features grouped into 31 classes.

Table 5. Results obtained for 100 different executions on the CMC dataset

Algorithm	Best solution	Worst solution	Average solution	Time(s)	F-measure
CA	1741.102	2811.556	2276.329	~46	0.4262
PSO-K	23690.88	28257.268	25974.074	~213	0.4055
PSO	24957.095	31579.183	28268.139	~100	0.4050
K-means	23707.72	24294.710	24001.216	~23	0.4102

We notice that the CA algorithm gives the best solution. It produces the most compact clusters since its fitness values are the lowest (Best, Worst, and Average solution).

5 Conclusion

In this paper we have presented a new clustering algorithm based on a new metaheuristic inspired from a strategy used by the militaries and the behavior of chameleon changing their color according to their state. We have implemented our approach and evaluated it on a set of well known datasets. The results show that the new algorithm produces the most compact clusters. The evaluation was made using some indexes well known in the domain of clustering data.

As future work, we plane to hybridize our approach with other clustering algorithms to improve the results.

References

1. Agrawal, R., Gehrke, J., Gunopulos, D., Raghavan, P.: Automatic subspace clustering of high dimensional data for data mining applications. In: International Conference on Management data, pp. 94–105. ACM SIGMOD (1998)
2. Nagesh, H., Goil, S.: Choudhary: MAFIA. Efficient and scalable subspace clustering for very large data sets. Technical Report CPDC-TR-9906-010 (1999)
3. Sheikholeslami, G., Chatterjee, S., Zhang, A.: Wave cluster: A multi-resolution clustering approach for very large spatial databases. In: 24th International Conference on Very Large Data Bases, New York City, USA, pp. 428–439 (1998)
4. Kaufman, L., Rousseeuw, P.J.: Finding groups in data. An Introduction to Cluster Analysis. John Wiley & Sons (1990)
5. Sneath, P.H.A., Sokal, R.R.: Numerical Taxonomy: the Principles and Practice of Numerical Classification. W. H. Freeman and Company, San Francisco (1973)
6. Vazirani, V.V.: Algorithmes d'approximation. Collection IRIS. Springer (2006)
7. Colanzi, T.E., Assunção, W.K.K.G., Pozo, A.T.R., Vendramin, A.C.B.K., Pereira, D.A.B., Zorzo, C.A., de Paula Filho, P.L.: Application of Bio-inspired Metaheuristics in the Data Clustering Problem. Clei Electronic Journal 14(3) (2011)
8. Merz, C.J., Blake, C.L.: UCI Repository of Machine Learning Databases, `http://www.ics.uci.edu/-mlearn/MLRepository.html`
9. Jain, A.K.: Data clustering: 50 Years beyond K-means. Pattern Recognition Letters 31, 651–666 (2010)
10. Xiaohui, C., Potok, T.E.: Document Clustering Analysis Based on Hybrid PSO+K-means Algorithm. Applied Software Engineering Research Group, Computational Sciences and Engineering Division, Oak Ridge National Laboratory, Oak Ridge, TN 37831- 6085, USA (2005)
11. Likas, A., Vlassis, M., Verbeek, J.: The global k-means clustering algorithm. Pattern Recognition 36, 451–461 (2003)
12. Milligan, G.W.: The validation of four ultrametric clustering algorithms. Pattern Recognition 12, 41–50 (1980)

13. Bradley, P.S., Fayyad, U.M.: Refining initial points for K-Means clustering. In: 15th International Conf. on Machine Learning, pp. 91–99. Morgan Kaufmann, San Francisco (1998)
14. Mirkin, B.: Clustering for data mining: A data recovery approach. Chapman and Hall, London (2005)
15. Kwedlo, W., Iwanowicz, P.: Using Genetic Algorithm for Selection of Initial Cluster Centers for the K-Means Method. In: Rutkowski, L., Scherer, R., Tadeusiewicz, R., Zadeh, L.A., Zurada, J.M. (eds.) ICAISC 2010, Part II. LNCS (LNAI), vol. 6114, pp. 165–172. Springer, Heidelberg (2010)
16. Alireza, A., Hamidreza, M.: Combining PSO and k-means to Enhance Data Clustering. In: International Symposium on Telecommunication, Tehran, vol. 1 & 2, pp. 688–691 (2008)
17. Saatchi, S., Hung, C.-C.: Hybridization of the Ant Colony Optimization with the K-Means Algorithm for Clustering. In: Kalviainen, H., Parkkinen, J., Kaarna, A. (eds.) SCIA 2005. LNCS, vol. 3540, pp. 511–520. Springer, Heidelberg (2005)
18. Taher, N., Babak, A.: An efficient hybrid approach based on PSO, ACO and k-means for cluster analysis. Applied Soft Computing 10, 183–197 (2010)
19. Davies, D.L., Bouldin, D.W.: A cluster separation measure. IEEE Transactions on Pattern Analysis and Machine Intelligence 1(2), 224–227 (1979)

A Modified Particle Swarm Optimization with Dynamic Particles Re-initialization Period

Chiabwoot Ratanavilisagul and Boontee Kruatrachue

Department of Computer Engineering, Faculty of Engineering,
King Mongkut's Institute of Technology Ladkrabang, Bangkok, Thailand
chaibwoot@hotmail.com, booontee@yahoo.com

Abstract. The particle swarm optimization (PSO) is an algorithm that attempts to search for better solution in the solution space by attracting particles to converge toward a particle with the best fitness. PSO is typically troubled with the problems of trapping in local optimum and premature convergence. In order to overcome both problems, we propose an improved PSO algorithm that can re-initialize particles dynamically when swarm traps in local optimum. Moreover, the particle re-initialization period can be adjusted to solve the problem appropriately. The proposed technique is tested on benchmark functions and gives more satisfied search results in comparison with PSOs for the benchmark functions.

Keywords: Particle Swarm Optimization, Particles Re-initialization, Mutation operator, Multi-start Particles.

1 Introduction

Kennedy and Eberhart [1, 2] introduced Particle Swarm Optimization (PSO) in 1995. It is motivated from the behavior of flying bird and their communication mechanism in solving optimization problems. In comparison with several other population-based Stochastic optimization methods, such as genetic algorithms (GAs) and evolutionary programming (EP), PSO performs better in solving a variety of optimization problems with fast and stable convergence rate [3-5]. In addition, PSO is capable of solving the function optimization problems [6-8], artificial neural network training [9, 10], as well as pattern classification and fuzzy system control [11, 12].

PSO has many advantages such as rapid convergence, simplicity, and little parameters to be adjusted. Its main disadvantage is trapping in local optimum and premature convergence. To overcome both problems, many researchers [13-19] increased searching diversity in the population of PSO by adding the particle re-initialization or the mutation operators in the process of PSO. The result showed that both methods can increase PSO optimization performance.

However, the particle re-initialization encounter parameters adjusting to suitably deal with any given problem. As inappropriate parameters could contribute to poorer

S. Boonkrong et al. (eds.), *Recent Advances in Information and Communication Technology*, Advances in Intelligent Systems and Computing 265,
DOI: 10.1007/978-3-319-06538-0_4, © Springer International Publishing Switzerland 2014

results vis-à-vis the standard PSO. The essential parameter of particle re-initialization is the particles re-initialization period (re-initialization period) which is the amount of time to be certain that the swarm trapped in local optimum and the reinitialized particles with PSO should be performed.

Researchers proposed various ways to define re-initialization period using the number of iteration with the best position unchanged [14], and the number of iteration with the consecutive neighborhood best position unchanged [15].

This paper investigates adjusting re-initialization period to suitably deal with any given problem applied with PSO. Normally, the reinitialized particles (resetting) should occur when swarm traps in local optima. Hence, this paper proposes the re-initialization period defined by the unchanged number of consecutive best particle which indicates the state of local optima trapping. Moreover, re-initialization period is automatically adjusted according to the difference of the best solution found after resetting and the previous best solution. A set of benchmark test functions is used to compare the standard PSO, the proposed PSO algorithm, PSO algorithm's [14], and PSO algorithm's [15]. The results show that the proposed PSO algorithm obtains the best results in all test functions.

The rest of this paper is organized as follows. Section 2 explains basic PSO and PSO with particles re-initialization. Section 3 explains PSO with dynamic particles re-initialization period. Section 4 explains the benchmark functions, the experiment setup and presents the experiment results. Section 5 concludes the paper with a brief summary.

2 Related Work

2.1 Particle Swarm Optimization

In standard PSO, each member of the population is called a "particle" with its own position and velocity. Each individual particle performs searching in the search space according to its velocity, the best position found in the whole swarm (GBEST) and the individual's best position (PBEST). The standard PSO algorithm starts with randomizing particle positions and their respective velocities, and the evaluation of the position of each particle is achieved using the objective function of the optimization problem. In a given iteration, each individual particle updates its position and velocity according to the expression below:

$$V_{id}' = \varpi V_{id} + \eta_1 rand()(P_{id} - X_{id}) + \eta_2 rand()(P_{gd} - X_{id}) \tag{1}$$

$$X_{id}' = X_{id} + V_{id}' \tag{2}$$

Where X_{id}' denotes the current positions of i particle and d dimension, X_{id} the previous positions of i particle and d dimension, V_{id} the previous velocity of i particle and d dimension, V'_{id} the current velocity of i particle and d dimension, P_{id} PBEST of i particle and d dimension, and P_{gd} GBEST of d dimension. $0 \le \varpi < 1$ is an inertia weight, η_1 and η_2 are acceleration constants, and rand() generates random number from interval [0,1]. A limit velocity calls V_{max}. If calculate velocity of a particle exceeds this value, it will replace value of V_{max}.

2.2 PSO with Particles Re-initialization

The particle re-initialization techniques prevent premature convergence and trapping in local optimum. They create a variation of a population. Many researchers proposed various techniques such as:

Den Bergh [13] proposed Multi-start PSO (MPSO). The concept of this algorithm can be summarized as follows: after some iteration with the maximum swarm radius less than some threshold value, all particles of swarm are initialized in order to increase the diversity of swarm and expand the searching space. This algorithm can jump out of the local minimum and converge to the new area. However, this technique destroys the current structure of the swarm with total re-initialization of particles which results in decreasing the convergence speed and the search accuracy [14, 15].

In fact, the swarm structure is not totally destroyed because GBEST is not reset. As a result, the re-initialized swarm is position very far from the GBEST and will rapidly converge to the GBEST. If swarm can locate a better position than the previous GBEST, the better position is then occupied as a new GBEST. However, MPSO has difficulty in adjusting suitably re-initialization period. Applying MPSO with general problems is complicated. In addition, the results from non-reset PBEST contribute to non-distributed searching with rapid convergence, so MPSO is rarely improve the searching performance of PSO.

Ning Li et.al. [14] proposed PSO with mutation by considering from the Euclidean distance and GBEST (PSOMEG). The concept of this algorithm can conclude as follows: The velocity and position of particles are initialized depend on mutation probability. The initialization occur when GBEST unchanged more than some the threshold value (MaxStep) and the maximum Euclidean distance between all the particles is less than the threshold value (BorderRadius). This technique improves the diversity of swarm without decreasing the convergent speed and the search accuracy. However, PSOMEG has three additional parameters from standard PSO. MaxStep must be in some range, otherwise the results is poor. BorderRadius is very difficult to set with any given problem. Mutation probability can be set to over-mutation which results in similar searching to MPSO. On the other hand, set to under-mutated results in remain trapping. The difficulty of parameters setting reduces the practicality of PSOMEG with general problems.

Xuedan Liu et.al. [15] proposed PSO with mutation by considering from the neighborhood's best position (PSOMN). In this paper, the velocity and position of particles is initialized when the unchanged numbers of the consecutive neighborhood best position are more than some threshold value (MaxStep). PSOMN increase the global search capability and avoid premature convergence. However, selection of MaxStep is recommended within some range to obtain good results.

3 PSO with Dynamic Particles Re-initialization Period

As previously mentioned, the main problem of applying particles re-initialization technique is parameters adjusting. To avoid this problem, the proposed algorithm automatically adjusts its parameters. In addition, the number of parameters is kept minimal.

The trapping in local optimum is possible for standard PSO, which leads to stagnation of the searching in which the solution (GBEST) obtained at the point of trapping, is repeatedly produced irrespective of the length of searching time. Therefore, the unchanged number of consecutive GBEST can indicate state of swarm when swarm traps in local optimum.

Normally, the particles re-initialization (resetting) should be used when swarm trap or trend to trap in local optimum in order to distribute particles to search in other area. The easy techniques to indicate the trapping state of the swarm is to monitor the unchanged number of consecutive GBEST. Thereby, this paper proposed the particles re-initialization period (re-initialization period) is considered from the unchanged number of consecutive GBEST.

If re-initialization period was set too small, particles will be reset before convergence. On the other hand, if re-initialization period was set too large, particles remain trapping in local optimum instead of search in other areas.

Therefore, this paper proposes the novel technique which can decrease problem of adjusting re-initialization period. The proposed technique can adjust the re-initialization period to solve any optimization problems appropriately. The concept of proposed technique is explained as follows: Before reset, if the best solution in current resetting round is worse than previous round GBEST, particles could not converge to the GBEST. Hence, the re-initialization period should be increased to enhance convergence time of particles to the GBEST in next resetting round. On the other hand, if the best solution in current resetting round equals GBEST, particles can converge to the GBEST. In this case, the re-initialization period should be decreased to reduce the searching time in local optimum. In addition, the proposed technique choose to initialize all particles to improve chance in obtain better solution than initializing some particles. Because the more the number of particles reset, the better chance in obtaining the new GBEST while particles return to GBEST. Furthermore, PBEST of all particles should be reset to enhance distribution of particles. The proposed technique is called PSO with dynamic particles re-initialization period (PSODR). Pseudo code of PSODR is shown below:

```
Initial position and velocity of each particle
Initial PBEST GBEST RBEST
While termination condition ≠ true do
  For each particle
    Evaluate particle fitness
    If fitness of each particle is better than that of PBEST
      Update PBEST (P_i = X_i)
    End If
    If fitness of each particle is better than that of GBEST
      Update GBEST (P_g = X_i)
    End If
    If fitness of each particle is better than RBEST
      Update RBEST (RBEST= X_i)
      Set the times of RBEST consecutive unchanged to 0
    End If
    Update particle position according to formula (1) and (2)
  End For
  Times of RBEST consecutive unchanged ++;
  If the times of RBEST consecutive unchanged >= PRP
    If fitness of RBEST is worse than fitness of GBEST
      PRP = PRP + NPRP
    Else
      PRP = PRP - NPRP
    End If
    If PRP is less than the MINPRP
      PRP = MINPRP
    Else If PRP is more than the MAXPRP
      PRP = MAXPRP
    End If
    Reset RBEST
    Reinitialize PBEST, velocity, and position of all particles
    Set the times of GBEST consecutive unchanged to 0
  End If
End while
```

PRP is the particle re-initialization period (re-initialization period). RBEST is the best solution in each resetting round. NPRP is adjusting round number of the particle re-initialization period (this is the dynamic adjustment of PRP). MINPRP is the minimum value of the particle re-initialization period. MAXPRP is the maximum value of the particle re-initialization period.

4 Experiments and Results

4.1 Benchmark Functions

The proposed algorithm is tested on nine well-known benchmark functions, listed in Table 1. These functions consists of six multimodal functions from function one to six. The remaining three functions are unimodal functions. The results of PSODR on the benchmark functions are compared with PSO [1, 2], PSOMEG [14] and PSOMN [15]. To guarantee fairness, the inertia weigh of PSOMN algorithm cannot be adapted as written in original paper. The inertia weigh of PSOMN algorithm is fixed as other algorithms.

4.2 Parameters Setting

Parameters are as follows for all experiments: acceleration constants of η_1 and η_2 are both set to be 1.496180 and inertia weight ω = 0.729844, as suggested by den Bergh [13], the number of population is 20. The number of experiments of each function is 100 runs. The maximum iteration is 5000 independently of number of resetting. Hence, the algorithm that locates local optimum faster has more number of trial runs.

The non-PSO parameters are as follows: For PSODR algorithm, MINPRP = 10, MAXPRP = 1000, NPRP = 50, PRP = 200. The compared algorithm parameters are set according to suggested by the original papers. For PSOMEG algorithm, the mutation probability = 0.04, MaxStep = 10, the divided number in calculation of BorderRadius is 1000. For PSOMN algorithm, MaxStep = 30, the neighborhood length = 2.

Table 1. Details of Benchmark Test Functions

Function name	Expression
Ackley	$f(x) = -20\exp(-0.2\sqrt{\frac{1}{n}\sum_{i=1}^{n} x_i^2}) - \exp(\frac{1}{n}\sum_{i=1}^{n}\cos(2\pi x_i)) + 20 + e$
Griewank	$f(x) = \sum_{i=1}^{n}\frac{x_i^2}{4000} - \prod_{i=1}^{n}\cos(x_i/\sqrt{i}) + 1$
Rastrigrin	$f(x) = 10n + \sum_{i=1}^{n}(x^2 - 10\cos(2\pi x_i))$
Rosenbrock	$f(x) = \sum_{i=1}^{n}[100(x_i^2 - x_{i+1}) + (x_i - 1)^2]$
Schwefel	$f(x) = 418.9829 \times n + \sum_{i=1}^{n}(x_i \times \sin(\sqrt{\lvert x_i \rvert}))$
Exponential	$f(x) = -\exp(-0.5\sum_{i=1}^{n} x_i^2) + 1$
Sphere	$f(x) = \sum_{i=1}^{n} x_i^2$
Parallel Ellipsoid	$f(x) = \sum_{i=1}^{n}(i \times x_i^2)$
Rotated Ellipsoid	$f(x) = \sum_{i=1}^{n}(\sum_{j=1}^{i} x_j^2)$

Table 1. *(continued)*

Function name	Search space $[X_{max}, X_{min}]$	Objective function value	Dim.	V_{max}
Ackley	$x \in [-32.768, 32.768]^n$	0	40	32.768
Griewank	$x \in [-300, 300]^n$	0	40	300
Rastrigin	$x \in [-5.12, 5.12]^n$	0	40	5.12
Rosenbrock	$x \in [-2.048, 2.048]^n$	0	40	2.048
Schwefel	$x \in [-500, 500]^n$	0	40	500
Exponential	$x \in [-1, 1]^n$	0	40	1
Sphere	$x \in [-5.12, 5.12]^n$	0	40	5.12
Parallel Ellipsoid	$x \in [-5.12, 5.12]^n$	0	40	5.12
Rotated Ellipsoid	$x \in [-65.536, 65.536]^n$	0	40	65.536

This research is conducted by a personal computer of Intel Core i5 2450 with a 2.5-GHz CPU and 8 GB RAM, and Visual C++ as the programming language. The measures of algorithm performance in the experiments are as follows: The mean best fitness value (MBF) is the mean of best fitness in the final iteration from all running (100 runs). MBF indicates the solution searching efficiency of an algorithm. All experiments benchmark functions have zero results as a minimum point. In Table 1, an entry less than 10^{-324} is given the value of zero. The closer the MBF to the zero point of a method, the better the method. The mean particles re-initialization period (MPRP) is the mean of particles re-initialization period in the final iteration from all running. MPRP is the result of the calculation of PSODR algorithm. The mean of time per mean of fitness (TPF) is the mean of time in running multiply the mean of fitness of GBEST in the final iteration from all running. The closer the TPF to the zero point of a method, the better the method. Because, this method loses a less time in the solution searching and obtains a good solution. SD is the standard deviation.

Table 2. The results of PSODR by fixed PRP on Benchmark Test Functions averaged over 100 runs

Techniques	PSO	PSODR	PSODR	PSODR	PSODR	PSODR	PSODR
PRP	-	2	10	100	200	500	1000
Problems	MBF	MBF	MBF	MBF	MBF	MBF	MBF
Ackley	5.87489	5.20859	2.23955	2.16283	2.22361	2.89651	3.23259
Griewank	0.15754	1.59019	0.044938	0.0445354	0.048823	0.046841	0.052307
Rastrigin	158.724	332.885	135.183	139.055	145.385	146.352	150.278
Rosenbrock	9.67E-22	31.8709	2.10E-12	3.95E-24	4.89E-23	2.08E-20	1.37E-20
Schwefel	3979.11	3429.55	3143.13	3175.25	3102.46	3338.6	3614.64
Exponential	2.56E-15	0.0126073	2.92E-15	2.59E-16	2.24E-16	2.22E-16	2.45E-16
Sphere	3.66E-37	0.698723	2.13E-18	7.98E-35	2.43E-36	4.77E-37	2.04E-37
Parallel Ellipsoid	4.18E-36	30.2192	2.76E-14	5.29E-37	1.23E-36	3.28E-34	4.35E-37
Rotated Ellipsoid	2.42E-37	4651.77	2.94E-12	9.04E-35	1.20E-34	5.28E-37	2.30E-36

Table 3. Comparative results of PSO, PSOMEG, PSOMN, and PSODR on Benchmark Test Functions averaged over 100 runs

Techniques	PSO	PSOMEG	PSOMN	PSODR	
Problems	MBF	MBF	MBF	MBF	MPRP
Ackley	5.87489	5.6928	3.2664	2.21957	103
Griewank	0.15754	0.133446	0.103264	0.04741	78
Rastrigin	158.724	153.939	148.766	138.981	67.8
Rosenbrock	9.67E-22	1.49E-19	2.60E-20	6.47E-23	192
Schwefel	3979.11	3910.7	3857.61	3091.22	66
Exponential	2.56E-15	1.15E-13	7.08E-14	2.71E-16	43.5
Sphere	3.66E-37	8.32E-36	1.51E-34	3.42E-37	200
Parallel Ellipsoid	4.18E-36	1.55E-34	1.04E-33	1.84E-36	200
Rotated Ellipsoid	2.42E-37	4.51E-36	9.08E-35	1.77E-37	200

Table 4. Comparative results of SD and TPF of PSO, PSOMEG, PSOMN, and PSODR on Benchmark Test Functions averaged over 100 runs

Techniques	PSO		PSOMEG		PSOMN		PSODR	
Problems	SD	TPF	SD	TPF	SD	TPF	SD	TPF
Ackley	2.7838	1.2855	2.6532	1.3634	3.104	0.8522	0.5379	0.5029
Griewank	0.39	0.03535	0.1375	0.03749	0.2525	0.02883	0.04894	0.0108
Rastrigin	29.4346	33.4701	29.1122	33.143	31.5726	32.2524	25.87	29.3666
Rosenbrock	7.46E-13	1.54E-22	1.52E-20	2.48E-20	9.69E-15	4.24E-21	3.34E-21	1.03E-23
Schwefel	1063.64	1001.14	975.27	1006.73	955.22	1006.0646	933.454	791.661
Exponential	2.78E-14	3.67E-16	7.40E-14	1.66E-14	1.63E-12	1.02E-14	7.44E-17	3.912E-17
Sphere	1.78E-31	5.08E-38	1.59E-36	1.18E-36	6.44E-33	2.17E-35	3.47E-38	4.8E-38
Parallel Ellipsoid	1.41E-33	5.91E-37	2.64E-35	2.24E-35	2.37E-26	1.4823E-34	1.42E-36	2.622E-37
Rotated Ellipsoid	1.07E-35	5.53E-38	2.10E-35	1.03E-36	1.13E-28	2.0893E-35	3.20E-36	3.9205E-38

From the experimental results of Table 2 show both unimodal functions and multimodal functions, if the setting re-initialization period is 2, swarm will be over-reset. Its searching solution is poorer than that of the standard PSO, such as Griewank function, Rastrigin function, Rosenbrock function, Exponential function, Sphere function, Parallel Ellipsoid function, and Rotated Ellipsoid function. This is due to particles hardly converge. In the case of, re-initialization period is more than or equal to 100 in unimodal functions, overall its searching solutions are similar to that of the standard PSO since the particles re-initialization algorithms is not executed. For multimodal problems, if the setting re-initialization period is 1000, swarm will be under-reset. Its searching solution is poorer than setting re-initialization period to 100. Since swarm wastes time searching in local optimum rather than starting new search in other areas.

For multimodal problems in Table 2, PSODR gets good results because of the re-initialization period setting ranged from 10 to 200. The PSODR algorithm can adjust re-initialization period in this range and it can obtain the good solutions as shown in Table 3. Therefore, the PSODR algorithm can dynamically adjust to suitably deal with the problem.

From the experimental results of Table 3 show that PSODR outperforms PSO, PSOMEG, and PSOMN in multimodal function. PSODR produces a better quality solution because of its lowest MBF in multimodal function. For unimodel function, PSO and PSODR have similar overall results. The MBF results of PSO and PSODR are higher than that of PSOMEG and PSOMN because PSOMEG and PSOMN could execute the particles re-initialization which will hinder the convergence of particles in unimodel function. For multimodal function, some MBF results of PSOMEG and PSOMN is poorer than the standard PSO such as Exponential function and Rosenbrock function. If adjusting re-initialization period to deal with problems appropriately, the solutions of PSOMEG and PSOMN can obtain better than that of the standard PSO. However, adjusting re-initialization period of PSOMEG and PSOMN to deal with any given problem appropriately is complicated.

From the experimental results of Table 4 show that the reliability of PSODR is better than that of PSO, PSOMEG, and PSOMN because of its lowest SD in all test functions. Moreover, PSODR obtains the best results when the results from the solution searching are compared with lost time in the solution searching. Because, TPF of PSODR is less than that of PSO, PSOMEG, and PSOMN in all test functions.

5 Conclusion

The particles re-initialization is applied with PSO to solve problems of trapping in local optimum and premature convergence. However, applying the particles re-initialization with general problems is complicated which is the cause of complicatedly adjusting the particle re-initialization period in order that the particles re-initialization can enhance the solution searching performance and can be easily applied to general problems. This paper proposes the re-initialization period that is automatically adjusted according to the difference of the best solution found after resetting and the previous best solution. The proposed technique is called PSODR. On benchmark functions, the proposed PSODR, PSO, PSOMEG and PSOMN are tested and the results are compared. From the experimental results show that PSODR can adjust the particle re-initialization period to solve the problem appropriately and PSODR outperforms PSO, PSOMEG, and PSOMN.

References

1. Kennedy, J., Eberhart, R.C.: Particle Swarm Optimization. In: IEEE International Conference on Neural Networks, pp. 1942–1948 (1995)
2. Kennedy, J., Eberhart, R.C.: A New Optimizer Using Particle Swarm Theory. In: Proceedings of the 6th International Symposium on Micro Machine and Human Science, pp. 39–43 (1995)
3. Parsopoulos, K.E., Plagianakos, V.P., Magoulas, G.D., Vrahatis, M.N.: Objective function stretching to alleviate convergence to local minima. In: Nonlinear Analysis TMA, vol. 47, pp. 3419–3424 (2003)

4. Eberhart, R.C., Shi, Y.: comparison between genetic algorithms and particle swarm optimization. In: Porto, V.W., Waagen, D. (eds.) EP 1998. LNCS, vol. 1447, pp. 611–616. Springer, Heidelberg (1998)
5. Kennedy, J., Eberhart, R.C.: Swarm Intelligence. Morgan Kaufmann, San Mateo (2001)
6. Pappala, V.S., Erlich, I., Rohrig, K., Dobschinski, J.A.: A stochastic modal for the optimal operation of a wind–thermal power system. IEEE Trans. Power Syst., 940–950 (2009)
7. Acharjee, P., Goswamj, S.K.: A decoupled power flow algorithm using particle swarm optimization technique. Energy Convers. Manage, 2351–2360 (2009)
8. Dutta, S., Singh, S.P.: Optimal rescheduling of generators for congestion management based on particle swarm optimization. IEEE Trans. Power Syst. (2008)
9. Kiranyaz, S., Ince, T., Yildirim, A., Gabbouj, M.: Evolutionary artificial neural networks by multi-dimensional particle swarm optimization. Neural Networks, 1448–1462 (2009)
10. Wei, H.L., Billings, S.A., Zhao, Y.F., Guo, L.Z.: Lattice dynamical wavelet neural networks implemented using particle swarm optimization for spatio-temporal system identification. IEEE Trans. Neural Network, 181–185 (2009)
11. Lin, C.J., Chen, C.H., Lin, C.T.: A hybrid of cooperative particle swarm optimization and cultural algorithm for neural fuzzy networks and its prediction applications. IEEE Trans. Syst. Man Cybernetics Part C, 55–68 (2009)
12. Zhao, L., Qian, F., Yang, Y., Zeng, Y., Su, H.: Automatically extracting T-S fuzzy models using cooperative random learning particle swarm optimization. Appl. Soft Comput., 938–944 (2010)
13. Van Den Bergh, F.: An Analysis of Particle Swarm Optimizers. PhD thesis, Department of Computer Science, University of Pretoria, South Africa (2002)
14. Li, N., Suan, D., Cen, Y., Zou, T.: Particle swarm optimization with mutation operator. Computer Engineering and Applications 17, 12–14 (2004)
15. Liu, X., Wang, Q., Liu, H., Li, L.: Particle Swarm Optimization with Dynamic Inertia Weight and Mutation. In: Third International Conference on Genetic and Evolutionary Computing, pp. 620–623 (2009)
16. Chiabwoot, R., Boontee, K.: Mutation Period Calculation for Particle Swarm Optimization. In: 1st International Symposium on Technology for Sustainability, pp. 213–216 (2011)
17. Andrews, P.S.: An investigation into mutation operators for particle swarm optimization. In: Proceedings of the IEEE Congress on Evolutionary Computation, pp. 1044–1051. IEEE, Vancouver (2006)
18. Lin, M., Hua, Z.: Improved PSO Algorithm with Adaptive Inertia Weight and Mutation. In: 2009 World Congress on Computer Science and Information Engineering, pp. 622–625 (2009)
19. Higashi, N., Iba, H.: Particle swarm optimization with Gaussian mutation. In: Proc. of the 2003 IEEE Swarm Intelligence Symphosium, pp. 72–79 (2003)

An Ensemble K-Nearest Neighbor with Neuro-Fuzzy Method for Classification

Kaochiem Saetern and Narissara Eiamkanitchat

Department of Computer Engineering, Faculty of Engineering,
Chiang Mai University, Thailand
kaochiem97@gmail.com, narisara@eng.cmu.ac.th

Abstract. This paper introduces an *ensemble k-nearest neighbor with neuro-fuzzy* method for the classification. A new paradigm for classification is proposed. The structure of the system includes the use of neural network, fuzzy logic and k-nearest neighbor. The first part is the beginning stages of learning by using 1-hidden layer neural network. In stage 2, the error from the first stage is forwarded to Mandani fuzzy system. The final step is the defuzzification process to create new dataset for classification. This new data is called "*transformed training set*". The parameters of the learning process are applied to the test dataset to create a "*transformed testing set*". Class of the transformed testing set is determined by using k-nearest neighbor. A variety of standard datasets from UCI were tested with our proposed. The fabulous classification results obtained from the experiments can confirm the good performance of *ensemble k-nearest neighbor with neuro-fuzzy* method.

Keywords: Ensemble k-nearest neighbor, Neuro-fuzzy, Classification, Transformed data.

1 Introduction

The development of classification techniques gets the tremendous interest among researchers. The main reason is this technique can be applied in many work areas. There are several popular classification algorithms such as decision tree, Naïve Bayesian, support vector machine, neural network and fuzzy logic. Each method has different strengths and weaknesses. The neural network algorithms are outstanding on the noise robustness. Using back-propagation learning is the main reason, because parameters are updated from errors of the individual data. Class prediction calculated from test data with parameters derived from the learning phase. The drawback of general neural networks are usually referred to as the black box that incomprehensible to humans [1-3]. In the fuzzy system also has a variety of structures such as Mamdani, Takagi-Sugeno, etc. Fuzzy logic commonly used to analyze the problems of computational intelligence because it has allowed boundaries of the decision. Support the creation of linguistic variables and processes using fuzzy rule based which are easy to understand. The weakness of the fuzzy system is a fuzzy rule is not robust and it gives equal priority to all factors.

S. Boonkrong et al. (eds.), *Recent Advances in Information and Communication Technology*, Advances in Intelligent Systems and Computing 265,
DOI: 10.1007/978-3-319-06538-0_5, © Springer International Publishing Switzerland 2014

Over the years to the present, many researchers have combined the advantages of these algorithms to propose several interesting methods of data analysis. In [4-5], researchers presented a new neuro-fuzzy model that yielded a high classification performance. The model uses fuzzy membership functions to create linguistic variables then sent to the learning phase by using neural network. A fuzzy evolutionary techniques for analysis unknown input with clustering by fuzzy technique on the decision process phase was proposed in [6]. Fuzzy technique is able to determine and search the final class of unknown input by aggregation on winner prototype and its neighbors. Aggregation fuzzy integral technique and SOM is successful to improve performance of the classification. In [7] the new Mamdani fuzzy classification is developed by combining with genetic algorithm. The genetic algorithm is performed to optimize the number of samples, classification error, a number of the features, and the number of fuzzy rules of the classifier by selecting significant features. This new method could generate better classification performance. Some research has been applied the fuzzy prediction with the actual operational data. The results showed that the fuzzy analysis is a good technique for many types of data in a typical scenario, such as financial markets [8], signature series [9] and others.

A variety of methods discussed previously developed using different measures. Such criteria, for example, speed, complexity, robustness, and interpretation. The most important is the accuracy of the classification. The method proposes in this paper based on two main criteria to determine the direction of algorithm development. We focus on improving a robust and highly accurate classification. An ensemble of k-nearest neighbor (KNN) with neuro-fuzzy method using neural network and fuzzy logic features are combined in the process of learning. This method is robust because the decision class of data derived from the calculation. In addition the new methods and principles are developed for the testing process by using parameters derived from the learning process to create "*transformed dataset*". Applications of KNN in deciding the classes of the information presented here is simple, uncomplicated and has very high accuracy. The next part of this paper is organized as follows. The concept and structure of the model described in section 2. Section 3 is the experimental methods and results of experiments to demonstrate the effectiveness of the model. Finally in the last section is the conclusion of the experimental results.

2 Structure of Ensemble K-Nearest Neighbor with Neuro-Fuzzy

The new paradigm presented here using neuro-fuzzy algorithms for learning. This approach has two-steps of sequence in learning process. The first learning step is using a neural network to update parameters and stored. In the second learning process, errors from the first step are used to create Mandani fuzzy rule based. Decision process presents a method to create transformed data and the application of KNN to identify classes of data. All procedures detailed explanation of each part, as shown in the following sections.

2.1 Learning Process of Neuro-Fuzzy

The First Step. This section is intended to apply the advantages of neural networks to learn and simultaneously improve the significance of each feature. The structure of 1 hidden layer with the concept of back propagation is used. Number of nodes in the hidden layer are set to 3 nodes, all nodes use sigmoid as activation function. Figure 1 shows the structure of the neural network in our proposed algorithm.

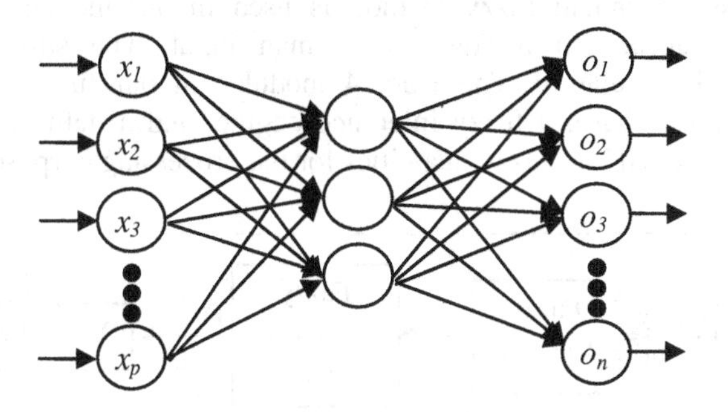

Fig. 1. The structure of neural network in learning phase

The first layer is the input layer, number of nodes in this layer equal to the dataset features. All data are fed forward to 3 nodes hidden layer. The results from hidden layer are fed to output layer which has nodes equal to the number of classes in the dataset. In the initial state all connection weights and biases are randomly defining. The learning rate (η) is set to 0.05. Considering output layer the error function of node j at iteration n is defined as

$$F_j(n) = d(n) - o_j(n) \tag{1}$$

The $F_j(n)$ is error of node j, $d(n)$ is desired output and $o_j(n)$ is the output from network. After completion of learning process, the errors $F_j(n)$ are used as input data for fuzzy phase. The total error energy is able to compute by equation (2).

$$E(n) = \frac{1}{2}\sum_j I_j^2(n) \tag{2}$$

The weights are updated with delta rule as

$$\Delta w_{ij}(n) = \eta \delta_j(n) o_j(n) . \tag{3}$$

The gradient ($\delta_j(n)$) value of output layer is defined by equation (4) where $\varphi'_j(o_j(n))$ is derivative of the associated activation function.

$$\delta_j(n) = I_j(n)\varphi'_j(o_j(n)) \tag{4}$$

For nodes in hidden layer the gradient values are derived from equation (5).

$$\delta_j(n) = \varphi'_j(o_j(n))\sum_k \delta_k(n) w_{kj}(n) \tag{5}$$

The $w_{kj}(n)$ is weight that consists of the weights associated with the considered node. The learning phase is stopped either maximum epoch or the defined total error energy value is reached.

The Second Step. Mandani fuzzy method is used in our model. It is intuitive, widespread acceptance, and it suited to human input. The structure is easily understanding and calculation. There are 4 modules in our fuzzy system, i.e., a fuzzifier, a fuzzy rule base, a fuzzy inference engine, and a defuzzifier. The crisp inputs are first converted into fuzzy quantities for the processing purposes.

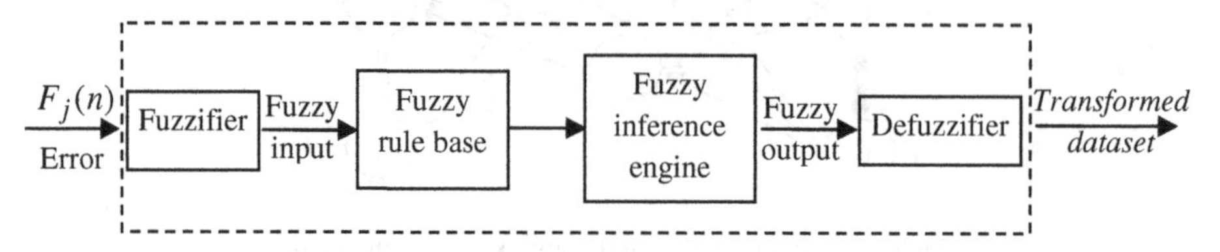

Fig. 2. The structure of Fuzzy system in learning phase

The errors from first step are used as input for this step. The numbers of errors are equal to the number of classes of the dataset, considered as features input. According to scatter plot analysis all errors can be separated into 2 groups {"high", "low"}. The scatter plots of "Lung cancer" errors are illustrated in Figure 3. There are 3 classes in Lung cancer row one in Figure 3 shows the distribution of errors from node 1 versus errors from itself, node 2 and node 3 respectively.

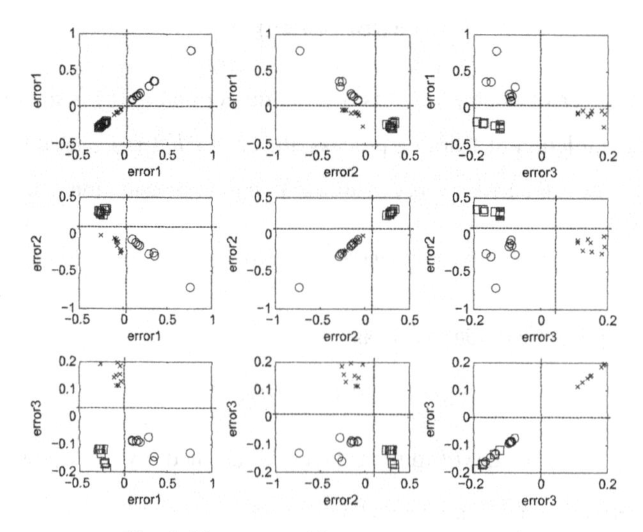

Fig. 3. The errors of Lung cancer scatter plot

The separation point of each error is defined in equation (6).

$$S_j = \frac{\max\{g_j\} + \min\{g_j\}}{2} \tag{6}$$

The value S_j is a threshold that calculates from the average maxim and minimum value in each group of error (g_j). As shown in Figure 3 and Figure 4, it can be interpreted as a high error group or low error group, regardless of the number of data classes. The Trapezoidal-shaped membership function is used in fuzzifier process. The Triangular-shaped membership function is used for fuzzifier output of training dataset. The graphs of membership function are shown in Figure 4.

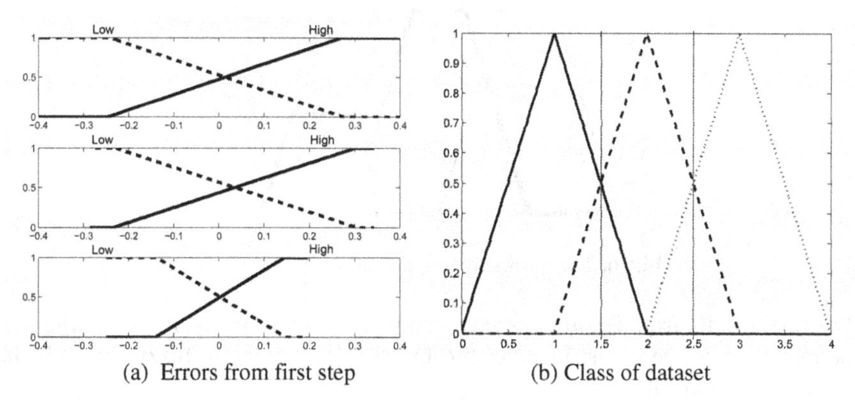

(a) Errors from first step (b) Class of dataset

Fig. 4. The membership function graphs of Lung cancer

The errors from learning process in first step are changed to fuzzy set according to fuzzy membership value. The new dataset is used to create a fuzzy rule relevant to classes. An example of the fuzzy rule base and classes are displayed in Figure 5. On the left hand side are the rules of the Lung cancer dataset. The membership graphs of each rule are shown on the right hand side.

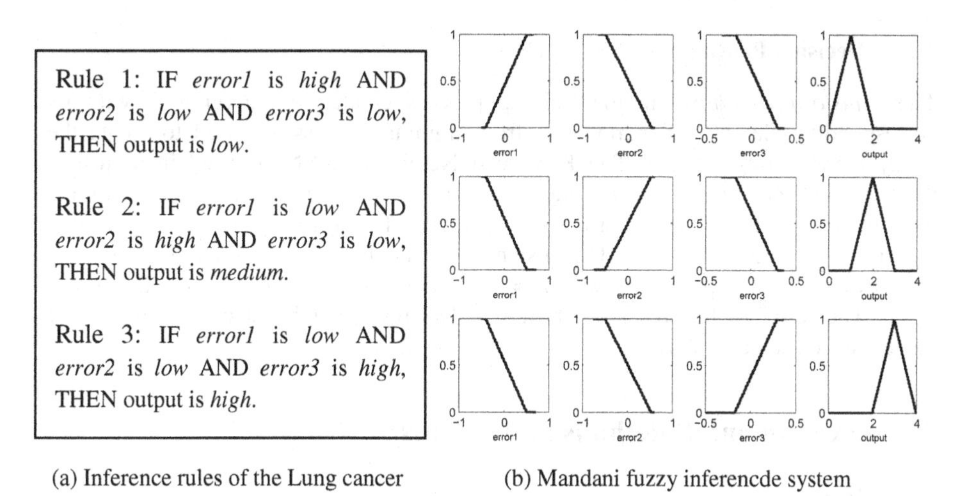

(a) Inference rules of the Lung cancer (b) Mandani fuzzy inferencde system

Fig. 5. The membership function graphs of Lung cancer

The *Max-Min* operation is used for the inference of the fuzzy output. The last step is the defuzzification method. The results from all rules are aggregated as displayed in Figure 6.

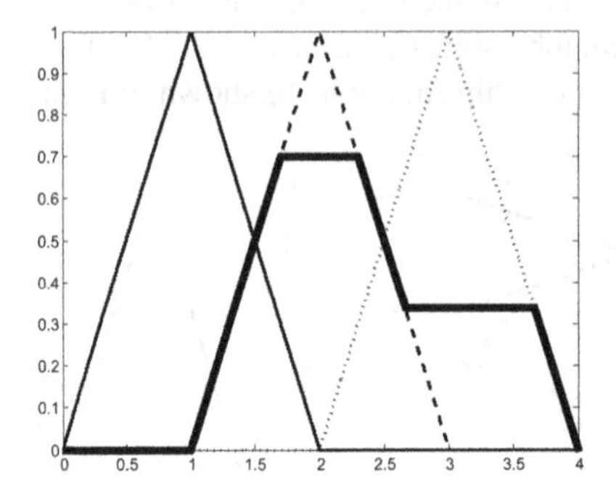

Fig. 6. The results from aggregation of all rules

The final results from learning process can be defined as an equation (8) where α_k is the maximum value of the membership function of the output fuzzy set, y_k is the corresponding crisp value.

$$T_i = \frac{\sum_{k=1}^{n} y_k \alpha_k}{\sum_{k=1}^{n} \alpha_k} \tag{8}$$

The result of T_i is used in the decision process to specify the class of the dataset.

2.2 Decision Process

The *transformed dataset* in the training process created by merging T_i with its corresponding classes. All parameters from training process are used to create the *transformed testing dataset*. The K-Nearest Neighbor (KNN) is used in to justify the classes of the *transformed testing dataset* in the decision phase. This algorithm is the simplest of all method algorithms for the classification and the accuracy of this algorithm is reasonable [10]. The distance is calculated using Euclidian distance measures. Several experiments with k = 3, 5, and 7 have done to define the best k value. The result of this research show that k=3 results higher accuracy than other values, which same as shown in [11].

3 Experimental Methods and Results

3.1 Dataset

The reliable algorithm is the most important thus several types of dataset are tested. The datasets from UCI Machine Learning Repository [12] are used in this paper.

Table 1. Details of dataset used in the experiment

Dataset	Size	Feature Type	Feature	Class
Car Evaluation	1728	Categorical	6	4
Contraceptive Method Choice	1473	Categorical, Integer	9	3
Credit Approval	690	Categorical, Integer, Real	14	2
Dermatology	366	Categorical, Integer	33	6
Glass	214	Real	9	7
Hayes-Roth	132	Categorical	5	3
Heart Disease (Cleveland)	303	Categorical, Integer, Real	13	5
Ionosphere	351	Integer, Real	34	2
Lung Cancer	32	Integer	56	3
Post-Operative Patient	90	Categorical, Integer	8	3

These datasets are considered the standard for verifying the classification algorithms which are consisted of several data types, and classes. The details of data in the experiments are shown in Table 1.

3.2 Experimental Results

The average accuracy results from the 10-fold cross validation of the Naïve Bayes, the 1-hidden layer neural network and the *ensemble k-nearest neighbor with neuro-fuzzy* are shown in Table 2. The column Ant-Clust-B is the accuracy results from Ant-Colony Cluster Bayesian proposed in [13].

Table 2. The comparison of classification results of the proposed method versus other methods

Dataset	Accuracy			
	Naïve Bayes	NN	Ant-Clust-B[13]	**Proposed Work**
Car Evaluation	85.53%	69.33%	93.30%	**96.18%**
Contraceptive	51.12%	49.75%	56.10%	**90.70%**
Credit	80.25%	61.87%	83.80%	**92.46%**
Dermatology	97.26%	51.89%	98.40%	**99.45%**
Glass	52.80%	48.30%	67.80%	**100%**
Hayes-Roth	76.51%	36.20%	85.60%	**94.67%**
Heart Disease	55.77%	54.12%	73.60%	**95.93%**
Ionosphere	35.89%	89.46%	**93.90%**	91.44%
Lung Cancer	59.37%	56.37%	95.80%	**96.66%**
Post-Operative	67.78%	71.23%	79.90%	**97.78%**

The experimental results show the high performance of our proposed method among other algorithms. The average accuracy values of 10-fold cross validation from the *ensemble k-nearest neighbor with neuro-fuzzy* are higher than the Naïve Bayes and the neural network in all datasets. The accuracy of 9 datasets are higher than the ant-colony cluster Bayesian algorithm [13].

Although there is no universal classifier, but SVM is one that has been widely accepted as effective classifier. The results in Table 3 is average accuracy results from the 10-fold cross validation of the *ensemble k-nearest neighbor with neuro-fuzzy* compared to pairwise adaptive support vector machine (*pa*-SVM) [14].

Table 3. The comparison of classification results of the proposed method versus *pa*-SVM [14]

Dataset	Accuracy	
	pa-SVM[14]	**Proposed Work**
Car Evaluation	**99.25%**	96.18%
Dermatology	98.35%	**99.45%**
Glass	77.93%	**100%**
Iris	98.00%	98.00%
Wine	99.44%	**100%**
Yeast	62.94%	**87.53%**

The experimental results show that the performance of the *ensemble k-nearest neighbor with neuro-fuzzy* is higher than pa-SVM. The accuracy results from our proposed of every dataset in Table 3 are high, moreover results from 4 datasets are higher than pa-SVM. Only 1 dataset that our method has lower accuracy, however the results still highest among other algorithms in Table 2.

4 Conclusion

The new paradigm of classification is proposed. The combination of strength of neural network, fuzzy logic and KNN algorithms are applied. The results from experimental clearly show the high efficiency of an *ensemble k-nearest neighbor with neuro-fuzzy* method for classification. The neuro-fuzzy with 2 learning steps is used to create transformed dataset. The simplest KNN with k=3 is used to decide the class of the transformed testing sets. The generalization of this method is verified by several datasets. The classification accuracy is very high and can be applied to use in application that need high classification reliability.

References

1. Han, J., Kamber, M., Pei, J.: Data mining: Concepts and Techniques, pp. 325–370. Morgan Kaufmann, San Francisco (2006)
2. Haykin, S.S.: Neural Networks and Learning Machines, 3rd edn. Prentice Hall, New York (2008)
3. Kriesel, D.: A Brief Introduction to Neural Networks, pp. 37–124 (2007) (retrieved August 15, 2011)
4. Eiamkanitchat, N., Theera-Umpon, N., Auephanwiriyakul, S.: A Novel Neuro-Fuzzy Method for Linguistic Feature Selection and Rule-Based Classification. In: The 2nd International Conference on Computer and Automation Engineering (ICCAE), pp. 247–252. IEEE Press (2010)
5. Eiamkanitchat, N., Theera-Umpon, N., Auephanwiriyakul, S.: Colon Tumor Microarray Classification Using Neural Network with Feature Selection and Rule-Based Classification. In: Zeng, Z., Wang, J. (eds.) Advances in Neural Network Research and Applications. LNEE, vol. 67, pp. 363–372. Springer, Heidelberg (2010)
6. Jirayusakul, A.: Improve the SOM Classifier with the Fuzzy Integral Technique. In: The 9th International Conference on ICT and Knowledge, pp. 1–4. IEEE Press (2011)
7. Weihong, Z., Shunqing, X., Ting, M.: A Fuzzy Classifier based on Mamdani Fuzzy Logic System and Genetic Algorithm. In: IEEE Youth Conference on Information Computing and Telecommunications (YC-ICT), pp. 198–201. IEEE Press (2010)
8. Bova, S., Codara, P., Maccari, D., Marra, V.: A Logical Analysis of Mamdani-type Fuzzy Inference, I theoretical bases. In: 2010 International Conference on Fuzzy System, pp. 1–8. IEEE Press (2010)
9. Tamás, K., Kóczy, L.T.: Selection from a Fuzzy Signature Database by Mamdani-Algorithm. In: 6th International Symposium on Applied Machine Intelligence and Informatics, pp. 63–68. IEEE Press (2008)
10. Juan, L.: TKNN: An Improved KNN Algorithm Based on Tree Structure. In: The 7th International Conference on Computational Intelligence and Security (CIS), pp. 1390–1394. IEEE Press (2011)
11. Liu, H., Zhang, S., Zhao, J., Zhao, X., Mo, Y.: A New Classification Algorithm Using Mutual Nearest Neighbors. In: The 9th International Conference on Grid and Cooperative Computing (GCC), pp. 52–57. IEEE Press (2010)
12. Bache, K., Lichman, M.: UCI Machine Learning Repository. University of California, School of Information and Computer Science, Irvine (2013), http://archive.ics.uci.edu/ml
13. Salama, K.M., Freitas, A.A.: Clustering-based Bayesian Multi-net Classifier Construction with Ant Colony Optimization. In: The IEEE Congress on Evolutionary Computation, pp. 3079–3086. IEEE Press (2013)
14. Hong, K., Chalup, S.K., King, R.A.: An Experimental Evaluation of Pairwise Adaptive Support Vector Machines. In: The 2012 International Joint Conference on Neural Networks (IJCNN), pp. 1–8. IEEE Press (2012)

A Modular Spatial Interpolation Technique for Monthly Rainfall Prediction in the Northeast Region of Thailand

Jesada Kajornrit, Kok Wai Wong, and Chun Che Fung

School of Engineering and Information Technology, Murdoch University,
South Street, Murdoch, Western Australia, 6150
j_kajornrit@hotmail.com,
{k.wong,l.fung}@murdoch.edu.au

Abstract. Monthly rainfall spatial interpolation is an important task in hydrological study to comprehensively observe the spatial distribution of the monthly rainfall variable in the study area. A number of spatial interpolation methods have been successfully applied to perform this task. However, those methods mainly aim at achieving satisfactory interpolation accuracy and they disregard the interpolation interpretability. Without interpretability, human analysts will not be able to gain insight of the model of the spatial data. This paper proposes an alternative approach to achieve both accuracy and interpretability of the monthly rainfall spatial interpolation solution. A combination of fuzzy clustering, fuzzy inference system, genetic algorithm and modular technique has been used. The accuracy of the proposed method has been compared to the most commonly-used methods in geographic information systems as well as previously proposed method. The experimental results showed that the proposed model provided satisfactory interpolation accuracy in comparison with other methods. Besides, the interpretability of the proposed model could be achieved in both global and local perspectives. Human analysts may therefore understand the model from the derived model's parameters and fuzzy rules.

Keywords: Monthly Rainfall, Spatial Interpolation, Modular Technique, Fuzzy Clustering, Fuzzy Inference System, Genetic Algorithm, Interpretable Model.

1 Introduction

Spatial continuous data (spatial continuous surface) play a significant role in planning, rick assessment and decision making in environmental management [1]. For countries located in the tropical area, rainfall spatial continuous surface is necessary for effective water management system leading to efficient irrigation planning, flood and drought prevention, and assessment of dam installation.

In geographic information systems (GIS), spatial interpolation is a key feature to create spatial continuous surface from discrete point data [2]. Presently, many spatial interpolation methods have been developed. The idiosyncrasy of those methods is

S. Boonkrong et al. (eds.), *Recent Advances in Information and Communication Technology*, Advances in Intelligent Systems and Computing 265,
DOI: 10.1007/978-3-319-06538-0_6,

different in terms of working mechanism and prior assumption. Those methods include deterministic, geostatistic and machine learning approaches [1], [3], [4].

Such methods have been successfully applied to several problems including hydrological [5] and environmental areas [1]. However, these methods mainly contribute towards the improvement of the interpolation accuracy and disregard the interpretability factor, which is an issue in data-driven models. The interpretability is important because interpretable models can provide insight knowledge of the model to human analysts when prior knowledge is unknown [6].

The fuzzy inference system (FIS) is a grey-box model that has been successfully applied to many disciplines such as system identification and control, fault diagnosis, classification and intelligent data analysis [6]. Compared to black-box nature models such as artificial neural network (ANN), fuzzy modeling formulates the system knowledge with rules in a transparent way for interpretation and analysis. However, establishing an efficient interpretable fuzzy system is not an easy task because interpretability and accuracy issues may exist as contradictive objectives [7].

To overcome the contrasting problem, many techniques have been proposed and summarized in [6]. Among those techniques, the *prototype-based fuzzy modeling* is recommended by [6] as the most efficient way to automatically create interpretable fuzzy system (i.e. fuzzy models are generated from a clustering technique and then optimized by an optimization algorithm). This technique has been successfully applied in [4]. However, the accuracy of the model proposed in [4] is still unsatisfactory because the interpretability of the fuzzy model will be lost when the number of model's parameters is increased in order to achieve better interpolation accuracy.

The main purpose of this paper is to enhance the outcomes from [4] by developing a fuzzy system that can provide better interpolation accuracy and interpretability of the fuzzy model. The proposed model will be applied to the problem of monthly rainfall spatial interpolation in the northeast region of Thailand. This study can be seen as an improvement from previous works. The techniques such as, fuzzy c-mean clustering (FCM) [8], Mamdani-type fuzzy inference system (MFIS) [9], genetic algorithm (GA) [10] are used and the concept of modular model is applied. The rest of the paper is organized as follows. In the next section, background information of the case study area and datasets are presented. Section 3 describes the proposed methodology and Section 4 shows the experimental results. In Section 5, experimental analysis will be discussed and the conclusion is finally presented in Section 6.

2 Case Study and Datasets

The case study area is located in the northeast region of Thailand as shown in Figure 1. Four selected case studies are *Aug 1998*, *May 1999*, *Aug 2000* and *Aug 2001*. These months have the highest rainfall of the year and also have small number of missing data. The dataset features comprise of information on the longitude (x), latitude (y) and amount of monthly rainfall (z). The datasets are normalized by linear transformation. General information of the datasets is showed in Table 1.

In this study, approximately 30 percent of the rain gauge stations are randomly selected to validate the interpolation accuracy. The correlation in Table 1 indicates the relationship between the amount of rainfall and the altitude of the rain gauge stations. Since the correlation values are close to zero, it could be considered that the orographic effect is not strong in the study area. So, this study does not use the altitude as one of the inputs to the model.

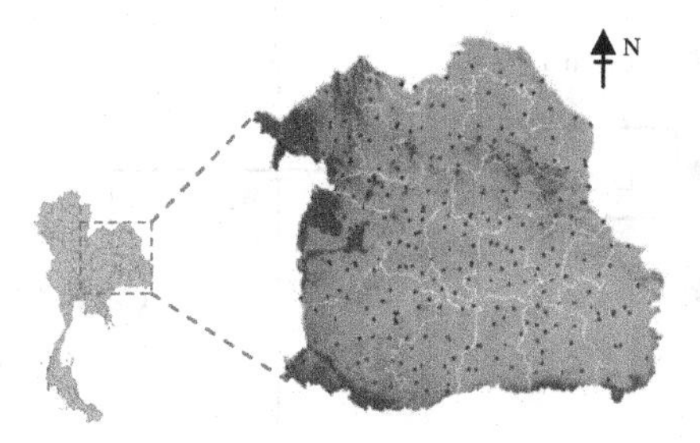

Fig. 1. The case study area locates in the northeast region of Thailand

Table 1. Information on the four datasets used in this study

Statistics	Case 1	Case 2	Case 3	Case 4
Mean (mm)	2317	2433	2756	3696
Calibration stations (Training data)	198	200	197	178
Validation stations (Testing data)	80	80	80	80
Correlation	-0.047	-0.001	-0.240	-0.046

3 The Proposed Methodology

The conceptual architecture of modular model used is similar to the multiple expert systems [11]. As shown in Figure 2, such model consists of a set of local modules and one gating module. Input data are fed into all local modules and gating module. The function of the local modules is to interpolate the rainfall value, whilst the gating module integrates results from the selected local modules into the final results. However, to simplify the interpretability of this proposed model, only one local module is selected to provide the final output. Therefore, this model could also be considered as a decision tree because the gating module works as the decision node and local module work as the leaves of decision tree. The proposed methodology consists of (1) localize the global area into local areas, (2) establish the local modules, and (3) establish the gating module.

3.1 Localize the Global Area

In the first step, localization, fuzzy c-means clustering technique is applied. FCM is not only capable of dealing with uncertainty of boundary, but it also provides the degree of membership values of the data that belong to other clusters. Calibration data (x, y, z) are clustered into n clusters. In general, the selected number of cluster in the prototype-based fuzzy modeling is important to fuzzy model interpretability [13],

[14]. If an appropriate clustering validation index for determining number of cluster is applied, fuzzy clustering can generate a *parsimonious* rule base [6]. In this study, the number of n is determined by the method proposed in [12], which is specifically developed for spatial data.

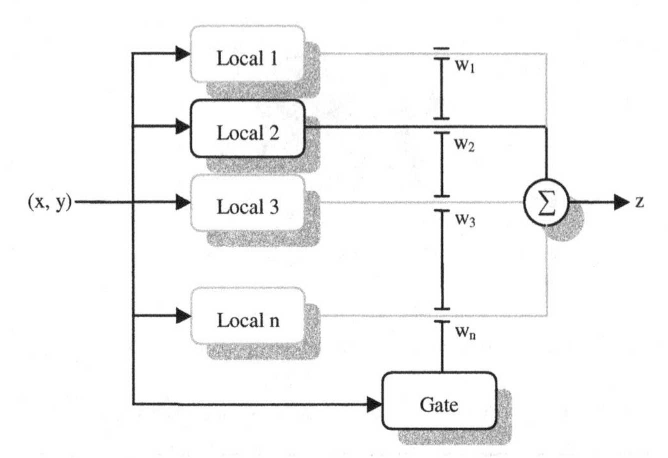

Fig. 2. An overview of the multiple expert system architecture [11]

3.2 Establish the Local Modules

In the second step, local modules establishment, the calibration data in each cluster are used to establish the MFIS model. To prevent the problem of extrapolation between the boundary of local areas, small overlaps between local areas are needed. Let matrix U is $m * n$ matrix, the additional information from FCM, where m is the number of cluster and n in the number of data. Let μ_{ij} , the member of matrix U, be the degree of membership values of the data j to cluster i and $\sum_{i=1}^{m} \mu_{ij} = 1$. Given x_j is the calibration data number j, x_j belongs to cluster i if (1) μ_{ij} in column j is maximum and (2) $\mu_{ij} \geq 0.5$ of maximum value of μ in column j. With this criterion, the overlapping datasets are created for the local modules.

To establish a local module, a prototype-based MFIS is generated from FCM and the membership function (MF) used is Gaussian function. The number of cluster is determined from Tumez's criterion [15], that is, *Minimize* n_c under, $Std[z(x)] \approx Std[z(c)]$, where n_c is the optimal number of cluster, Std is the standard deviation, $z(x)$ are the rainfall values of the dataset and $z(c)$ are the rainfall values at the cluster centers. The numbers of cluster are plotted against $Std[z(c)]$. The number of cluster that the show minimum distance between $Std[z(x)]$ and $Std[z(c)]$ is selected as optimal number.

One constrain of this criterion is that $Std[z(c)] \leq Std[z(x)]$. Rarely, it is possible that $Std[z(c)] > Std[z(x)]$ for all numbers of cluster. Consequently, all numbers of clusters do not satisfy this condition. In this case, a default value must be defined. This study set the default values, n_{df}, as: $n_{df} = Floor(\ (n_{max} + n_{min})\ /\ 2\ )$ and selected $n_{min} = 2$ and $n_{max} = 4$. This study selected the number of cluster as small as possible because the number of calibration data for each local module is not enough to optimize the MFIS's parameters.

Once the initial MFIS is generated, the MFIS's parameters (sigma and center) are then optimized by GA. The chromosome of the algorithm consists of the sequence of input 1, input 2 and output respectively. In turn, the input and output are the sequence of MFs which consists of two parameters (sigma and center). The fitness function is the minimized sum square error between observed value (z) and interpolated value (z') of calibration data in the cluster and it is given as: $SSE = \sum_{i=1}^{S}(z_i' - z_i)^2$.

In this process, the MFIS parameters are allowed to be searched in a certain range in order to prevent *indistinguishability* of the MFs [6]. Let α and β be user-defined control parameters, the center parameters (c) are allowed to be searched within $[c - \alpha, c + \alpha]$ and the sigma parameters (σ) are allowed to be searched within $[\sigma - \beta, \sigma + \beta]$.

3.3 Establish the Gating Modules

Two methods of gating are proposed for the gating module. In the first method, *generic gating module*, the gating module selects one local module by determining the distance between the input data and the center of clusters. The local module that gives the minimum distance is activated and is used to derive the final results. The formal expression of this method is $z = \sum_{i=1}^{n} w_i z_i$ and $w_i = 1,\ if\ d_i < d_{i'}\ where\ i \neq i'\ and\ 0,\ elsewhere$. In another word, the gating module uses a lookup table to calculate the geographic distance between the cluster center of local modules and the interpolated point (x_i, y_i).

In the second method, *fuzzy gating module*, the decision is derived from the antecedent part of the fuzzy system. The functional mechanism of fuzzy gating module is the same as a general FIS, except the consequent part is not used. The number of fuzzy rules is associated with the number of local modules (Figure 3) and, in turn, each fuzzy rule is associated with one MF in each input dimension. For example, if there are seven local modules, the number of MFs of input 1, input 2 and fuzzy rule are seven. Fuzzy rule$_1$ is associated with MF$_1$ in input 1 and input 2.

IF *longitude (x)* = *cls$_1$* AND *latitude (y)* = *cls$_1$* THEN *sub-module$_1$ is activated*
IF *longitude (x)* = *cls$_2$* AND *latitude (y)* = *cls$_2$* THEN *sub-module$_2$ is activated*
IF *longitude (x)* = *cls$_n$* AND *latitude (y)* = *cls$_n$* THEN *sub-module$_n$ is activated*

Fig. 3. An example of fuzzy rules in fuzzy gating module

When the new data is fed into the fuzzy gating module, the fuzzy inference process evaluates the degree of MFs of each *rule$_i$*. The firing strength is the algebraic product of a_i and b_i, (or $a_i * b_i = r_i$) where a_i and b_i is the degree of MF of input 1 and input 2 of *rule$_i$* respectively. The local module activated is associated to the rule that has the maximum firing strength. The formal expression of this method is $z = \sum_{i=1}^{n} w_i z_i$ and $w_i = 1,\ if\ r_i > r_{i'}\ where\ i \neq i'\ and\ 0, elsewhere$.

The process to establish the fuzzy gating module consists of two steps. Firstly, create and initialize the fuzzy gating module. For each local module, the center, minimum and maximum values of calibration data's input are used to initialize the MFs. The MF used is Gaussian function. The centers of the MFs are set to the center of the clusters. For example, if a cluster center (x,y) is (0.3,0.7), the center of input 1's MF is 0.3 and the center of input 2's MF is 0.7 and the sigma parameter of input 1's and input 2's MF are set to $0.1424*(x_{max}-x_{min})$ and $0.1424*(y_{max}-y_{min})$ respectively.

This constant makes the width of the MF approximately equal to the width of the data.

Secondly, optimize fuzzy gating module, parameters of the initial fuzzy gating modules are optimized by GA. The chromosome consists of MFs' parameters of input 1 and input 2. Again, the α and β parameters are used to control the search space in optimization process. The fitness function is to minimize sum square error between observed value (z) and interpolated value (z') of all the calibration data.

In summary, the methodology consists of localizing the global area into local areas by FCM. In each local area, MFIS local modules are generated by prototype-based method. Two gating modules are proposed. First method is based on geographic distance and another method used the concept of fuzzy system. In the next section, evaluation of this method will be discussed.

4 Experiments and Results

To evaluate the proposed method, two commonly used deterministic and geostatistic methods were used for comparison purposes. These methods are Inverse Distance Weighting (IDW), Thin Plate Spline (TPS), Ordinary Kriging (OK) and Universal Kriging (UK). These methods were performed on ArcGIS application, which is widely adopted in GIS area. Furthermore, the results of the model proposed in the previous study (GAFIS) are also compared [4]. Four qualitative measures have been used, that is, Mean Error (ME), Mean Absolute Error (MAE), Root Mean Square Error (RMSE) and Correlation Coefficient (R). These error measures will be normalized by the mean of the datasets for comparison purposes.

From now on, *Mod* FIS and *Mod* FIS-*FS* are referred to as the modular models with *generic* and *fuzzy gating module* respectively. The number of control points used in all GIS methods is six as recommended by [16]. For IDW, k parameter was automatically optimized by ArcGIS [17] and it is used as the control method [3], or standard benchmark. TPS with *tension* is selected rather than *regularized* since it provided more accurate results. Spherical semivariogram is used for OK and UK and the optimal number of lag is 12, which is enough to fit the experimental semivariogram with spherical model. Liner trend is used for UK method.

The numbers of cluster determined from case 1 to case 4 are 7, 7, 8 and 8 respectively. The α and β parameter of the local modules are both set to 0.05. The α parameter value is 5 percent of universe of discourse (UoD). This is set to a half of α parameter in GAFIS method [4] because the size of data in each cluster is decrease approximately more than half of each dimension. The value 0.05 should be large enough for search space. For β parameter, this setting allow the flank of MFs vary approximately in 10-15 percents of UoD. However, if α and β parameter are set too small, the optimal solution may not be met since the fuzzy model cannot handle the uncertainty in rainfall data well under this condition.

For α and β parameter of fuzzy gating module, they are set to 0.05 and 0.025. The α parameter is set to 0.05 because, generally, the centroid of cluster is not at the geometric center of cluster, so, the parameter allows small space for the centroid to move to the geometric center of the cluster in order to let MF covers the entire cluster. The β parameter is set at half of the setting in sub-module because it allows to move in smaller space in overlapped area between local modules. The GA's population is set to 150 and GA's generation is set to 30, where the best and average fitness values are met. The experimental results are shown in Table 2 to Table 5. Please note that the row "*avg*" is the average values from the four cases and the row "*imp*" is the percentage improvement of the average values based on IDW method.

Table 2. Normalized Mean Error

Case	IDW	TPS	OK	UK	GAFIS	Mod FIS	Mod FIS-FS
1	0.0049	0.0036	0.0152	0.0007	0.0133	0.0242	-0.0093
2	0.0354	0.0400	0.0405	0.0422	0.0490	0.0019	0.0202
3	-0.0177	-0.0163	-0.0121	-0.0121	-0.0130	-0.0120	-0.0037
4	-0.0083	-0.0086	-0.0246	-0.0118	-0.0234	-0.0340	-0.0323
Avg	0.0166	0.0171	0.0231	0.0167	0.0247	0.0180	0.0164
Imp	-	-3.41	-39.46	-0.85	-49.04	-8.85	1.17

Table 3. Normalized Mean Absolute Error

Case	IDW	TPS	OK	UK	GAFIS	Mod FIS	Mod FIS-FS
1	0.2871	0.2825	0.2897	0.2898	0.2843	0.2578	0.2484
2	0.2006	0.2002	0.2054	0.2168	0.2238	0.1975	0.1978
3	0.2703	0.2707	0.2750	0.2717	0.2667	0.2597	0.2490
4	0.2293	0.2106	0.2220	0.2028	0.2328	0.2055	0.2044
Avg	0.2468	0.2410	0.2480	0.2453	0.2519	0.2301	0.2249
Imp	-	2.37	-0.48	0.64	-2.05	6.77	8.88

Table 4. Normalized Root Mean Square Error

Case	IDW	TPS	OK	UK	GAFIS	Mod FIS	Mod FIS-FS
1	0.3570	0.3549	0.3718	0.3603	0.3656	0.3351	0.3177
2	0.2747	0.2771	0.2794	0.2885	0.2904	0.2697	0.2610
3	0.3583	0.3566	0.3575	0.3576	0.3528	0.3526	0.3340
4	0.3204	0.2969	0.3086	0.2840	0.3257	0.2950	0.2890
Avg	0.3276	0.3214	0.3293	0.3226	0.3336	0.3131	0.3004
Imp	-	1.90	-0.53	1.52	-1.84	4.43	8.29

Table 5. Correlation Coefficient

Case	IDW	TPS	OK	UK	GAFIS	Mod FIS	Mod FIS-FS
1	0.4249	0.4360	0.3433	0.3998	0.3700	0.5282	0.5913
2	0.7705	0.7649	0.7564	0.7418	0.7490	0.7830	0.8039
3	0.5080	0.5082	0.4931	0.4935	0.5130	0.5147	0.5806
4	0.7682	0.8051	0.7937	0.8257	0.7604	0.8111	0.8196
Avg	0.6179	0.6285	0.5966	0.6152	0.5981	0.6592	0.6989
Imp	-	1.72	-3.45	-0.44	-3.21	6.69	13.10

5 Analysis and Discussion

In terms of ME, in case 1, *Mod* FIS-*FS* provided negative bias, whilst the others showed positive. It was possibly caused by the fuzzy gating modules because *Mod* FIS-*FS* used the same local modules as *Mod* FIS. When the activated local modules are different, small changing of bias are possible. In the other cases, the sign directions are the same, therefore, no uncommon condition is found. Based on average values, the quality of interpolators can be ordered as: *Mod* FIS-*FS* > IDW > UK > *Mod FIS* > TPS > OK > GAFIS.

In terms of MAE, RMSE and R, in deterministic GIS methods, TPS provided better interpolation accuracy than IDW. It is because only geographic distance may not be sufficient to reflect the spatial variation in the area. In geostatistic methods, UK showed better results than OK. It is possible that the linear trend appear rather strong in this spatial rainfall data. Overall, it seems that spatial variation is not strong enough. Consequently, TPS tends to provide better interpolation than UK, except for the case 4.

Mod FIS and *Mod* FIS-*FS* consistently show satisfactory interpolation accuracy in general. They provided better interpolation accuracy than the best of GIS method in three cases. Furthermore, it showed a large improvement from GAFIS. This means that modular technique is a promising method for spatial interpolation problem when fuzzy system is applied.

Between both modular models, *Mod* FIS-*FS* has improved from *Mod* FIS. This is because the shape of local area is not spherical. Hence, only geographic distance from cluster center is not sufficient to select the local module (i.e. *Mod* FIS). However, this problem can be solved by using fuzzy concept, consequently, the accuracy is improved (i.e. *Mod* FIS-*FS*). Overall, based on the average values, the accuracy of all interpolators can be ranked as: *Mod* FIS-*FS* > *Mod* FIS > TPS > UK > IDW > OK > GAFIS.

The quantitative results showed above indicated that *Mod* FIS-*FS* could be a good alternative method. However, the key objective of the system is the interpretability of interpolators. When interpolators are being in the form of a model, human analysts can gain insights from data to be modeled when a prior knowledge are unknown. In *Mod* FIS and *Mod* FIS-*FS* the model is transparent to human analysts through the fuzzy parameters and fuzzy rules. Furthermore, due to this transparency, human analysts can use their knowledge to manipulate the models.

However, not all fuzzy models can be counted as the interpretable model. Large number of fuzzy rule or MFs can make interpretability of the model deteriorate. In previous work, GAFIS, the model is good interpretable since it is rather satisfied to the criterion proposed in [6] but such model still lack of accuracy when compared to those GIS methods. *Mod* FIS makes use of modular technique to increase the accuracy. Not only the accuracy was improved, the model can also be interpretable in more details in each of the local areas.

Mod FIS-*FS*, in turn, makes use of fuzzy concept to improve *Mod* FIS. With the fuzzy concepts the model is interpretable in the global and local layers. The parameters in fuzzy gating module provide the global view of the model, whereas the parameters in the local modules provide details of the rainfall data. Figure 4 and 5 show an example of the fuzzy rules and associated MFs in the fuzzy gating module and one local module. The MFs and fuzzy rules are not complicated and easy to understand.

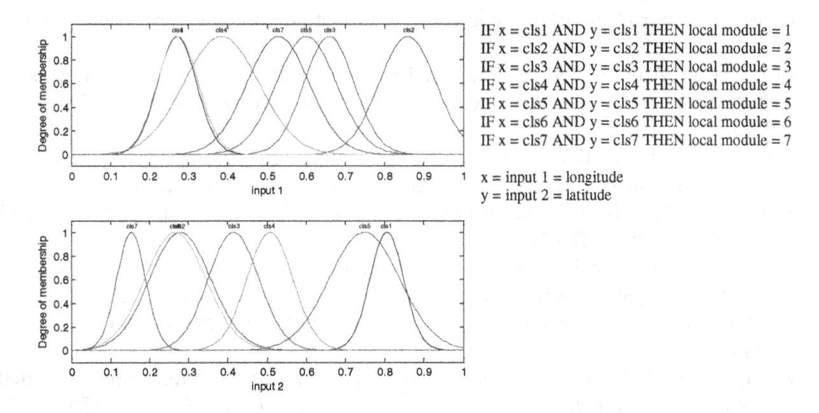

Fig. 4. An example of fuzzy sets and fuzzy rules in the fuzzy gating module (Case 1)

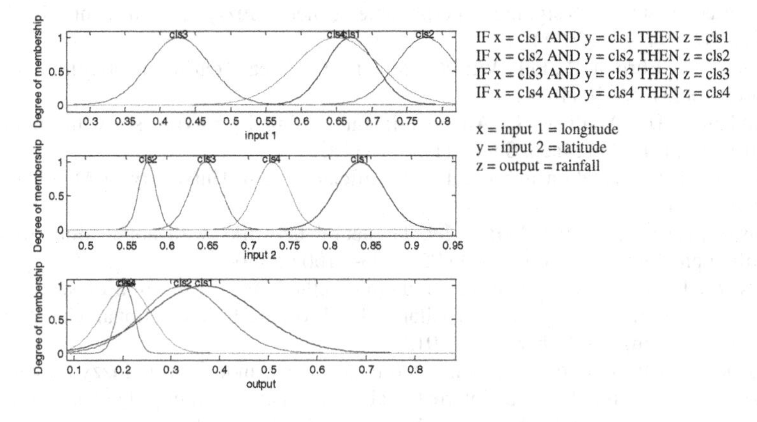

Fig. 5. An example of fuzzy sets and fuzzy rules in one of the local modules (Case 1)

6 Conclusion

This study proposed the use of fuzzy system and modular technique to address the spatial interpolation problem and to be applied to the monthly rainfall data in the northeast region of Thailand. Four datasets were used to evaluate the proposed methods. The experiments showed that the proposed models, especially, *Mod* FIS-*FS*, provided satisfactory interpolation accuracy in comparison with GIS methods and showed improvement from previous work. By making use of fuzzy rule, the model is interpretable, in which human analysts can gain insight knowledge from the data being modeled; and by using modular technique, the knowledge gained can be provided in more details in a modular way.

References

1. Li, J., Heap, A.D.: A review of spatial interpolation methods for environmental scientists. Geoscience Australia, record 2008/23 (2008)
2. Chang, T.: Introduction to geographic information systems, 3rd edn. McGraw-Hill, Singapore (2006)
3. Li, J., Heap, A.D., Potter, A., Daniell, J.J.: Application of machine learning methods to spatial interpolation of environmental variables. Environ Modell. Softw. 26, 1647–1659 (2011)
4. Kajornrit, J., Wong, K.W., Fung, C.C.: An integrated intelligent technique for monthly rainfall spatial interpolation in the northeast region of Thailand. In: Lee, M., Hirose, A., Hou, Z.-G., Kil, R.M. (eds.) ICONIP 2013, Part II. LNCS, vol. 8227, pp. 384–391. Springer, Heidelberg (2013)
5. Di Piazza, A., Lo Conti, F., Noto, L.V., Viola, F., La Loggia, G.: Comparative analysis of different techniques for spatial interpolation of rainfall data to create a serially complete monthly time series of precipitation for Sicily, Italy. Int. J. Appl. Earth OBS 13, 396–408 (2011)
6. Zhou, S., Gan, J.Q.: Low level interpretability and high level interpretability: a unifed view of data-driven interpretable fuzzy system modelling. Fuzzy Set Syst. 159, 3091–3131 (2008)
7. Cordon, O.: A history review of evolutionary learning methods for Mamdani-type fuzzy rule-based system: Designing interpretable genetic fuzzy system. Int. J. Approx. Reason 52, 894–913 (2011)
8. Bezdek, J.C., Ehrlich, R., Full, W.: FCM: the fuzzy c-means clustering algorithm. Comput. Geosci. 10(23), 16–20 (1984)
9. Mamdani, E.H., Assilian, S.: An experiment in linguistic synthesis with fuzzy logic controller. Int. J. Man Mach. Stud. 7(1), 1–13 (1975)
10. Holland, J.H.: Adaptation in natural and artificial system. University of Michigan Press, Ann Arbor (1975)
11. ChrisTseng, H., Almogahed, B.: Modular neural networks with applications to pattern profiling problems. Neurocomputing 72, 2093–2100 (2009)
12. Kajornrit, J., Wong, K.W.: Cluster validation methods for localization of spatial rainfall data in the northeast region of Thailand. In: Proc. 12th International Conference on Machine Learning and Cybernetics (2013)
13. Hoppner, F., Klawonn, F.: Obtaining interpretable fuzzy models from fuzzy clustering and fuzzy regression. In: Proc. 4th International Conference on Knowledge-based Intelligent Engineering Systems and Allied Tech (KES), pp. 162–165 (2000)
14. Wang, X.Z., Yeung, D.S., Tsang, E.C.C.: A comparative study on heuristic algorithms for generating fuzzy decision trees. IEEE Trans. Systems Man Cybernet, Part B 31(2), 215–226 (2001)
15. Tutmez, B., Tercan, A.E., Kaymak, U.: Fuzzy modeling for reserve estimation based on spatial variability. Mathematical Geology 39(1) (2007)
16. Zimmerman, D., Pavlik, C., Ruggles, A., Armstrong, M.P.: An experimental comparison of ordinary and universal kriging and inverse distance weighting. Mathematical Geology 31, 375–390 (1999)
17. Luo, W., Taylor, M.C., Parkerl, S.R.: A comparison of spatial interpolation methods to estimate continuous wind speed surfaces using irregularly distributed data from England and Wales. Int. J. Climatol. 28, 947–959 (2008)

On N-term Co-occurrences

Mario Kubek and Herwig Unger

FernUniversität in Hagen, Faculty of Mathematics and Computer Science, Hagen, Germany
{dr.mario.kubek,herwig.unger}@gmail.com

Abstract. Since 80% of all information in the World Wide Web (WWW) is in textual form, most of the search activities of the users are based on groups of search words forming queries that represent their information needs. The quality of the returned results -usually evaluated using measures such as precision and recall- mostly depends on the quality of the chosen query terms. Therefore, their relatedness must be evaluated accordingly using and matched against the documents to be found. In order to do so properly, in this paper, the notion of n-term co-occurrences will be introduced and distinguished from the related concepts of n-grams and higher-order co-occurrences. Finally, their applicability for search, clustering and data mining processes will be considered.

Keywords: keyword, co-occurrence, search engine, context, clustering.

1 Motivation

Search words and groups of search words are the usual input data of the big powerful search engines like Google to filter suitable information out of the approx. 785 million websites [1]. Their pages offer approx. 80% of their information in textual form [2], but in average only 2 words are used as input [3] to the search engines to address the wanted content. It is clear that groups of 4 to 6 search words may return more precise search results due to a better context disambiguation which leads to a reduced number of search results containing less unwanted information.

Several publications address the question, how users may identify the right sets of search words for their information needs. Big search engines use powerful statistics to suggest frequently entered keywords by many users along with the initial query [4]. A personalised user account beyond that enables a significant refinement of the search. By doing so, the search engine can store a bigger amount of information of the respective user and therefore learns about his or her special search behavior and personal interests. Other approaches try to analyse locally stored documents [5, 6] or knowledge of a group of collaboratively working users [7, 8] for this purpose.

Mostly, simple but fast methods like the determination of characteristic word frequency vectors [9] are used to determine the similarity (or dissimilarity) of documents and their terms. More complex methods such as Latent Semantic Indexing (LSI) [10] require complex programming, as well as more time and/or computational resources.

In this paper, the authors concentrate on an extension of the use of term co-occurrences, i.e. the appearance of (so far) two words in a sentence, paragraph or whole text T, and the justification of their usage in (interactive) search applications.

S. Boonkrong et al. (eds.), *Recent Advances in Information and Communication Technology*, Advances in Intelligent Systems and Computing 265,
DOI: 10.1007/978-3-319-06538-0_7, © Springer International Publishing Switzerland 2014

Co-occurrence relations can be easily processed by scanning the documents and can also be easily visualized using graphs, too. In doing so, the words w_i of the text build the set of nodes W of a graph while an edge connects two nodes, if and only if the respective two words w_1 and w_2 appear together in a sentence (paragraph, text) [11] of T with a given significance σ (i.e. at least σ times).

The most simple measure to determine the significance of a co-occurrence is the weight $c(w_1, w_2)$ of the respective co-occurrence/edge and denotes, how often this co-occurrence has been observed in the given text T. Well-known other measures to gain co-occurrence significance values on sentence level are for instance the mutual information measure [12], the Dice [13] and Jaccard [14] coefficients, the Poisson collocation measure [15] and the log-likelihood ratio [16]. The resulting co-occurrence graphs are generally undirected which is suitable for the flat visualisation of term relations (Fig. 1) and for applications like query expansion via spreading activation techniques [6].

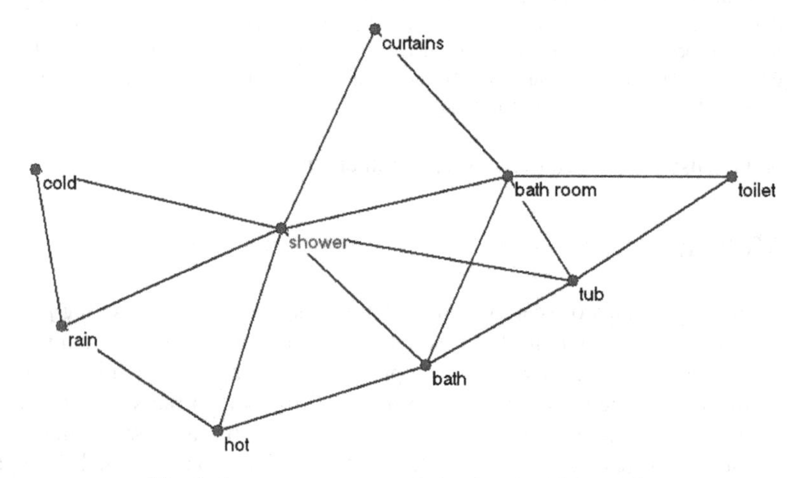

Fig. 1. A co-occurrence graph for the word "shower"

Co-occurrences benefit from the fact that they do not consider single terms only, but built a relation between two words allowing a much clearer disambiguation. Especially the consideration within a sentence or paragraph (i.e. text subunits) presents much more information than simple, globally counted word vectors. Nevertheless, so far there is no broader use of co-occurrence relations in the search process of the big search engines, e.g. for suggesting related query terms or the grouping of similar search results. However, tools such as the Google Ngram Viewer [17] or Google Autocomplete [4] extract phrases (a special form of co-occurrences) from books, web contents and user queries in order to visualise trends in the usage of n-grams over the years and to suggest query terms that often appear in the immediate neighbourhood of entered search words.

However, it might be sensible to also consider co-occurrences for this purpose that do not only contain terms which are immediate neighbours and that are not limited to two words only. The herein described extension and generalization of co-occurrences therefore allows them to consist of n terms ($n \geq 2$) that do not need to appear alongside each other in a text fragment (the order of their occurrence does not play a role) and may be used like word frequency vectors in order to determine content differences in large corpora, to group related documents with similar n-term co-occurrences as well as to support a fine-grained determination of related query terms in order to semantically narrow down a search context.

The remaining paper is structured as follows: the next section explains the notion of n-term co-occurrences in detail before section three provides results of conducted experiments using three large text corpora. Section four discusses the application field of web search for n-term co-occurrences and outlines a suitable method to determine their overall importance in a corpus. In section five, the interactive search application "DocAnalyser" is introduced that implements the gained results from the experiments. Section six summarises the paper and provides a look at future research directions to evaluate the overall significance of n-term co-occurrences in single texts using large well-balanced or topically oriented corpora for the purpose of finding their most characteristic term combinations and to topically classify these texts.

2 N-term Co-occurrences

Considering the above facts and techniques, the question arises, why should co-occurrences consist of only two words appearing together in a given text environment? An n-term co-occurrence may be easily defined as a set of n terms ($n \geq 2$) appearing together in a sentence, paragraph or whole text. In an analogous manner, a σ-significant co-occurrence requires that its set of words appears at least σ times together in a given text. The notion of n-term co-occurrence, however, has another meaning than higher-order co-occurrences [18] that rely on paradigmatic relations of word forms that can be derived by comparing the semantic contexts (e.g. vectors of significant co-occurring terms) of word forms. Thus, higher-order co-occurrences cannot be directly extracted by parsing the mentioned text fragments. N-term co-occurrences, however, represent important syntagmatic relations between two or more word forms and can be easily extracted in the process of parsing these text fragments.

With this new definition, a set of questions immediately arises:

1. What does the distribution of n-term co-occurrences in a given set of texts look like?
2. What is the absolute frequency of n-term co-occurrences?
3. How significant are those n-term co-occurrences for tasks such as text classification and search/topic determination? Can they be considered to be a fingerprint of a text consisting of the most prominent term relations?
4. How high is the computational effort to calculate n-term co-occurrences?

In order to answer these questions, a set of experiments have been conducted using different corpora with topically grouped articles on German politics ("Politik-Deutschland"), cars ("Auto") and humans ("Mensch") of the German news magazine "Der Spiegel". The politics corpus consists of 45984 sentences, the corpus on cars has 21636 sentences and the corpus on topics related to humans contains 10508 sentences. For this purpose, stop words have been removed, only nouns and names as the only allowed parts-of-speech have been extracted and a baseform reduction has been applied prior to the co-occurrence extraction.

3 Experiments

In order to answer the first question, the first diagram (Fig. 2) shows the power-law distribution of significant n-term co-occurrences (consisting of 2, 3, 4, 5, 6, 7 and 8 terms) for the corpus "Politik-Deutschland" about German politics. These co-occurrences must have appeared at least twice in the corpus (in two different sentences) in order to be called significant. It can be clearly seen, that the most frequent co-occurrence is a 2-term co-occurrence and was found 476 times. This co-occurrence ("Euro, Milliarde") appeared more than three times more often than the most frequent 3-term co-occurrence ("Euro, Jahr, Milliarde") that was found 130 times. This observation is very interesting as it shows that by taking into account just one additional term (in this case "Jahr", German for "year") the semantic context of a 2-term co-occurrence can be specified much more precisely. This is especially true for co-occurrences that appear very frequently and therefore in many different contexts in the used corpus.

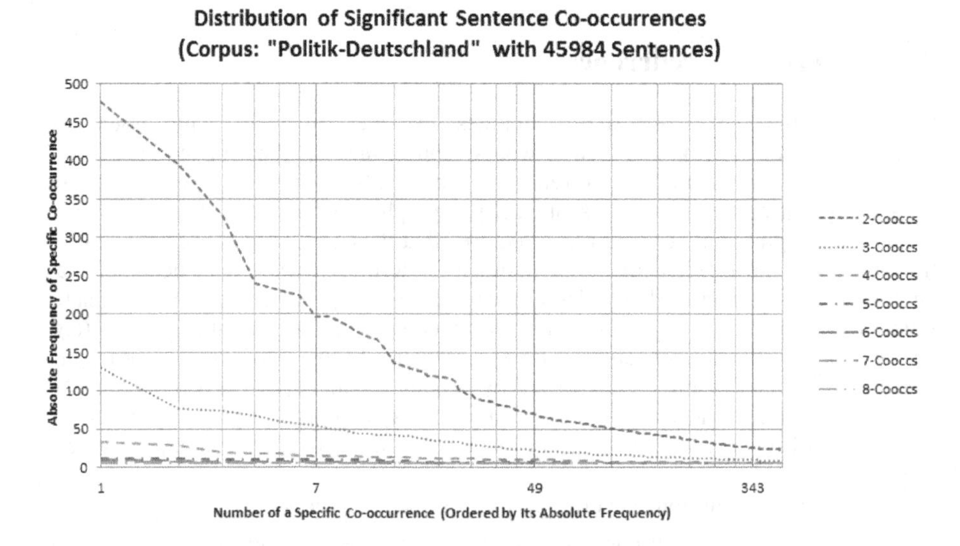

Fig. 2. Distribution of n-term co-occurrences (logarithmic scale on X-axis)

By adding another co-occurring term such as "Bund" to the most frequent 3-term co-occurrence, the absolute frequency of this 4-term co-occurrence is again drastically reduced to 12. However, the frequency of specific 5-term, 6-term, 7-term or 8-term co-occurrences is generally much lower than the frequency of 2-term, 3-term or 4-term co-occurrences. That means that by adding another term, the semantic context can be defined even more precisely to a very little extent only. When using five or more terms in a query, it is therefore likely that only a very low number of results will be returned, or even no results at all, when improper terms are used. However, the effort to determine such types of n-term co-occurrences will rise significantly due the increased number of possible term combinations. For the refinement of queries, 3-term or 4-term co-occurrences hold, as mentioned before, valuable expansion terms.

Nevertheless, in order to cluster (group) related terms in texts, this approach to find n-term co-occurrences is a very natural way for doing so and does not involve calculations of term similarities [11] at all.

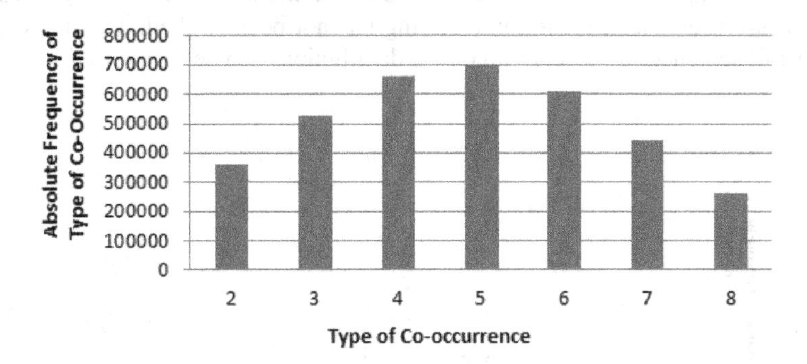

Fig. 3. Absolute frequencies of the types of co-occurrences

Even, when the absolute of frequency of specific 2-term co-occurrences will be much higher than the frequency of specific 3-term, 4-term or 5-term co-occurrences, in general, there are more 3-term til 7-term co-occurrences (cumulated) due to the increased number of possible term combinations as it can be seen in Fig. 3 again for the politics corpus. The number of 8-term co-occurrences, however, is significantly lower than the number of the other types as a lower number of sentences contained eight nouns or names.

Another feasible way to filter sets of co-occurrences (so far, a co-occurrence must have appeared at least twice to be regarded) is by adapting the significance threshold (the number of times a co-occurrence must be found at least for further considerations). Fig. 4 shows the influence of this parameter. A simple increase from 2 to 5 reduces the number of remaining co-occurrences for all types drastically.

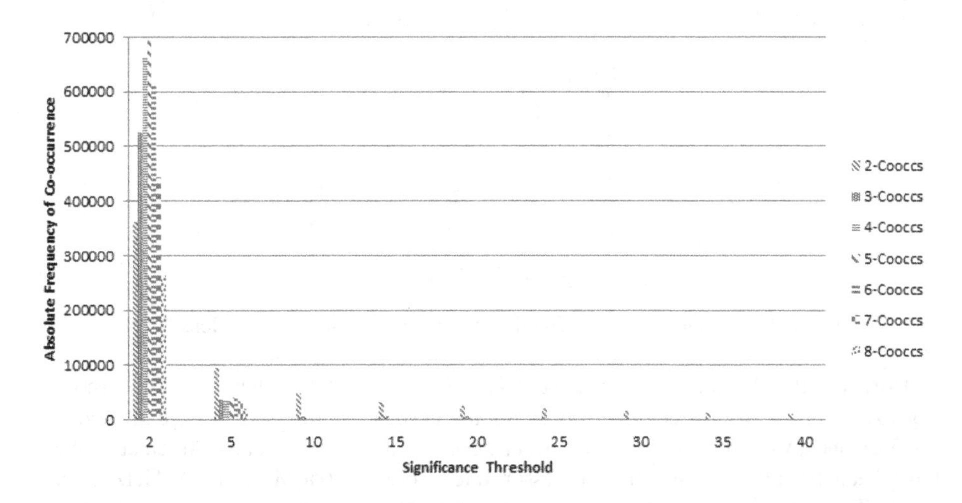

Fig. 4. Absolute frequencies of types of co-occurrences grouped by the significance threshold

Similar distributions of n-term co-occurrences have been observed for smaller corpora, too. This can be seen in Fig. 5, whereby "Corpus 1" stands for the corpus "Politik -Deutschland", "Corpus 2" for the corpus "Auto" and "Corpus 3" for the corpus "Mensch". The diagram shows an extract of the distributions of their most significant 4-term co-occurrences. Here, logarithmic scaling has not been applied. The influence of the number of sentences in the corpora on the distributions can be easily seen.

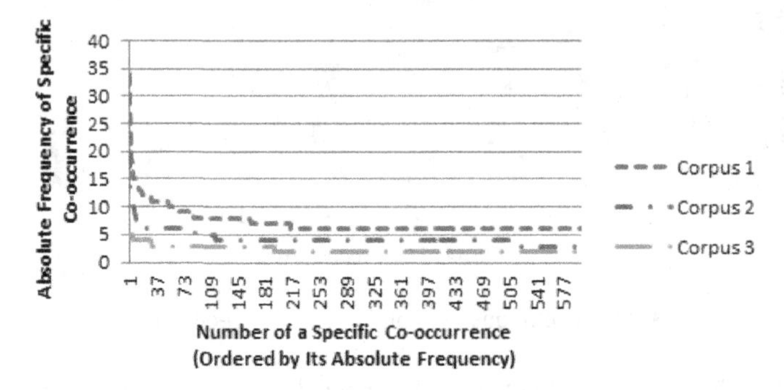

Fig. 5. Comparison of the distributions of significant 4-term co-occurrences for 3 corpora

As an example, in the smallest corpus 3, the absolute frequency of specific 4-term co-occurrences is much lower. Also, for all corpora, this number drops drastically in the range of the first 100 (most frequent) specific co-occurrences.

Another interesting factor to consider is the computational effort to calculate n-term co-occurrences. For the corpus "Mensch", in Fig. 6, the time in seconds needed to do so is denoted for the considered seven types of n-term co-occurrences. Here, the calculations for the 3-term and 4-term co-occurrences needed most of the time. A similar relation of the processing times to each other could be observed for the two corpora "Politik-Deutschland" and "Auto".

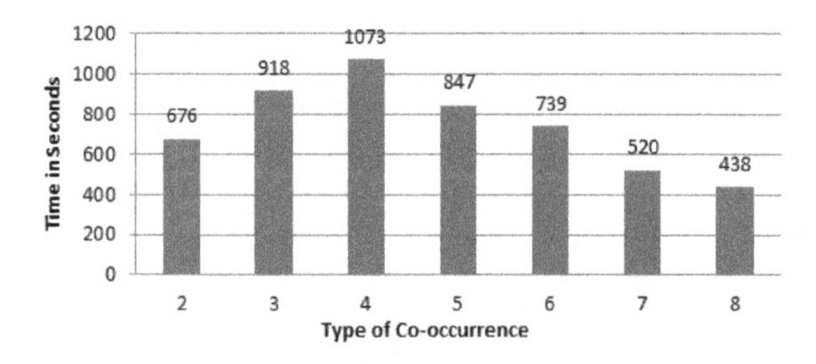

Fig. 6. Distribution of processing time in seconds for the corpus "Mensch"

However, the absolute number of seconds needed was much higher for these corpora. For example, the 4-term co-occurrences for the corpus "Auto" have been extracted in 5815 seconds, whereby the 3-term co-occurrences needed 4683 seconds. All calculations have been performed using the processor Intel Core i7-2640M with 2.8 GHz clock speed. These results indicate that the computation of n-term co-occurrences cannot be sensibly performed for large corpora during interactive search sessions, e.g. for the purpose of calculating query expansion terms. For single texts, however, the extraction of classic 2-term co-occurrences is a feasible option in order to generate co-occurrence graphs for specific applications such as keyword and search word extraction and query expansion. This option has been used in the search application "DocAnalyser" which will be introduced in section five.

4 Discussion

In the application field of web search (as an example), it is known that by using more than two query terms, the number of search results in the WWW can be greatly reduced and they will likely contain more proper information, too. Query expansion based on n-term co-occurrences (not only n-grams) is therefore a good choice to refine queries. Not only will those suggestions reduce the result set, the result set will not be empty at all as the co-occurrences are extracted from textual contents that already exist and can therefore be found with the co-occurring terms they comprise.

However, it is still an open question of how to determine the overall importance of an n-term co-occurrence in a corpus? A method to do so would be of importance when it comes to determine the significance of n-term co-occurrences in single texts, e.g. by comparing their relative frequencies with the respective ones from large reference corpora. This approach bears some resemblance to the so-called differential analysis [19] which measures the deviation of word frequencies in single texts from their frequencies in general usage. To solve this problem, the method proposed by Luhn [20] to extract words with a high entropy could be applied. He argues that neither the most frequent (too common) nor the least frequent words can be used for filtering out the characteristic words. For this purpose, he sets high-frequency and low-frequency cutoff lines. In between those two lines, the most important words can be found. The problem with this approach, however, is to establish "confidence limits" using statistical methods for these lines. Even so, it can be generalised to filter out too common or too infrequent n-term co-occurrences. The applicability of Luhn's method, adapted for n-term co-occurrences, is currently being evaluated.

5 "DocAnalyser" – Searching with Web Documents

The search application DocAnalyser [21] is a new service that offers a novel way to interactively search for similar and related web documents and to track topics without the need to enter search queries manually. The user just needs to provide a web content to be analysed. This is usually the web page (or a specific part of it) currently viewed in the web browser. DocAnalyser then extracts its main topics and their sources (important inherent, influential aspects / basics) [22] and uses them as search words. The returned search words and web search results are generally of high quality. This is mainly due to the implemented state-of-the-art graph-based extraction of keywords of the texts provided. Usually, the currently analysed web document is found again among the Top-10 search results which underlines and confirms this statement. Therefore, this tool can also be used to some extent to detect plagiarism in the WWW.

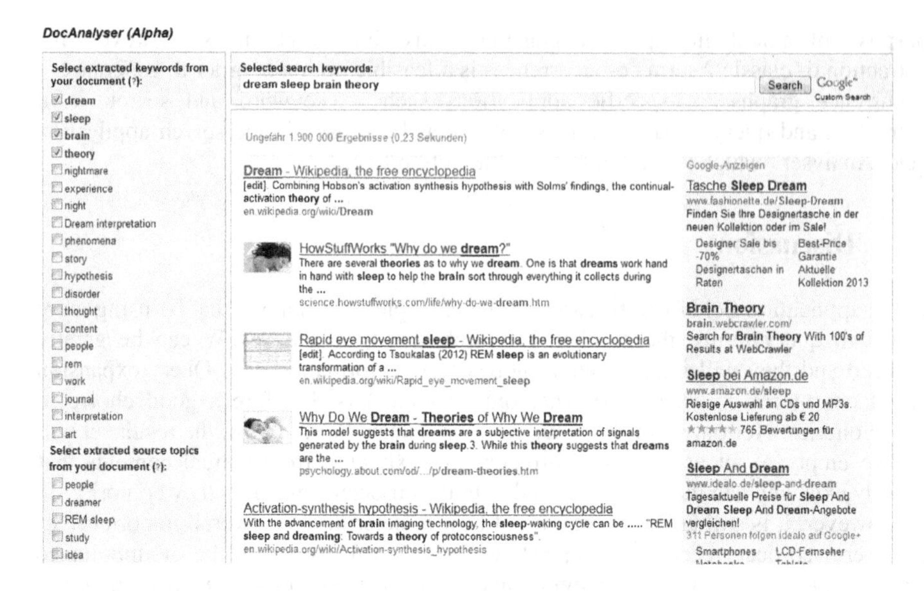

Fig. 7. Screenshot of "DocAnalyser" with extracted search words from the Wikipedia-article "Dream" and Google's returned web results

As indicated before, for the purpose of fast keyword/search word extraction, "DocAnalyser" creates directed 2-term co-occurrence graphs of the provided texts using a special approach to determine directed term associations [22] and analyses them with an extended version of the HITS algorithm [23] that takes into account the strength of the term relations. This algorithm returns two lists of terms: the characteristic terms (authorities) and the source topics (hubs) that strongly influence a text's main topics, whereby authorities are nodes (terms) often linked by many other nodes, and hubs are nodes (terms) pointing to many other nodes and, therefore, topically influence them. These term lists (can be regarded as term clusters, too) can also show, however, an overlap when analysing directed co-occurrence graphs. Hence, this is a soft graph clustering technique. In each case, these lists are ordered according to their terms' centrality scores.

The idea behind this approach is that especially the determined source topics can lead users to documents that cover important aspects of their analysed web documents and presented search results. This goes beyond a simple search for similar documents as it offers a new way to search for related documents and to track topics to their roots by using these source topics as search words. This functionality can be regarded as a useful addition to Google Scholar (http://scholar.google.com/), which offers users the possibility to search for similar scientific articles.

Furthermore, in order to provide instant results, the four most important keywords/key phrases from the authorities list are automatically selected as search words which are sent to Google. The authors chose four terms because, as discussed in section two, 4-term co-occurrences (when used as search words) provide the best trade-off between a precise search context description and a useful amount of search results. In this case, however, the 4-term co-occurrence is constructed for the complete web content provided, not only for a text fragment such as a sentence or paragraph. Although this approach yields useful results, the user still has the option to easily modify the preselected query by selecting and/or deselecting entries from the term lists or by entering own query terms.

6 Conclusion

The notion of n-term co-occurrence has been introduced herein and an in-depth discussion on the characteristics and the applicability of n-term co-occurrences based on empirical results from conducted experiments has been provided. Also, the interactive search application "DocAnalyser" has been presented that implements the gained results from those experiments. Ongoing research concentrates on providing means and methods to evaluate the significance of n-term co-occurrences in single texts using n-term co-occurrences from large and topically well-balanced text corpora. Here, modified approaches that measure the importance of terms such as TF-IDF [9] and differential analysis [19] will be useful. Also, it is evaluated if these methods can be used to classify single texts using smaller, but topically oriented corpora. The results of this research will be presented in future contributions.

References

1. November 2013 Web Server Survey (2013), `http://news.netcraft.com/archives/2013/11/01/november-2013-web-server-survey.html` (last retrieved on March 01, 2014)
2. Grimes, S.: Unstructured Data and the 80 Percent Rule (2008), `http://breakthroughanalysis.com/2008/08/01/unstructured-data-and-the-80-percent-rule` (last retrieved on March 01, 2014)
3. Agrawal, R., Yu, X., King, I., Zajac, R.: Enrichment and Reductionism: Two Approaches for Web Query Classification. In: Lu, B.-L., Zhang, L., Kwok, J., et al. (eds.) ICONIP 2011, Part III. LNCS, vol. 7064, pp. 148–157. Springer, Heidelberg (2011)
4. Website of Google Autocomplete, Web Search Help (2013), `http://support.google.com/websearch/bin/answer.py?hl=en&answer=106230` (last retrieved on March 01, 2014)
5. Xu, J., Croft, W.B.: Query expansion using local and global document analysis. In: Frei, H.-P., Harman, D., Schäuble, P., Wilkinson, R. (eds.) Proc. of the 19th Annual International ACM/SIGIR Conference on Research and Development in Information Retrieval, SIGIR 1996, Zurich, pp. 4–11 (1996)
6. Kubek, M., Witschel, H.F.: Searching the Web by Using the Knowledge in Local Text Documents. In: Proceedings of Mallorca Workshop 2010 Autonomous Systems. Shaker Verlag, Aachen (2010)
7. Keiichiro, H., et al.: Query expansion based on predictive algorithms for collaborative filtering. In: Proc. of the 24th Annual International ACM/SIGIR Conference on Research and Development in Information Retrieval, SIGIR 2001, pp. 414–415 (2001)
8. Han, L., Chen, G.: HQE: A hybrid method for query expansion. Expert Systems with Applications Journal 36, 7985–7991 (2009)
9. Salton, G., Wong, A., Yang, C.S.: A vector space model for automatic indexing. Communications of the ACM 18(11), 613–620 (1975)

10. Deerwester, S., et al.: Indexing by latent semantic analysis. Journal of the American Society of Information Science 41(6), 391–407 (1990)
11. Heyer, G., Quasthoff, U., Wittig, T.: Text Mining: Wissensrohstoff Text: Konzepte, Algorithmen, Ergebnisse. W3L-Verlag, Dortmund (2006)
12. Büchler, M.: Flexibles Berechnen von Kookkurrenzen auf strukturierten und unstrukturie-ten Daten. Master's thesis, University of Leipzig (2006)
13. Dice, L.R.: Measures of the Amount of Ecologic Association Between Species. Ecology 26(3), 297–302 (1945)
14. Jaccard, P.: Étude Comparative de la Distribution Floraledansune Portion des Alpeset des Jura. Bulletin de la SociétéVaudoise des Sciences Naturelles 37, 547–579 (1901)
15. Quasthoff, U., Wolff, C.: The Poisson Collocation Measure and its Applications. In: Proc. of the Second International Workshop on Computational Approaches to Collocations, Wien (2002)
16. Dunning, T.: Accurate methods for the statistics of surprise and coincidence. Computational Linguistics 19(1), 61–74 (1994)
17. Michel, J., et al.: Quantitative Analysis of Culture Using Millions of Digitized Books. Science 14 331(6014), 176–182 (2011)
18. Biemann, C., Bordag, S., Quasthoff, U.: Automatic Acquisition of Paradigmatic Relations using Iterated Co-occurrences. In: Proc. of the 4th International Conference on Language Resources and Evaluation (LREC 2004), Lisbon, Portugal, pp. 967–970 (2004)
19. Witschel, H.F.: Terminologie-Extraktion - Möglichkeiten der Kombination statistischer und musterbasierter Verfahren. Ergon-Verlag (2004)
20. Luhn, H.P.: Automatic Creation of Literature Abstracts. IBM Journal of Research and Development 2(2), 159–165 (1958)
21. Website of DocAnalyser (2014), `http://www.docanalyser.de` (last retrieved on March 01, 2014)
22. Kubek, M., Unger, H.: Detecting Source Topics by Analysing Directed Co-occurrence Graphs. In: Proc. 12th Intl. Conf. on Innovative Internet Community Systems, GI Lecture Notes in Informatics, vol. P-204, pp. 202–211. Köllen Verlag, Bonn (2012)

Variable Length Motif-Based Time Series Classification

Myat Su Yin[1], Songsri Tangsripairoj[1], and Benjarath Pupacdi[2]

[1] Faculty of ICT, Mahidol University, Bangkok, Thailand
myat.su@student.mahidol.ac.th, songsri.tan@mahidol.ac.th
[2] Chulabhorn Research Institute, Bangkok, Thailand
benjarath@cri.or.th

Abstract. Variable length time series motif discovery has attracted great interest in the community of time series data mining due to its importance in many applications such as medicine, motion analysis and robotics studies. In this work, a simple yet efficient suffix array based variable length motif discovery is proposed using a symbolic representation of time. As motif discovery is performed in discrete, low-dimensional representation, the number of motifs discovered and their frequencies are partially influenced by the number of symbols used to represent the motifs. We experimented with 4 electrocardiogram data sets from a benchmark repository to investigate the effect of alphabet size on the quantity and the quality of motifs from the time series classification perspective. The finding indicates that our approach can find variable length motifs and the discovered motifs can be used in classification of data where frequent patterns are inherently known to exist.

Keywords: Variable length time series motif discovery; suffix array; repeated pattern discovery, time series classification.

1 Introduction

Time series mining using patterns has received much attention recently as time series data are best understood and interpreted by their representational structures. Repeated patterns, also known as motifs, are extensively considered to represent time series due to repetition behavior of a pattern is considered to carry consistent context information of the underlying process.

Motif patterns discovery is often the data preprocessing step in mining association rule mining [1], k-mean clustering [2], as well as in classification [3], information visualization and summarization [4] and visualization of medical data time series [5]. With its wide range of applications, motif discovery and mining tasks increasingly become an important research area.

A majority of existing works on motif discovery in time series relies on the predefined length of a motif. However, frequent pattern of different lengths might be present in time series data [5]. Consequently, a pattern having a length different from the predefined length might be overlooked during motif discovery process. A naïve approach to find motifs of different lengths is to repeat the motif discovery algorithm for every possible length.

S. Boonkrong et al. (eds.), *Recent Advances in Information and Communication Technology*, Advances in Intelligent Systems and Computing 265,
DOI: 10.1007/978-3-319-06538-0_8,

Recently, several algorithms have been proposed to address the variable length motif discovery. A recent work [6] presented motifs of different length enumeration in real-valued time series efficiently. Since probability of observing any real-number is zero, working with real-valued time series might restrict further data analysis choices. It is well accepted that appropriate high-level representation may be best captured trends, shapes and patterns contained within the time series data [7]. One of such representation is symbolic representation and variable length motif discovery using it are found in [5, 8]. Known issues surrounding these methods are their reliance on input parameters, requirement for post processing, poor scalability with large datasets, and whole sequence discretization not capturing enough local details in data compared to the subsequence discretization [5]. Existing greedy grammar induction approach [5] is time and space efficient and tackles the above shortcomings. Although a greedy approach is efficient, an additional step is required to improve the quality of induced grammar.

In this study, we propose a novel motif discovery approach for extracting variable length time series motifs. It is a simple algorithm which looks for frequent motif patterns directly from the data. Apart from the parameters required for representation transformation and minimum support threshold, no additional parameter is required. Using suffix array and an auxiliary array, variable length motifs are extracted linearly with the input data. Motivated by recent progress in motif based time series classification, the usefulness of variable length motifs is further demonstrated in time series classification task. A statistical test is applied to measure the discriminative quality of motifs. We also examine the effect of the alphabet size chosen during time series representation transformation stage on the motif quality in a classification task.

Our contributions are as follows. Using suffix array, our approach can find a set of variable length motifs without requiring an input for a predefined motif length. No additional step is required to refine motif results.

The remainder of this paper is organized as follows: Section 2 describes the related works on motif discovery, interestingness of motif and time series classification. We discussed our proposed approach in Section 3. Section 4 contains experimental evaluation and discussion. Finally, Section 5 draws conclusion and gives suggestions for future work.

2 Related Works

Recently, a few methods have been proposed for variable length time series motif discovery in [3,5,8]. Except the parameters required for SAX transformation, the motif length is not a mandatory parameter in both approaches [3] and [5]. However, requirement for post processing, scalability, and adopted discretization methods which do not capture enough local details in data are considered as limitations of existing methods.

The number of variable length motifs discovered from a given time series dataset is generally large. It is essential to identify patterns that can both qualitatively and quantitatively characterize the underlying time series classes. Castro and Azevedo [8] incorporates a statistical significance (p-value) method to evaluate time series motifs. To assess the statistical significance of a motif, [4] computes information gain which measure the information content and log-odd significant measure which determines the degree of surprise of a motif.

In this study, we assessed the usability of variable length motifs in time series classification. Buza and Schmidt-Thieme [9] constructs feature vectors using motif time series classification. With SAX representation, a prefix tree like data structure performs classification with Bayesian and Support Vector Machine approach with reported accuracy values of 70.51% and 65.07% (66.28% with SVM-logistic) respectively. Specifically, an extended version of the Apriori algorithm is used to extract motifs. Similar to their work, we construct the motif feature vector using binary values as attribute values.

3 Suffix Array-Based Variable Length Time Series Motif Discovery

The overview of suffix array-based variable length motif discovery and time series classification is illustrated in Fig. 1. Major steps in our approach are Symbolic representation transformation, Suffix array-based motif discovery and Motif selection. They are followed by Classifier training and evaluation.

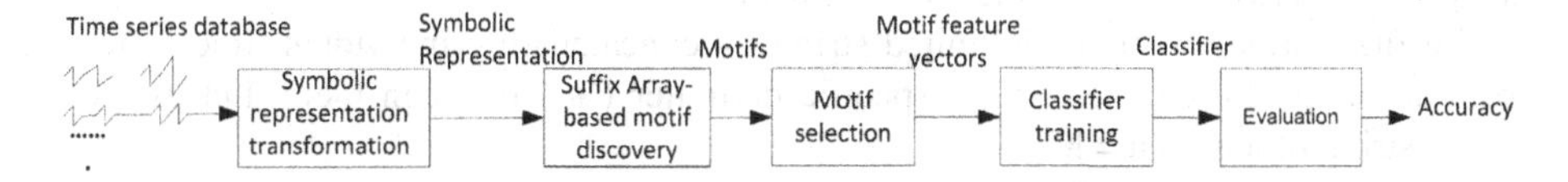

Fig. 1. Suffix array-based variable length time series motif discovery and classification

3.1 Symbolic Representation Transformation

We begin by introducing definitions for time series and subsequence.

Definition 1. *Time Series*: A time series T of length n is an ordered sequence of n real-valued variables, i.e., $T = (t_1..t_n), t_i \in \mathbb{R}$. T_{SAX} is a time series represented as sequence of discrete symbols of Symbolic Aggregate Approximate (SAX) representation [10].

Definition 2. *Subsequence*: Given T_{SAX} of length n, subsequence T_{seq} of T_{SAX} is a subsection of length m contiguous time instances starting from p, i.e., $T_{seq} = t_p..t_{p+m-1}$for $1 \leq p \leq n - m + 1$ [11].

Suffix array is originally designed for discrete data. Therefore, we convert real-valued time series data into symbolic representation. Among all available symbolic representations, SAX is widely used in time series motif discovery researches [4, 10, 11]. SAX transformation consists of two-stages. The process started by obtaining time series subsequences via a sliding window. Each window is divided into equal length segments and the mean value of each segment is computed. In the subsequent discretization stage, the calculated mean values from the previous stage are mapped to corresponding symbols (alphabets) via a statistical lookup table. Throughout the paper, symbols and alphabets will be used interchangeably.

The recommended alphabet size for the SAX representation is between 5 to 8 [10]. The choice of the alphabet sizes a affects the quantity and quality of motifs found in later stages. With small a, the level of information captured in each alphabet is low. Consequently, a less variety in the symbol pool to represent the subsequences also leads to increase in trivial matches. On the other hand, with a large a, more information is captured by each alphabet and would result in fewer trivial matches. The concatenation of symbols representing the segments forms a word w. The number of segments in SAX first stage determines the length of each word.

In our study, the sliding window length of 20, word length of 5 (characters) and the numbers of alphabet size is varied between 5 and 8. We adopted numerosity reduction option provided by [10] to avoid over representing of trivially matched pattern in the later stages.

3.2 Suffix Array-Based Motif Discovery

The suffix array data structure [12], introduced in 1990 is one of the most widely used data structures in bioinformatics due to its efficiency and efficacy to compute the frequency, and positions of a substring in a long sequence [13-14]. Suffix array works well for a character sequence without space/delimiter such as DNA biological sequences or protein sequences. For the delimited symbolic string, the word suffix array (WSA) [13] is an appropriate data structure.

We first formed a single delimited string by concatenating individual time series from an entire training set with a specific delimiter ('#') between them. Let GS be such a string of length n = |GS|.

Definition 3. *Word suffix array* [13] of *GS* is an array of integer array of integer in the range of 0 to |GS|, representing the lexicographic ordering of n word suffixes of GS, i.e., for a given string GS, $GS_{WSA[0]}, GS_{WSA[1]}, \dots, GS_{WSA[n]}$ is the sequence of suffixes of GS in ascending lexicographic order, where $GS_i = GS[i..|GS|]$ denotes the ith nonempty suffix of the string GS, $0 < i \le |GS|$.

Definition 4. *Longest Common Prefix (LCP)* [14] array for a string *GS* is an integer array with each cell indicating how many words a particular suffix has in common with the consecutive suffix in a word suffix array, i.e., $lcp[i]$ is the length of *lcp* of $GS_{WSA[i-1]}$and $GS_{WSA[i]}$, for $1 \le i \le n$. As an example, consider a string *S* = bac-cab-abb-bbc-bac-cab-abb-bac-bbc-bac, a time series in symbolic representation with word length *w*=3 and alphabet size a =3 with symbols {a,b,c}. The word suffix array and lcp-array for *S* are shown in Fig. 2.

Definition 5. *l-interval* Given *lcp* of length *n*, lcp-interval, denoted by (*l-interval* or $l[i..j]$) is an interval $[i..j]$, $0 \le i \le n$ with a lcp-value *l* which satisfies the following set of conditions [15].

- $lcp[i] < l$,
- $lcp[k] \ge l$ for all *k* with $i + 1 \le k \le j$,
- $lcp[k] = l$ for at least one k with $i + 1 \le k \le j$
- $lcp[j + 1] < l$.

To discover the frequent patterns from a word suffix array, an lcp-interval tree [15] is constructed. A lcp-interval tree is rooted with an lcp-interval, 0-interval ($0[1..n]$). Each suffix of the string is represented as a singleton interval [1..1] in lcp-interval tree. The parent interval of such a singleton interval is the smallest lcp-interval $[i..j]$ with $l \in [i..j]$. The parent child relationship forms the lcp-interval tree. For our running example string S, the lcp-interval tree is also shown in Fig 2.

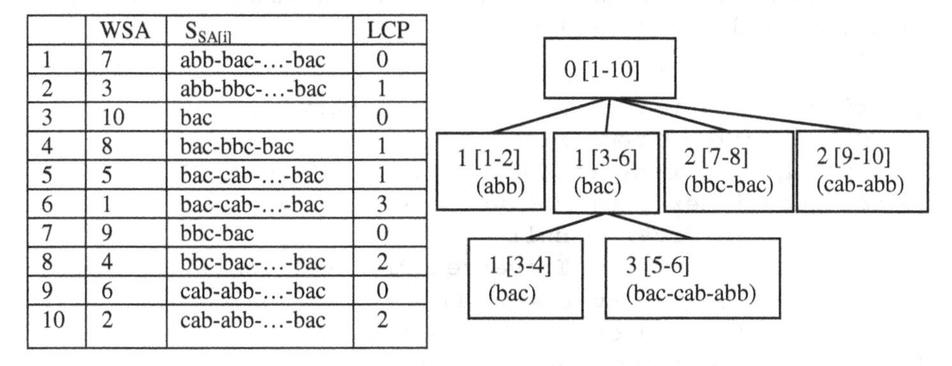

	WSA	$S_{SA[i]}$	LCP
1	7	abb-bac-…-bac	0
2	3	abb-bbc-…-bac	1
3	10	bac	0
4	8	bac-bbc-bac	1
5	5	bac-cab-…-bac	1
6	1	bac-cab-…-bac	3
7	9	bbc-bac	0
8	4	bbc-bac-…-bac	2
9	6	cab-abb-…-bac	0
10	2	cab-abb-…-bac	2

Fig. 2. Word suffix and lcp-array and the corresponding lcp-interval tree for S

Each interval node *l-interval* $l[i..j]$ carries essential information of a motif pattern. Specifically, l denotes the length of a pattern occurred between index range $i..j$ of WSA. $i..j$ represents the position(s) of the pattern in S. The number of repetition of a pattern is derived from the absolute difference between the interval lower and upper bounds $|i-j|+1$. For example, consider the node $1[3..6]$ in Fig.2; it can be interpreted as there is a substring of length *1*-word commonly occurs at WSA indexes between 3 and 6. A *1*-word string can be encoded back using $S(WSA[i]..WSA[i]+l-1)$. Therefore, pattern string for the *l*-interval $1[3..6]$ is 'bac'. Likewise, position indexes of a pattern (10,8,5,1) is obtained from $WSA[i..j]$ and frequency (4) is obtained by $|i-j|+1$.

As soon as an *l*-interval node is identified, its frequency is compared with minimum support value. Since each l-interval from lcp- interval tree compactly describes the patterns of arbitrary length along with their frequency, variable length repeated patterns within the given symbolic sequence are discovered directly from it. Finally, we define the motif as follows.

Definition 6. *Motif*: Given a string GS and minimum support threshold value called *minsup*, a symbolic subsequence P of length *q*, *q*<*|GS|* in GS is considered as a motif if the count of all occurrences of P in GS is at least *minsup*.

We describe the pseudo code of variable length motif discovery using suffix array in Algorithm 1. The algorithm is relatively straightforward. It takes a time series database in symbolic representation *DSAX* and minimum support threshold *minsup* as inputs. Firstly, a delimited string *GS* is formed from time series instances in *DSAX* and is subsequently used as an input into the word suffix array. From line 3 through 15, lcp- interval tree is traversed in a bottom up fashion as described in [15], and *l*-intervals are obtained. Each *l*-interval is a triple *(lcp, lb, rb)* where *lcp* is the lcp-value of the interval, *lb* is its left boundary and *rb* is its right boundary. Number of indexes that shares the same lcp-value is computed from lb and rb and is compared against the minimum support threshold *minsup*. Only the l-intervals that are having frequency at least *minsup* are taken to remap *l*-interval into the pattern string back from *GS*. Respective patterns occurrence positions in *GS* are obtained via suffix array WSA for further process. lcp-interval tree traversal and minimum support checking is performed for all elements in lcp-table.

```
Algorithm 1: VariablelengthMotifDiscovery(D_SAX, minsup)
1: GS ← FormDelimitedString(D_SAX)
2: (WSA, LCP) ← ConvertoWordIndex(charSA, charLCP)
3: for each index i in LCP do
4:         l_interval ← null
5:         l_interval ← TraverseLCPInteral(lcp[i])
6:         if (l_interval ≠ null)
7:          if(|l_interval_lb-l_interval_rb|> minsup)
8:            startIndex ←SA[l_interval_lb]
9:            endIndex  ← SA[l_itnerval_lb]+l_interval_lcp -1
10:           pattern ← GS_startIndex .. GS_endIndex
11:           positions  ← WSA[l_interval_lb-l_interval_rb]
12:           VarLenMotifs  ←Append(pattern,positions)
13:         end if
14:        end if
15: end for
16: return VarLenMotifs
```

As a sanity check, *winding* dataset [16] is used to extract variable length motifs. The length of the dataset is 2500. We fixed the SAX parameters at number of symbols = 3, word length = 4 and sliding window length =100 as in [5]. The motifs discovered are approximately similar to the findings of [5] in terms of locations and lengths. Examples of variable length motif are shown in Fig. 3. In this section, variable length motif discovery approach using suffix array and lcp-array is presented. The discovered motifs are ready for further analysis.

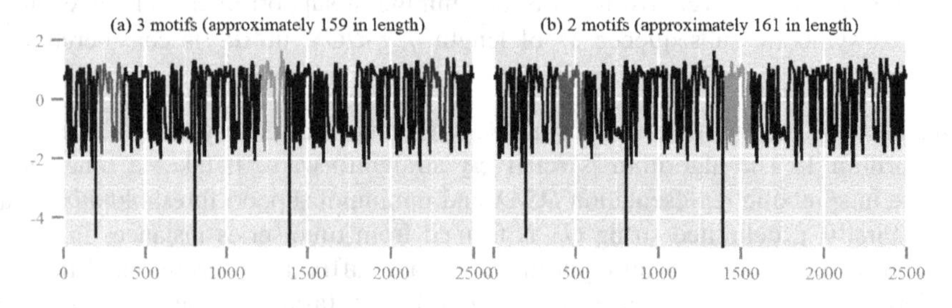

Fig. 3. Examples motif discovered from the *winding* dataset

3.3 Motif Selection and Motif Based Time Series Classification

In the previous stage, a pattern is considered to be frequent if its frequency is greater than a pre-determined minimum support threshold. Even with the minimum support threshold in place, the number of variable length frequent patterns will still result in a huge feature space. To obtain the discriminative patterns that can help to classify the objects more accurately while reducing the feature space, the discriminative quality of a pattern is determined using Chi-Square (Chi).

A supervised time series classification task involves the process of assigning a label to a time series based on an abstraction that is generated by learning through a set of training time series. The classification performance attaches a great importance to the patterns that are highly discriminative with respect to the class label. Through feature selection process, the quality-motifs are selected and time series are described with motif feature vectors. Any vector-based learning algorithm could be used in constructing the classification model.

4 Experiment

To investigate the effects of the alphabet size on the number of motifs discovered and its subsequent effect on classifier performance, an experiment is conducted using four ECG time series datasets from [16]. Selected datasets and their characteristics are shown in Table 1. Original training and testing sets are merged together and later partitioned into 70% and 30% subsets for training and testing set with a random sampling method. Minimum support threshold for pattern frequency is fixed at 3% of training set size. In our experiment, we use Larsson and Sadakane algorithm [17] to build suffix arrays and the algorithm of Kasai et al. [18] for lcp-array construction.

Table 1. Characteristics of datasets

No.	Name	Classes	Length	Training	Testing
#1	Two Lead ECG	2	82	23	1139
#2	ECG	2	96	100	100
#3	ECG Five Days	2	136	23	861
#4	CinC_ECG_Torso	4	1639	40	1380

Support Vector Machines (SVM) and k-nearest neighbor method with k=1 (1-NN) are selected as classification methods due to their superior performance among the best available classification methods. We use libsvm implementation in weka using RBF kernel with default parameters and parameter tuning function. To have a stable comparison result, 10-fold cross validation is used for training the classifiers. The performance of classifiers with motif-based feature vectors is evaluated from the accuracy value aspect. We compared the accuracy of motif feature vector based classifiers with nearest neighbor (1-NN) classifier with Euclidean distance using *real-valued* time series.

4.1 Findings from Experiment

Accuracy values of classifiers obtained from each dataset using different alphabet size (5-8) during transformation representation and Chi feature selection measure are presented in Fig. 4.

As shown in Fig. 4, in Dataset #1, both 1-NN (motif) and SVM (motif) achieved near perfect classification as 1-NN classifier in all alphabet size. Similarly, in Dataset #3, the 1-NN (motif) and SVM (motif) classifiers performance reaches closet to the 1-NN classifier result using motifs represented with alphabet size 8. Classifiers built with the variable length motifs represented with either lowest alphabet size of 5 or highest alphabet size of 8 give lower performance than middle range alphabet size (6-7) in Datasets #2 and # 4. Among two classifiers, SVM (motif) achieves the higher accuracy independently from alphabet size than 1-NN (motif) classification.

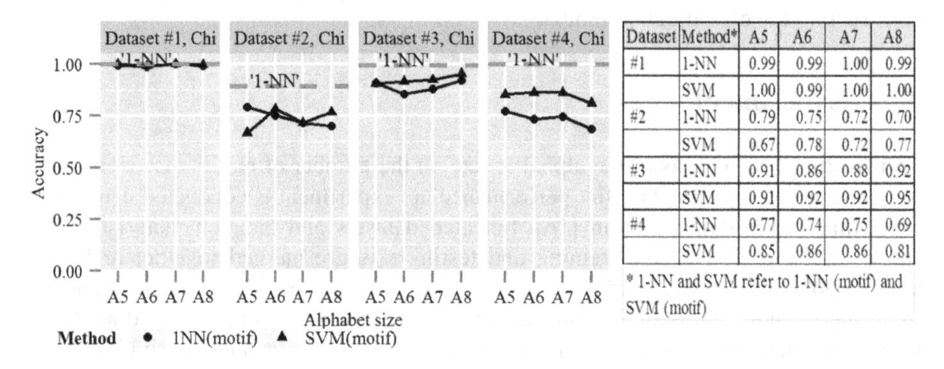

Dataset	Method*	A5	A6	A7	A8
#1	1-NN	0.99	0.99	1.00	0.99
	SVM	1.00	0.99	1.00	1.00
#2	1-NN	0.79	0.75	0.72	0.70
	SVM	0.67	0.78	0.72	0.77
#3	1-NN	0.91	0.86	0.88	0.92
	SVM	0.91	0.92	0.92	0.95
#4	1-NN	0.77	0.74	0.75	0.69
	SVM	0.85	0.86	0.86	0.81

* 1-NN and SVM refer to 1-NN (motif) and SVM (motif)

Fig. 4. Classification accuracy w.r.t alphabet size. x- axis (A5-A8) represents number of symbols used to represent motifs while accuracy is presented in y-axis. 1-NN (dashed-line) is an accuracy value of 1-NN classification performed on *real-valued* time series, 1-NN (motif) and SVM (motif) are accuracy values of 1-NN and SVM classifications performed on variable length motifs based feature vectors.

4.2 Discussion

Our approach gained advantages directly contributed from suffix array and lcp-array. The construction of both arrays is linear with input symbolic database size. Input raw time series dataset is accessed once during symbolic representation transformation stage. Only a linear scan of lcp-array [15] is required to construct the lcp-interval tree. After motif patterns are identified via lcp-interval tree, a second pass to symbolic time series sequences is required to retrieve the pattern strings. It is carried out via suffix indexes associating with each pattern. The overall performance of this process depends on the number of arbitrary length patterns obtained after pruning with a minimum support threshold. Another advantage of lcp-interval tree is the ease in determining the frequency and the pattern length easily from an l-interval node. With this feature, we can optionally filter out top k-frequent patterns or l-length patterns, which iterative and incremental algorithms do not support.

Based on the accuracy values, 1-NN classifier using real-valued time series outperformed the classifiers using feature vectors constructed from variable length motifs. However, for 1-NN classifier, there is a tradeoff between the high accuracy with high storage requirement and computational complexity in testing process. Motif feature vectors can also be regarded as a set of features that are representative of the underlying data and it also allows us to work with any vector-learning algorithms.

For our work, a limitation lies in direct correctness evaluation of discovered motifs. It is due to training and testing dataset adjustment steps we have taken. However, we demonstrated that our motif results are of a high quality via our classification results. Even though, comparable works on variable length motif discovery exist in literature, differences in experimental data and inaccessibility to existing programs for approximate representation hindered the evaluation process.

5 Conclusion and Future Work

In this work, we have presented an approach to find approximate variable length motif discovery for time series analysis using suffix array. Our approach does not require a predefined motif length or any additional step to refine the motif results. The usefulness of extracted variable-length motifs is examined by incorporating selected motifs into a feature vector and subsequently performed classification with two classification methods.

Findings indicate that suffix array-based can find variable length motifs and discovered patterns can be effectively used in time series classifications. Since frequent patterns are used in constructing feature vectors, our approach could be beneficial to data which are inherently exhibiting frequent characteristics, in particular, medical data time series. The discovery and study of such frequent patterns can provide valuable clues relating to the medical data under examination.

There are many directions to extend the current work. In our work, the motifs are considered from an occurrence frequency aspect, the exact ordering of the patterns is not taken into account. It is possible to extend the current work by considering the sequential or temporal orders of patterns which can be efficiently obtained from suffix array.

References

1. Das, G., Lin, K., Mannila, H.: Rule Discovery from Time Series. In: Proceedings of Fourth Int'l Conf. Knowledge Discovery and Data Mining, New York, NY, USA, pp. 16–22 (1998)
2. Phu, L., Anh, D.T.: Motif-Based Method for Initialization the K-Means, pp. 11–20 (2011)
3. Hegland, M., Clarke, W., Kahn, M.: Mining the MACHO dataset. Computer Physics Communications 142, 22–28 (2001)
4. Lin, J., Keogh, E., Lonardi, S.: Visualizing and discovering non-trivial patterns in large time series databases. Information Visualization 4, 61–82 (2005)
5. Li, Y., Lin, J., Oates, T.: Visualizing Variable-Length Time Series Motifs. In: Proceedings of the 2012 SIAM International Conference on Data Mining, Anaheim, CA, pp. 895–906 (2012)

6. Mueen, A.: Enumeration of Time Series Motifs of All Lengths. In: ICDM. IEEE (2013)
7. Ratanamahatana, C.A., Lin, J., Dimitrios, G., Keogh, E.J., Vlachos, M., Das, G.: Mining Time Series Data. In: The Data Mining and Knowledge Discovery Handbook, pp. 1069–1103 (2005)
8. Castro, N., Azevedo, P.: Significant Motifs in Time Series. Statistical Analysis and Data Mining 5(1), 35–53 (2012)
9. Buza, K., Schmidt-Thieme, L.: Motif-based Classification of Time Series with Bayesian Networks and SVMs. In: Fink, A., Lausen, B., Seidel, W., Ultsch, A. (eds.) Advances in Data Analysis, Data Handling and Business Intelligence, pp. 105–114 (2010)
10. Lin, J., Keogh, E., Wei, L., Lonardi, S.: Experiencing SAX: A Novel Symbolic Representation of Time Series. Data Mining and Knowledge Discovery 15, 107–144 (2007)
11. Chiu, B., Keogh, E., Lonardi, S.: Probabilistic Discovery of Time Series Motifs. In: Proceedings of the Ninth ACM SIGKDD International Conference on Knowledge Discovery and Data Mining, KDD 2003, pp. 493–498. ACM Press, New York (2003)
12. Manber, U., Myers, G.: Suffix Arrays: A New Method for On-Line String Searches. SIAM Journal on Computing 22, 935–948 (1993)
13. Ferragina, P., Fischer, J.: Suffix Arrays on Words. In: Ma, B., Zhang, K. (eds.) CPM 2007. LNCS, vol. 4580, pp. 328–339. Springer, Heidelberg (2007)
14. Fischer, J.: Inducing the LCP-Array. In: Dehne, F., Iacono, J., Sack, J.-R. (eds.) WADS 2011. LNCS, vol. 6844, pp. 374–385. Springer, Heidelberg (2011)
15. Abouelhoda, M.I., Kurtz, S., Ohlebusch, E.: The Enhanced Suffix Array and Its Applications to Genome Analysis. In: Guigó, R., Gusfield, D. (eds.) WABI 2002. LNCS, vol. 2452, pp. 449–463. Springer, Heidelberg (2002)
16. Keogh, E., Zhu, Q., Hu, B., Hao, Y., Xi, X., Wei, L., Ratanamahatana, C.A.: The UCR Time Series Classification/Clustering Homepage, `http://www.cs.ucr.edu/~eamonn/time_series_data/`
17. Larsson, N.J., Sadakane, K.: Faster suffix sorting. Theoretical Computer Science 387, 258–272 (2007)
18. Kasai, T., Lee, G., Arimura, H., Arikawa, S., Park, K.: Linear-Time Longest-Common-Prefix Computation in Suffix Arrays and Its Applications. In: Amir, A., Landau, G.M. (eds.) CPM 2001. LNCS, vol. 2089, pp. 181–192. Springer, Heidelberg (2001)

Adaptive Histogram of Oriented Gradient for Printed Thai Character Recognition

Kuntpong Woraratpanya and Taravichet Titijaroonroj

Faculty of Information Technology,
King Mongkut's Institute of Technology Ladkrabang, Bangkok, Thailand
{Kuntpong,bank2533}@gmail.com

Abstract. A similarity of printed Thai characters is a grand challenge of optical character recognition (OCR), especially in case of a variety of font types, sizes, and styles. This paper proposes an effective feature extraction, adaptive histogram of oriented gradient (AHOG), for overcoming the character similarity. The proposed method improves the conventional histogram of oriented gradient (HOG) in two principal phases, which are (i) adaptive partition for gradient images and (ii) adaptive binning for oriented histograms. The former is implemented with quadtree partition based on gradient image variance so as to provide for an effective local feature extraction. The later is implemented with non-uniform mapping technique, so that the AHOG descriptor can be constructed with minimal errors. Based on 59,408 single character images equally divided into training and testing samples, the experimental results show that the AHOG method outperforms the conventional HOG and state-of-the-art methods, including scale space histogram of oriented gradient (SSHOG), pyramid histogram of oriented gradient (PHOG), multilevel histogram of oriented gradient (MHOG), and HOG column encoding algorithm (HOG-Column).

Keywords: Printed Thai Character Recognition, Pattern Recognition, Histogram of Oriented Gradient (HOG), Adaptive Histogram of Oriented Gradient (AHOG), Feature Extraction.

1 Introduction

An evolution of printed Thai fonts has continually developed and created for a variety of print medias, such as magazines, books, brochures, newspapers, and so on. These are appealing to readers, but are useless to visually impaired persons. In addition, a variety of font types, sizes, and styles is a key factor to degrade recognition rate of optical character recognition (OCR), which is a process of converting document images to editable text and Braille books. Therefore, the improvement of OCR based on a variety of print medias in Thai language is essential for visually impaired persons.

Although the printed Thai character recognition has been continually researched over the past two decades [1], it still requires the performance improvement for applying to real applications [2–4]. One of the main problems is the diversity of new

S. Boonkrong et al. (eds.), *Recent Advances in Information and Communication Technology*, Advances in Intelligent Systems and Computing 265,
DOI: 10.1007/978-3-319-06538-0_9, © Springer International Publishing Switzerland 2014

Thai character fonts, i.e., the more new fonts are created, the more recognition errors are increased. Furthermore, a similarity of printed Thai characters becomes a grand challenge of the Thai OCR. In order to achieve the higher performance of character recognition, the effective feature extraction method is required. Based on observation regarding a Thai character structure, its shape is a significant feature. For this reason, this paper investigates on the effective shape feature extraction. One of the successful methods for shape recognition is the histogram of oriented gradient (HOG) introduced by Navneet D. et al. [5]. Since then, many research papers have proposed modified HOGs, for example, scale space histogram of oriented gradient [6], pyramid histogram of oriented gradient [7], and multilevel histogram of oriented gradient [8]. Most of these have been successful in high efficiency for object recognition, but few papers [9] studied on HOG for character recognition.

Even though the conventional HOG and modified HOG methods have been successful for object recognition, these fail to printed Thai character recognition. The main obstacle is the similarity of Thai characters. Therefore, this paper proposes an AHOG method to overcome this problem. This method improves the conventional HOG algorithm in two principal phases, (i) adaptive partition of gradient images and (ii) adaptive binning of oriented histograms. The former phase is implemented with quadtree decomposition based on gradient image variance so as to provide for effectively extracting local shape features. The later phase is implemented with non-uniform quantization, so that the AHOG descriptor can be constructed with minimal errors. In this way, the recognition rate of Thai characters is significantly improved.

The rest of this paper is organized as follows: Section 2 reviews backgrounds of the conventional HOG and modified HOG algorithms, and points out their drawbacks. In Section 3, an AHOG method is proposed. Section 4 shows experimental results and discussions. Finally, conclusions are presented in Section 5.

2 Related Works

In this section, the important issues of conventional HOG and state-of-the-art methods, including SSHOG, PHOG, MHOG, and HOG-Column, are reviewed and pointed out their drawbacks when applied to Thai character recognition.

A similarity of printed Thai characters becomes a critical issue in recognition as shown in Fig. 1. In this case, one way to classify similar characters effectively is using local shape features. This paper investigates on an improvement of feature extractions based on shape recognition. One of the effective methods for shape recognition is HOG, which was first proposed by Navneet D. et al. [5] and was successfully applied to human detection. The HOG algorithm is composed of five procedures, (i) gamma and color normalization, (ii) gradient computation, (iii) spatial and oriented binning, (iv) normalization and descriptor blocks, and (v) detector window and context. Then the features extracted from the HOG algorithm are classified by a linear support vector machine (SVM) to identify persons or non-persons. Although the HOG method achieves in human detection with high performance, it is not suitable for Thai characters with high similarity. The main reason is that the HOG approach was designed for coarse scale of object recognition, thus making it difficult to classify similar shape objects.

Fig. 1. An example of three groups of printed Thai characters with high similarity

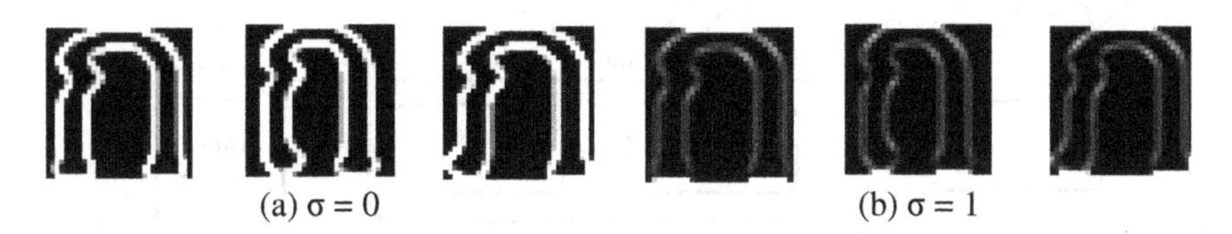

Fig. 2. An effect of applying two scale spaces, $\sigma = 0$ and $\sigma = 1$, to gradient images

SSHOG was introduced by Ning H. et al. [6] to solve complex objects and human detections which can be better to perceive at different scales. The main idea of this method is applying Gaussian kernel with multiple scales as demonstrated in Fig. 2. The conventional HOG differs from SSHOG in that it has only uni-scale ($\sigma = 0$). The experimental results show that the SSHOG method outperforms the conventional HOG method. Nonetheless, the higher scale of the SSHOG method filters out fine shape information. This leads to the misclassification when the SSHOG method applies to Thai characters. Furthermore, the more increased scales construct the larger feature size, thus consuming the computational time.

Representing shape with a spatial pyramid kernel was introduced by Anna B. et al. [7]. Its main idea is to extend a resolution of the conventional HOG method to multi-resolutions. In this method, the resolution implies a number of sub-region images divided from the original image. The descriptor made from the multi-resolutions is called PHOG. Its efficiency is better than the traditional HOG method. However, each resolution of the PHOG method is allocated with a fixed size. This technique works well with coarse shape objects, but it leads to the less recognition rate when applied to Thai characters with high similarity.

MHOG was proposed by Subhransu M. et al. [8]. Due to this method derived from the PHOG algorithm, it cannot extract the fine shape information to generate effective features. Furthermore, feature vectors are increasingly larger size.

Andrew J.N. et al. [9] introduced HOG column encoding algorithm (HOG-Column) and showed that the descriptor constructed from this algorithm is robust to character recognition. This approach is an extension of the conventional HOG descriptor by including features extracted from multiple scales, a base scale, σ_{base}, and a coarser scale, $r\sigma_{base}$, where r is a scale ratio. It is successful to recognize imperfect characters acquired from natural scenes. However, in case of a higher scale, there is no difference when compared to the SSHOG method.

Based on an analysis of Thai character patterns and conventional HOG algorithms, it can be summarized into two issues. (i) The Thai characters require fine shape features for classification, whereas (ii) the conventional HOG method needs to be improved for fine shape recognition. In order to meet the requirements, the critical issue of the HOG algorithm is regarded as three factors, (i) a fixed partitioning, (ii) a fixed binning, and (iii) a scalar magnitude sum. Fortunately, such an issue can be solved by the proposed method which is described in details in the next section.

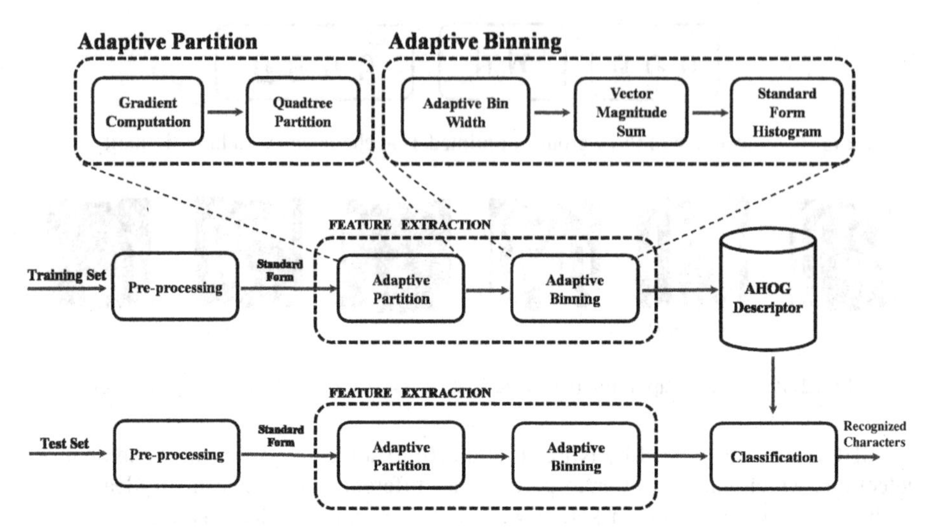

Fig. 3. A framework of AHOG for printed Thai character recognition

3 Proposed Method

In this paper, AHOG, which is an effective feature extraction, is proposed for overwhelming Thai character recognition with high similarity. A framework of the proposed method is schematically depicted in Fig. 3 and the manipulation details of each phase are described as follows.

3.1 Adaptive Partition

In general, printed Thai character recognition can be viewed as object recognition. Features extracted from printed Thai characters are represented in shape, which has a certain pose with a single view point. The best way to achieve in extracting the effective descriptor is using adaptive partition. In this procedure, there are two important phases, gradient computation and quadtree partition, to extract effective features.

Gradient Computation. Gradient images can be calculated by several methods. The mostly used technique is the first order derivative of Gaussian filters, such that S_x = [-1 0 1] and S_y = $[-1\ 0\ 1]^T$, where superscript T is a matrix transpose operator. After that, two features, magnitude, M, and orientation, θ, of the gradient images are calculated by using equations (1) and (2) for each pixel, respectively, and the results are illustrated in Fig. 4.

$$M = \sqrt{I_x^2 + I_y^2} \tag{1}$$

$$\theta = \tan^{-1}\left(\frac{I_y}{I_x}\right); \ -\pi \le \theta \le \pi \tag{2}$$

where I_x and I_y are gradient images.

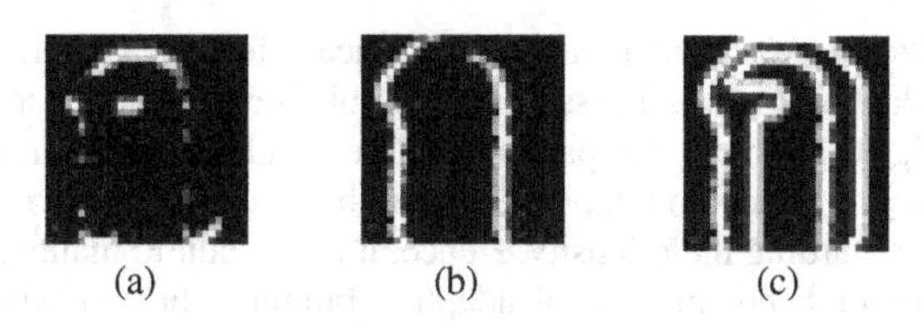

Fig. 4. (a) gradient image convolved by S_x, (b) gradient image convolved by S_y, and (c) a combination of gradient images from (a) and (b)

Quadtree Partition. In this phase, magnitudes of an image gradient are divided into sub-images or cells whose sizes are a square. The conventional HOG algorithm uses a fixed partition technique, i.e., all cells are equal sizes. Nevertheless, this technique has a disadvantage. As illustrated in Fig. 5, two similar shape character images are partitioned by means of an adaptive decomposition based on variance of the gradient images and are partitioned by means of a fixed partition. It is noticed that in the middle-left side of Fig. 5(a) and 5(b), the fixed partition does not provide an effective local shape feature to distinguish between two similar characters. In order to improve this, the AHOG algorithm makes use of adaptive partition based on gradient image variance as depicted in Fig. 5(c) and 5(d). It is evident that the adaptive partition is able to provide a better local shape feature.

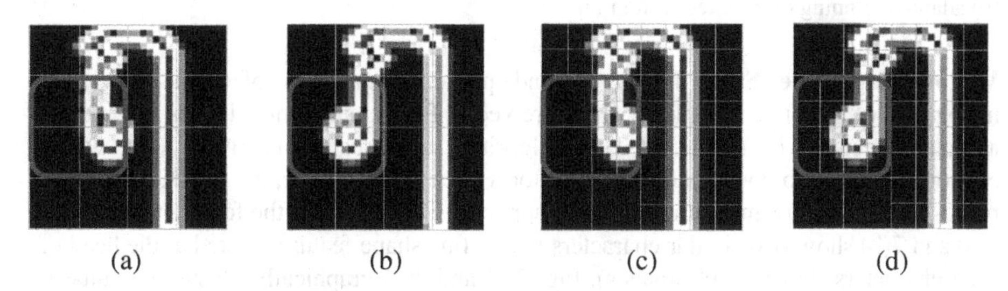

Fig. 5. A comparison of fixed and adaptive partitions

This paper implements the cell size with two levels, 4×4 pixels and 8×8 pixels, since the template of printed Thai characters is designed with size 32×32 pixels. If cell size is smaller than 4×4 pixels, some bins have zero data. This leads to those bins having no vectors in orientation. On the other hand, it is impossible to use cell size greater than 8×8 pixels, because the cell's data is coarser for classification.

3.2 Adaptive Binning

An adaptive binning is an important procedure to minimize error of feature extraction. This procedure is composed of three phases as explained in the following subsections.

Adaptive Bin Width. A bin width is an important factor having an impact on the quality of an AHOG descriptor. In this paper, the bin width assignment makes use of a non-uniform quantization technique. Such a technique is a many-to-few mapping, thus the bin width becomes a key factor to maintain the significant feature as well as possible. The main idea of an adaptive method is assigning the suitable bin width for the high density of information. Fig. 6 shows a comparison of fixed and adaptive bin-ning of oriented histograms. In case of fixed binning, the bin width is equally defined for the oriented histogram without regarding their density. Hence, it is difficult to maintain the significant features. On the other hand, in case of adaptive binning, the bin width is adaptively defined for those features with regarding the density of the oriented histogram. The higher density of features, the finer bin width is defined to preserve significant features. Here, the suitable bin width can be obtained by using this criterion 2.5σ, where σ is a standard deviation of oriented histograms of a training set.

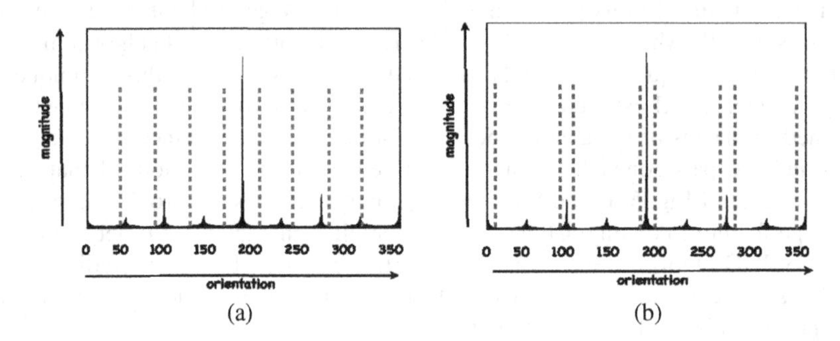

Fig. 6. A comparison of bin width assignments: (a) fixed binning of oriented histogram and (b) adaptive binning of oriented histogram

Vector Magnitude Sum. In the second phase, magnitudes of each partitioned histograms of orientated gradient, which are vector forms, are summed so as to generate a feature vector. The traditional HOG algorithm uses scalar magnitude sum, which cannot provide a more accurate feature vector. Hence, in this paper, the AHOG algorithm makes use of a vector magnitude sum to improve the efficiency of the feature vector. Fig. 7(a) and 7(b) show two similar characters with a fine shape feature located at the head of such characters (left-top sub-images). Fig. 7(c) and 7(d) graphically illustrate results of the fine shape feature by using vector and scalar magnitude sums, respectively. The results from both techniques are different. The vector magnitude sum presents a more accurate result; therefore, it is more appropriate for printed Thai characters. The outcome of this phase is a two-level AHOG descriptor as depicted in Fig. 8(a).

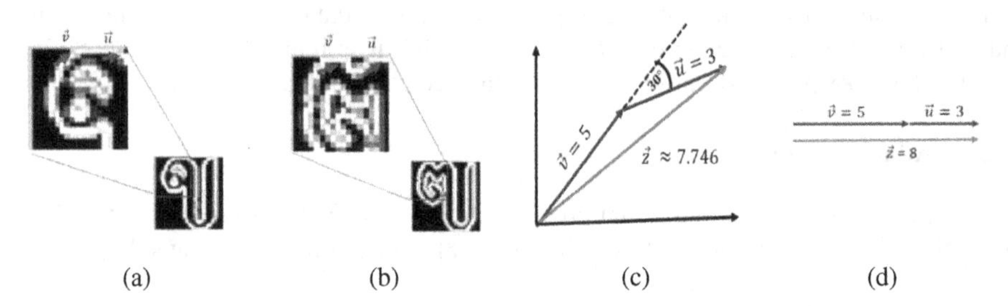

Fig. 7. (a) a sub-image of 'ข', (b) a sub-image of 'ฆ', (c) a vector addition, and (d) a scalar addition

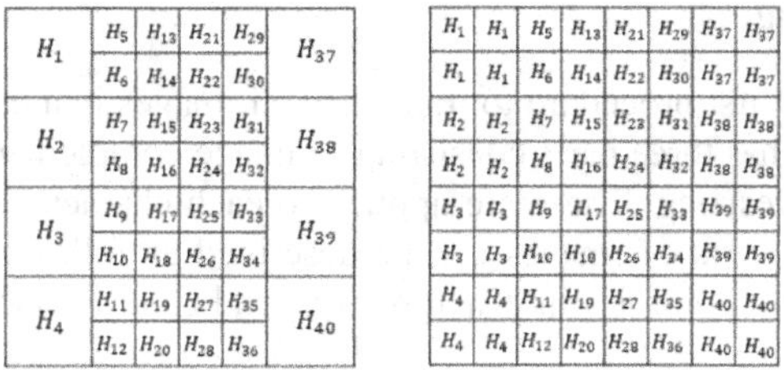

H_1	H_5	H_{13}	H_{21}	H_{29}	H_{37}
	H_6	H_{14}	H_{22}	H_{30}	
H_2	H_7	H_{15}	H_{23}	H_{31}	H_{38}
	H_8	H_{16}	H_{24}	H_{32}	
H_3	H_9	H_{17}	H_{25}	H_{33}	H_{39}
	H_{10}	H_{18}	H_{26}	H_{34}	
H_4	H_{11}	H_{19}	H_{27}	H_{35}	H_{40}
	H_{12}	H_{20}	H_{28}	H_{36}	

(a)

H_1	H_1	H_5	H_{13}	H_{21}	H_{29}	H_{37}	H_{37}
H_1	H_1	H_6	H_{14}	H_{22}	H_{30}	H_{37}	H_{37}
H_2	H_2	H_7	H_{15}	H_{23}	H_{31}	H_{38}	H_{38}
H_2	H_2	H_8	H_{16}	H_{24}	H_{32}	H_{38}	H_{38}
H_3	H_3	H_9	H_{17}	H_{25}	H_{33}	H_{39}	H_{39}
H_3	H_3	H_{10}	H_{18}	H_{26}	H_{34}	H_{39}	H_{39}
H_4	H_4	H_{11}	H_{19}	H_{27}	H_{35}	H_{40}	H_{40}
H_4	H_4	H_{12}	H_{20}	H_{28}	H_{36}	H_{40}	H_{40}

(b)

Fig. 8. Standard form histogram construction: (a) two-level AHOG and (b) one-level AHOG, standard form histogram

Standard Form Histogram. In the last phase, the standard form histogram is formed by splitting the larger cell size into smaller cell size, so that the AHOG descriptor has the same dimensionality of features. Fig. 8 demonstrates that a two-level AHOG with 8×8 and 4×4 pixels is decomposed into a one-level AHOG with 4×4 pixels. For instance, one $H_{1(8\times8)}$ feature cell is separated into four $H_{1(4\times4)}$ cells. The standard form histogram is the AHOG descriptor as shown in Fig. 8(b).

4 Experimental Results

In order to evaluate the efficiency of the proposed method, AHOG, in terms of recognition accuracy and computing time, the experiments are set up as follows.

4.1 Dataset Preparation

The dataset used in all experiments is a printed Thai character image corpus made from a variety of Thai medias. It is composed of consonants, vowels, and tonal marks. A resolution of such an image dataset is a 400 dpi. Font types are composed of AngsanaUPC, BrowalliaUPC, CordiaUPC, DilleniaUPC, EucrosiaUPC, FreesiaUPC, IrisUPC, and JasmineUPC, and font sizes are composed of 8, 10, 12, 14, 16, 18, 20, and 22. Font styles are regular, bold, italic, and bold-italic. There are totally 59,408 samples equally divided into training and testing sets. All of font types, sizes, and styles have three levels: upper, middle, and lower levels. Each level contains a number of characters as shown in Table 1.

Table 1. Printed Thai characters divided into three levels

Level	Member
Upper level	◌ิ ◌ี ◌ึ ◌ื ◌่ ◌้ ◌๊ ◌๋ ◌็ ◌ั ◌์ ◌ํ
Middle level	ก ข ฃ ค ฅ ฆ ง จ ฉ ช ซ ฌ ญ ฎ ฏ ฐ ฑ ฒ ณ ด ต ถ ท ธ น บ ป ผ ฝ พ ฟ ภ ม ย ร ล ว ศ ษ ส ห ฬ อ ฮ ๐ ๑ ๒ ๓ ๔ ๕ ๖ ๗ ๘ ๙ ะ า โ ไ ใ เ ฤ ฦ ๆ ฯ
Lower level	◌ุ ◌ู

4.2 Pre-processing

Pre-processing is the first procedure to transform all images in a dataset to a standard form. The procedure has three parts consisting of image complementation, image zero padding, and image resizing. These are applied to each character image as shown in Fig. 9. The standard form of images can be described as follows: the image size is 32×32 pixels, character color is white, and background color is black.

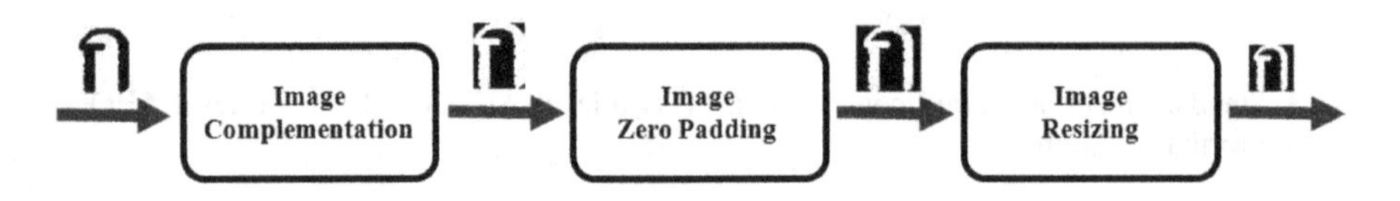

Fig. 9. Pre-processing procedure: image complementation, image zero padding, and image resizing

4.3 Performance Evaluation

This paper focuses on the feature extraction for constructing the effective descriptor. Thus, the descriptor performance is evaluated by means of the simple measurement, Euclidean distance, for classification.

Two experiments are set up in order to evaluate the performance of the AHOG algorithm. The first experiment aims to test the accuracy of the proposed method in terms of recognition rate. There are four state-of-the-art methods—including SSHOG, PHOG, MHOG, and HOG-column—and one conventional HOG, which are implemented as baseline algorithms. For all algorithms, the number of bins of oriented histograms is set to 9 as recommended in [5]. The experimental results are shown in Table 2. It is found that the recognition rate of the AHOG algorithm outperforms all of state-of-the-art algorithms.

In upper level, there are totally 1,128 character images for test. 1.00% of upper level characters are approximately 11 images. In this case, the maximum and minimum improvements of recognition rate in regular style are 7.10% and 0.09%, when compared to the HOG-Column and SSHOG methods. That is, the number of characters correctly recognized in the maximum improvement case is increasing about 80 character images. On the other hand, at least one character is correctly recognized in the minimum improvement case. In bold style, the maximum and minimum improvements of the recognition rate are 5.67% and 1.41%, when compared to the HOG-Column and SSHOG methods, respectively. It means that there are 64 characters greatly improved for recognition when compared to the HOG-Column method whereas there are at least 16 characters improved for recognition when compared to SSHOG. In the same way, in italic style, the maximum and minimum recognition rates of improvements are 6.56% and 0.10% when compared to the HOG-Column and SSHOG methods, respectively. In bold-italic style, the maximum and minimum recognition rates of improvements are 7.62% and 1.06% when compared to the HOG-Column and MHOG methods, respectively.

Middle level has 6,110 character images for test. 1.00% of middle level is approximate 61 character images. The great improvements of recognition rate in regular, bold, italic, and bold-italic styles are 5.61%, 4.91%, 5.06%, and 4.13%, respectively, when compared to the HOG-Column method. Simultaneously, when compared to the SSHOG method, the AHOG method outperforms with the better recognition rate, 0.50%, 0.64%, 0.59%, and 0.28% for regular, bold, italic, and bold-italic styles, respectively.

In lower level, there are 188 characters for test. The experimental results illustrate that all methods yield the same results, 100%.

Table 2. A comparison of recognition accuracy of HOG, SSHOG, PHOG, MHOG, HOG-Column, and AHOG methods

Method	Upper Level				Middle Level				Lower Level			
	Regular	Bold	Italic	Bold-Italic	Regular	Bold	Italic	Bold-Italic	Regular	Bold	Italic	Bold-Italic
HOG	96.54	94.77	96.90	95.48	97.50	98.67	98.28	98.89	100.00	100.00	100.00	100.00
SSHOG	97.07	95.66	97.42	95.92	98.58	98.77	98.79	99.25	100.00	100.00	100.00	100.00
PHOG	94.24	94.59	94.33	93.44	95.01	95.99	95.52	96.45	100.00	100.00	100.00	100.00
MHOG	96.19	95.66	96.63	96.01	97.07	97.23	96.73	96.87	100.00	100.00	100.00	100.00
HOG Column	90.06	91.40	90.96	89.45	93.47	94.50	94.32	95.40	100.00	100.00	100.00	100.00
AHOG	**97.16**	**97.07**	**97.52**	**97.07**	**99.08**	**99.41**	**99.38**	**99.53**	100.00	100.00	100.00	100.00

Table 3. A comparison of feature size and computing time of HOG, SSHOG, PHOG, MHOG, HOG-Column, and AHOG methods

Method	HOG	SSHOG	PHOG	MHOG	HOG Column	AHOG
Feature Size	1,764	7,056	765	16,992	720	**576**
Computing Time	0.008 s	0.031 s	0.012 s	0.111 s	0.018 s	0.034 s

The second experiment aims to test the proposed method in terms of computing time and feature size. The experimental results are shown in Table 3. The proposed method has the minimal feature, 576 feature vectors. However, the computing time of proposed method is in an average, 0.034 sec, when compared with all methods. Based on the experimental results, it is summarized that the proposed method, AHOG, achieves the higher recognition rate in comparison with the state-of-the-art methods. In addition, the AHOG descriptor is constructed with the smallest size of feature vectors.

5 Conclusions

In this paper, an AHOG method has been proposed to improve the efficiency of extracting local shape features and to increase recognition rate of printed Thai characters with high similarity. In order to achieve these purposes, the proposed method enhances the conventional HOG algorithm in two principal phases, i.e., (i) using an adaptive partition for gradient images to increase the efficiency of a local shape feature extraction, and (ii) applying an adaptive binning for oriented histograms to reduce the error of the local shape feature extraction. Based on these improvements, the AHOG descriptor is minimal, 576 features, when compared with baseline methods, whereas the computing time of feature extraction is in an average, 0.034 sec. Furthermore, the AHOG algorithm is evaluated in terms of recognition accuracy. Based on 59,408 single character images equally divided into training and testing samples, the experimental results show that the AHOG algorithm outperforms the conventional HOG and state-of-the-art algorithms, including SSHOG, PHOG, MHOG, and HOG-column.

References

1. Kimpan, C., Itoh, A., Kawanishi, K.: Fine Classification of Printed Thai Character Recognition Using the Karhunen-Loeve Expansion. IEE Proceedings 134, 257–264 (1987)
2. Tanprasert, C., Sae-Tang, S.: Thai type style recognition. In: 1999 IEEE International Conference on Circuits and Systems, pp. 336–339. IEEE Press, New York (1999)
3. Thammano, A., Duangphasuk, P.: Printed Thai Character Recognition Using the Hierarchical Cross-correlation ARTMAP. In: the 17th IEEE International Conference on Tools with Artificial Intelligence, pp. 695–698. IEEE Press, New York (2005)
4. Woraratpanya, K., Titijaroonrog, T.: Printed Thai Character Recognition Using Standard Descriptor. In: Meesad, P., Unger, H., Boonkrong, S. (eds.) IC^{2}IT2013. AISC, vol. 209, pp. 165–173. Springer, Heidelberg (2013)
5. Navneet, D., Bill, T.: Histograms of Oriented Gradients for Human Detection. In: IEEE International Conference on Computer Vision and Pattern Recognition, pp. 1–8. IEEE Press, New York (2005)
6. Ning, H., Jiaheng, C., Lin, S.: Scale Space Histogram of Oriented Gradients for Human Detection. In: 2008 International Symposium on Information Science and Engineering, pp. 167–170. IEEE Press, New York (2008)
7. Anna, B., Andrew, Z., Xavier, M.: Representing Shape with a Spatial Pyramid Kernel. In: The 6th ACM International Conference on Image and Video Retrieval (2007)
8. Subhransu, M., Alexander, C.B., Jitendra, M.: Classification Using Intersection Kernel Support Vector Machines is Efficient. In: IEEE International Conference on Computer Vision and Pattern Recognition, IEEE Press, New York (2008)
9. Andrew, J.N., Lewis, D.G.: Multiscale Histogram of Oriented Gradient Descriptors for Robust Character Recognition. In: 2011 International Conference on Document Analysis and Recognition, pp. 1085–1089. IEEE Press, New York (2011)

A Comparative Machine Learning Algorithm to Predict the Bone Metastasis Cervical Cancer with Imbalance Data Problem

Kasama Dokduang[1], Sirapat Chiewchanwattana[1], Khamron Sunat[1], and Vorachai Tangvoraphonkchai[2]

[1] Department of Computer Science, Faculty of Science, Khon Kaen University, Thailand
kasama.d@kkumail.com, sunkra@kku.ac.th, khamron_sunat@yahoo.com
[2] Department of Radiology, Faculty of Medicine, Khon Kaen University, Thailand
vorachai@kku.ac.th

Abstract. This paper attempted to develop and validate a tool to predict the immediate results of radiation on bone metastasis in cervical cancer cases. Cases of bone metastasis in cervical cancer are based on radiation treatment data, which is imbalanced. This imbalanced data is a challenge among the researchers in data mining, called class imbalance learning (CIL) and has lead to difficulties in machine learning and a reduction in the classifier performance. In this paper, we compared several algorithms to deal with the data imbalance classification problem using the synthetic minority over-sampling technique (SMOTE) used to drive classification models: Ant-Miner, RIPPER, Ridor, PART, ADTree, C4.5, ELM and Weighted ELM using Accuracy, G-mean and F-measure to evaluate performance. The results of this paper show that the RIPPER algorithm outperformed the other algorithms in Accuracy and F-measure, but weighted ELM outperformed other algorithms by G-mean. This may be useful when evaluating clinical assessments.

Keywords: cervical cancer, classification algorithm, radiotherapy, imbalance data, machine learning, metastasis.

1 Introduction

Cervical cancer is a malignant neoplasm of cells originating from cervix uteri. It is caused by an infection of human papilloma virus (HPV). This cancer is the second most common cancer in Thai woman after breast cancer [1], 6243 new cases and 2620 deaths were reported in 2002 [2], and the rate of incidence is 24.7 per 100,000 the female population annually (Khuhaprema, et al., 2007). The treatment options of this cancer in early stages are monitored closely by the Pap smear, loop electrosurgical excision, and therapeutic conization while the treatment regimen of metastasis stages is surgical therapy, radiotherapy and chemotherapy.

The study on the prediction of cervical cancer by using statistical techniques is primarily carried out through multiple logistic regression analysis [3] which establishes the models from the data of cervical cancer patients who have been

S. Boonkrong et al. (eds.), *Recent Advances in Information and Communication Technology*, Advances in Intelligent Systems and Computing 265,
DOI: 10.1007/978-3-319-06538-0_10, © Springer International Publishing Switzerland 2014

diagnosed and confirmed through pathological results as well as admitted for radiation treatment. The application of artificial neuron networks was utilized in the prediction of patients with uterine cervical cancer treated by radiation therapy. The Artificial Neuron Networks (ANNs) were evaluated by comparing a receiver operating characteristic (ROC) curve with the area under the ROC curve to analyze the prognosis factors for cervical cancer proposed by (Takashi Ochi et. al, 2003)[4].

The information taken from cases of patients diagnosed with cervical cancer was imbalanced. In supervised classification, the imbalance dataset is a challenge for the research community in data mining and is referred to as Class Imbalance Learning (CIL). The classification tends to be biased towards the majority class and the minority class is more likely to be misclassified. Many techniques have been developed for imbalanced data, the techniques used to solve to the data imbalance include the Over-sampling and Under-sampling.

In under-sampling, the majority examples are removed randomly, until a particular class ratio is met. There are many ways to apply a method such as Tomek Link or TLink which was proposed by Ivan Tomek [5], One-Sidied Selection (OSS) proposed by Kubat and Matwin [6]. On the other hand, the Synthetic minority over-sampling technique (SMOTE) [7] is an oversampling method where new synthetic examples are generated in the neighborhood of the existing minority examples rather than directly duplicating them. SMOTE are performed using C4.5, Ripper and a Naive Bayes classifier. Chawla proposed SMOTE-NC (Synthetic Minority Over-sampling Technique Nominal Continuous) and SMOTE-N (Synthetic Minority Over-sampling Technique Nominal), the SMOTE can also be extended for nominal features [18]. However, over-sampling may lead to over-fitting [8] while under-sampling may loss of important data.

In our research groups, Chaowanan and co-workers [9] have applied ANNs, Logistic regression and Bayesian network with SMOTE imbalance techniques and cost-sensitive learning to compare the performance in the prediction of cervix cancer treated by radiotherapy in Srinakarin Hospital, Khon Kaen, Thailand. However, imbalanced data is a significant problem for data mining classification

The aim of this study is to select the appropriate model to predict the bone metastasis of cervix cancer after being treated by radiotherapy in Srinakarin Hospital, Khon Kaen, Thailand. We use the SMOTE technique in preprocessing order to modify the class distributions in the training data through over-sampling of the minority class. Moreover, classification algorithms in data mining are also considered. Methods such as Extreme Learning machine, weighted ELM, Ant-Miner, RIPPER, PART, Ridor, ADTree and C4.5 and using Accuracy, G-mean and F-measure to evaluates imbalanced data.

2 Bone Metastasis of Cervical Cancer

This Analytical study is aimed at building and testing a model for predicting the bone metastasis of cervix cancer. In this study patient with cervical cancer were diagnosed and verified through pathological results and then admitted in the Radiation unit, Srinakarin Hospital, Faculty of Medicine. This was approved by the human research ethics of Khon Kaen University (HE561461). We analyzed applying confirmation from the patients treated with radiation. The data used in the model were collected from 1994 to 2012. This dataset has 5 continuous attributes and 14 nominal attributes. We considered two-class metastasis (class yes) with 148 samples and no metastasis (class no) with 3716 samples, which is imbalanced data.

Variables used in the study consisted of age, menstruation, gravidity, abortion, parity, pathological groups, keratinizing, stage of cervix cancer, Tumor size, aim of treatment, type of irradiation Anemia, level of hemoglobin, Interval (duration of the detected tumor cells). The dependent variables were effective treatment for patients with cervical cancer after being treated by radiotherapy consisting of metastasis and no metastasis (yes or no). We consider metastasis (class yes) with 148 samples and no metastasis (class no) with 3716 samples.

3 Evaluation Matrices

We can estimate the performance of the classifier using a test data set rather than training data [19]. An imbalanced data set mainly focuses on a two-class problem, the class label of the minority class is positive; while the class label of the majority class is negative. Table 1 illustrates a confusion matrix of the two-class problem. TP and TN denote the number of positive and negative examples that are classified correctly, while FP and FN denote the number of misclassified positive and negative examples. In this paper we use Accuracy, G-mean and F-measure as performance evaluation measures.

Table 1. Confusion matrix for two-class

		Predictive	
		Positive	Negative
Actual	Positive	True Positive (TP)	False Negative (FN)
	Negative	False Positive (FP)	True Negative (TN)

$$\text{Accuracy} = (\text{TP} + \text{TN})/\ (\text{TP} + \text{FN} + \text{TN} + \text{FP}) \tag{1}$$

$$\text{Sensitivity} = \frac{\text{TP}}{\text{TP} + \text{FN}} \tag{2}$$

$$\text{Specificity} = \frac{\text{TN}}{\text{TN} + \text{FP}} \tag{3}$$

$$\text{G} - \text{mean} = \sqrt{\text{Sensitivity} \times \text{Specificcity}} \tag{4}$$

$$\text{Precision} = \text{TP}/\ (\text{TP} + \text{FP}) \tag{5}$$

$$\text{Recall} = \text{TP}/\,(\text{TP+FN}) \tag{6}$$

$$\text{F} - \text{measure} = 2 \times \frac{\text{precision} \times \text{recall}}{\text{precision} + \text{recall}} \tag{7}$$

The above defined evaluation metrics can reasonably evaluate the algorithms for imbalanced data sets because their formulae are relative to the minority class.

4 Methodology

Figure 1 illustrates the overview in the classification algorithms using the SMOTE technique

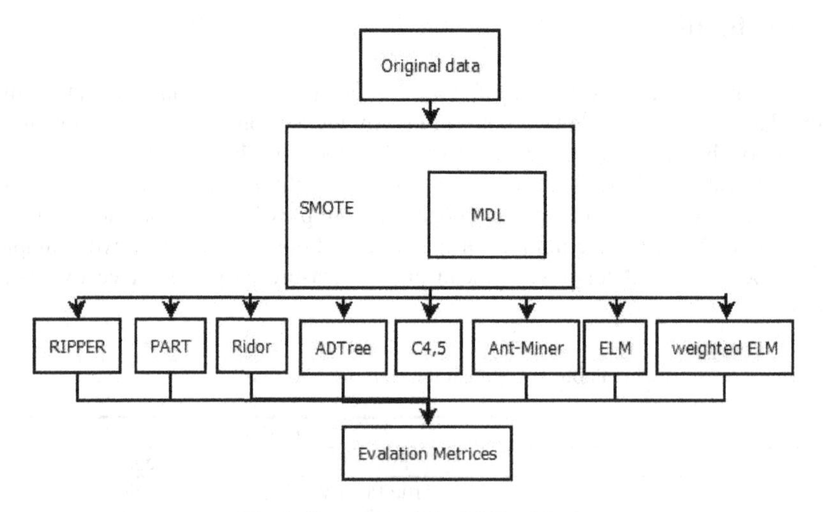

Fig. 1. Overview of the Methodology

4.1 Over-Sampling Dataset with SMOTE

SMOTE: Synthetic Minority Over-sampling Technique [7] is an approach to the construction of classifiers from imbalanced data sets. The minority class is over-sampled by creating "synthetic" samples rather than by over-sampling with replacement. Depending upon the amount of over-sampling required neighbors from the k nearest neighbors are randomly chosen. We use of the SMOTE in the preprocessing method.

In this paper, SMOTE applied to the bone metastasis in cases of cervical cancer. In table 2 SMOTE was used to determine the parameters nearestNeighbors = 5 and increasing percentage of oversampling is 100%, 500%, 1000%, 1500%, 2000% and 2400%. In the experiment design, the positive examples were over-sampled by 2400% so that the size of the positive class was roughly equal to the size of the negative class.

4.2 Algorithms for Comparison and Parameters

The experiments have been carried out using Ant-Miner, Extreme Learning Machine (ELM), weight ELM, and the WEKA learning environment [19] with C4.5(J48), ADTree, PART, Ridor, RIPPER (JRip), and the synthetic minority oversampling technique (SMOTE), whose parameter values used in the experiments are given in Table 2.

Ant-Miner. An Ant colony optimization based algorithm in data mining used to generate classification rule discovery called, Ant-Miner [10], proposed by R. Parpinelli, H. Lopes. The incremental construction is a classification rule of the form:

IF <conditions> THEN <class>

The rule antecedent is a conjunction of terms. Each term is a triple (attribute, operator, value) as Sex = female. Based on simulating the behavior of ants to leave pheromone on the path and converge on the best solution.

RIPPER (JRip). Repeated Incremental Pruning to Produce Error Reduction –a rule based classifier proposed by Cohen W. [11] that integrated REP algorithm (Reduced Error Pruning) with the separate and conquer rule learning algorithm, called IREP (Incremental Reduced Pruning). This IREP build up a rule set in greedily adding rule (one rule at time) until all positive examples are covered by rules.

Ridor. An algorithm which is the implementation of a RIpple-DOwn Rule learner [12], proposed by Brian R. Gaines and Paul Compton. It generates a default rule first and then the exceptions for the default rule with the least (weighted) error rate. Then it generates the "best" exceptions for each exception and iterates until pure. Thus it performs a tree-like expansion of exceptions. The exceptions are a set of rules that predict classes other than the default. IREP is used to generate the exceptions.

PART. A separate-and-conquer rule learner [13], proposed by Eibe Frank and Ian H. Witten. The algorithm produces sets of rules called "decision lists" which are an ordered set of rules. PART builds a partial C4.5 decision tree in each of the iteration and makes the "best" leaf into a rule. The algorithm is a combination of C4.5 and RIPPER rule learning.

ADTree. The alternating decision tree [14] learning algorithm is a natural generalization of decision trees proposed by Freund, Y. and Mason L. This algorithm consists of decision nodes and prediction nodes. Decision nodes specify a predicate condition and prediction nodes contain a single number. ADTrees always have prediction nodes at both root and leaves.

C4.5 (J48)**.** C4.5 [15] is a top-down decision tree learner extension of the ID3 algorithm proposed by J. Ross Quinlan. At each node predictive and split node based on information gain as well as ID3.

ELM. Extreme Learning Machine [16] is a new single hidden layer feed forward neuron network (SLFNs) used in classification and regression proposed by G.B. Huan g, D.H. Wang and Y. Lan. This algorithm consists in k hidden nodes, randomly assigned input weight, hidden layer bias, and activation function. Applied to the learning data for N arbitrary distinct samples (x_i, ti), $i = 1, 2, \ldots, N$, where $xi = (x_{i1}, x_{i2}, \ldots, x_{in})^T \in R^N$ and $t = (t_{i1}, t_{i2}, \ldots, t_{im})^T \in R^m$. ELM is performed by using a linear system, where:

$\boldsymbol{\beta}: j \times m$, the output weight matrix

$\mathbf{T}: N \times m$, target

$$\mathbf{H} = \begin{bmatrix} g(\mathbf{w}_1\mathbf{x}_1 + b_1) & \cdots & g(\mathbf{w}_j\mathbf{x}_1 + b_j) \\ \vdots & \ddots & \vdots \\ g(\mathbf{w}_1\mathbf{x}_N + b_1) & \cdots & g(\mathbf{w}_j\mathbf{x}_N + b_j) \end{bmatrix}$$

$$\mathbf{H}\boldsymbol{\beta} = \mathbf{T} \tag{8}$$

$$\min_{\boldsymbol{\beta}} \|\mathbf{H}\boldsymbol{\beta} - \mathbf{T}\| \tag{9}$$

$$\boldsymbol{\beta} = (\mathbf{H}^T\mathbf{H})^{-1}\mathbf{H}^T\mathbf{T} \tag{10}$$

Weighted Extreme Learning Machine. This algorithm used for imbalanced learning proposed by Weiwei Zong [17]. A weighted ELM is based on the original ELM for binary and multiclass classification tasks. To deal with data with imbalanced class distribution, a weighted ELM is proposed which is able to generalize to balanced data. Given a set of training data (xi, ti), i = 1, 2,…, N, belonging to two classes, we define an NxN diagonal matrix W associated with every training sample xi. The optimization problem mathematically written as: $\mathbf{W} = \begin{bmatrix} w_{11} & \ldots & 0 \\ \vdots & \ddots & \vdots \\ 0 & \ldots & w_{NN} \end{bmatrix}$

$$\mathbf{WH}\boldsymbol{\beta} = \mathbf{WT} \tag{11}$$

$$\min_{\boldsymbol{\beta}} \|\mathbf{WH}\boldsymbol{\beta} - \mathbf{WT}\| \tag{12}$$

$$\boldsymbol{\beta} = ((\mathbf{WH})^{\mathrm{T}}\mathbf{WH})^{-1}(\mathbf{WH})^{\mathrm{T}}\mathbf{WT} \tag{13}$$

Table 2. Number of minority class after over-sampling

Dataset	#cases	#categ. attributes	#cont. attributes	#class yes	#class No
Original Cervix cancer	3862	14	5	148	3716
Cervix cancer100%	4012	14	5	296	3716
Cervix cancer500%	4604	14	5	888	3716
Cervix cancer1000%	5344	14	5	1628	3716
Cervix cancer1500%	6084	14	5	2368	3716
Cervix cancer2000%	6824	14	5	3108	3716
Cervix cancer2400%	7416	14	5	3700	3716

Table 3. Parameters used in the classifiers

Algorithm	Parameters
Ant-Miner	Number of ants = 100 Min case per rule = 10 Max uncovered case = 10 Rule for convergence = 10
ELM	Number of hidden nodes = 500 Activation function = 'sig' (sigmoidal)
Weighted ELM	Number of hidden nodes = 500 Activation function = 'sig' (sigmoidal) Trade-off constant C = 2^8 (256)
C4.5	Prune tree; Confidence factor = 0.25 Minimum of instances per leaf = 2
RIPPER	Number Of Boosting iterations = 10
PART	Confidence factor = 0.25 Minimum of instances per leaf = 2
Ridor	minimum total weight of the instances in a rule = 2 number of nearest neighbors = 5
SMOTE	percentage of SMOTE instances to create =100, 500, 1000, 1500, 2000,and 2400

5 Results and Discussion

We evaluated the performance on a number of classification problems with SMOTE in bone metastasis of cervical cancer datasets. The goal of this paper is to determine the learning framework to achieve better accuracy and other evaluation matrices.

In all experiments we evaluate Accuracy, G-mean and F-measure using 10- fold cross validation. We experimented with the bone metastasis of cervix cancer datasets from Srinakarin Hospital, Khon Kaen, Thailand; this dataset is imbalanced data.

Experiments on this dataset were compared with the classification performance of class imbalance learning methods. We used SMOTE to preprocess the dataset used in this paper to get a balanced distribution of classes with a percentage = 2400. In classifications with imbalance, SMOTE has proved to be the appropriate preprocessing step to improve most learning algorithms.

Table 4 and 6 show the summary of accuracy and F-measure. For the rule algorithm, it was observed that the RIPPER algorithm was the most accurate classifier (having an average of 95.16% and 96.69%) compared with other algorithms which had an output average ranging from 86.75% to 94.14% and 61.45% to 96.59%. Ant-Miner had the lowest average accuracy when compared with other algorithms, with an average of 86.75%. For tree algorithm, it was observed that the C4.5 had the highest accuracy (having an average of 94.99% and 96.43%) followed by ADTree (having an average of 91.96% and 93.99%). For ELM and weighted ELM, we determined 500 nodes. It was observed that ELM had the highest accuracy (having an average of 92.81% and 74.28%) followed by weight ELM (having an average of 82.27% and 66.14%).

Table 4. Summary of 10 fold cross validation performances for Accuracy on all the datasets

Dataset	RIPPER	PART	Ridor	Ant-Miner	ADTree	C4.5	ELM	Weighted ELM
without SMOTE	96.17	95.52	96.17	96.18	96.17	96.17	95.94	59.94
SMOTE 100%	94.34	93.69	94.24	93.57	94.09	94.44	93.49	75.27
SMOTE 500%	94.05	94.18	93.53	86.48	90.86	93.61	92.42	84.75
SMOTE 1000%	95.08	95.06	94.52	87.11	89.82	94.14	91.30	87.41
SMOTE 1500%	94.99	95.43	94.81	64.52	89.91	95.07	91.15	88.69
SMOTE 2000%	95.28	95.79	95.35	90.32	91.29	95.59	92.01	89.65
SMOTE 2400%	96.18	96.29	95.69	89.1	91.56	95.90	93.34	90.21
Average	**95.16**	94.14	94.90	86.75	91.96	94.99	92.81	82.27

Table 5. Summary of 10-fold cross validation performances for G-mean on all the datasets

Dataset	RIPPER	PART	Ridor	Ant-Miner	ADTree	C4.5	ELM	Weighted ELM
without SMOTE	0	8.19	0	0	0	0	2.58	59.04
SMOTE 100%	54.39	60.67	48.94	60.64	49.23	56.48	45.44	70.42
SMOTE 500%	85.62	88.03	83.94	80.38	77.76	86.8	85.22	86.48
SMOTE 1000%	92.56	93.16	91.76	79.49	85.76	92.1	89.85	84.67
SMOTE 1500%	94.45	94.81	93.72	45.67	88.46	94.5	91.96	89.32
SMOTE 2000%	95.51	95.65	95	82.87	91.11	95.44	93.08	89.87
SMOTE 2400%	96.18	96.29	95.69	88.38	91.56	95.9	93.31	90.15
Average	74.1	76.69	72.72	62.49	69.13	74.46	71.63	**81.42**

Table 5 shows the G-mean summary. This weight ELM has the most accurate classifiers (having average of 81.42%) compared with ELM (having 71.63%) and other algorithms. The weighted ELM is able to deal with data imbalance while maintain its performance good with balanced data, as well as unweighted ELM.

The results obtained by eight algorithms using the SMOTE in imbalanced data of bone metastasis cervix cancer by radiation therapy are followed as: the RIPPER algorithm outperformed the other two evaluated algorithms (Accuracy and F-measure). A comparison of the intuitive neuron network ELM and weighted ELM found that the RIPPER rule based classification had better results. However, the size of the dataset analyzed in WEKA may be limited by the computer memory available and need to be considered.

Table 6. Summary of 10-fold cross validation performances for F-measure on all the datasets

Dataset	RIPPER	PART	Ridor	Ant-Miner	ADTree	C4.5	ELM	Weighted ELM
without SMOTE	98.05	97.71	98.05	0	98.05	98.05	12.70	13.26
SMOTE 100%	97.02	96.65	96.98	30.13	96.90	97.07	42.93	30.62
SMOTE 500%	96.40	96.44	96.10	70.67	94.52	96.11	94.43	69.26
SMOTE 1000%	96.54	96.50	96.15	81.23	92.87	95.84	89.54	81.74
SMOTE 1500%	96.34	96.31	95.85	66.58	91.95	96.00	92.12	86.88
SMOTE 2000%	96.22	96.18	95.84	89.84	92.90	95.99	93.75	89.47
SMOTE 2400%	96.27	96.34	95.80	91.68	91.57	95.95	94.43	90.83
Average	**96.69**	96.59	96.40	61.45	93.99	96.43	74.28	66.14

6 Conclusion

In this paper, we have discussed the used classification data of Bone metastasis cervix cancer with the imbalanced data problem by using SMOTE technique. The experiments were carried out by using open source Machine Learning tools. We have compared the performance of Ant-Miner, RIPPER, PART, Ridor, ADTree, C4.5, ELM and weight ELM. The results show that, concerning predictive accuracy, RIPPER obtained a higher accuracy rate than other algorithms. On the other hand, weight ELM found G-mean better than other classifiers. In addition, the ELM and weight ELM algorithms are computationally inexpensive to run. Finally, the rule based classifiers are suitable to be used to generate classification rules for new patients and may be labeled as metastasis or not metastasis and aid an Oncology Doctor in decisions about cancer in a short time by utilizing the Graphical User Interface.

As a next step, the algorithm RIPPER has the most potential to be the standard analysis method with other over-sampling, under-sampling and combined over-sampling and under-sampling techniques. We will use the classification rules that are applied to predict the symptoms that indicate the patients with cervical cancer.

References

1. Nartthanarung, A., Thanapprapasr, D.: Comparison of Outcomes for Patients With Cervical Cancer Who Developed Bone Metastasis After the Primary Treatment With Concurrent Chemoradiation Versus Radiation Therapy Alone. Int. J. Gynecol. Cancer 20(8), 1386–1390 (2010)
2. Thanapprapasr, D., Nartthanarung, A., Likittanasombut, P., Na Ayudhya, N.I., Charakorn, C., Udomsubpayakul, U., Subhadarbandhu, T., Wilailak, S.: Bone Metastasis in Cervical Cancer Patients over a 10-Year Period. Int. J. Gynecol. Cancer 20(3), 373–378 (2010)
3. Kamsa-ard, S., Tangvorapongchai, V., Krusun, S., Sriamporn, S., Suwanrungruang, K., Mahaweerawat, S., Pomros, P.: A model to predict the immediate results of radiation on cervix cancer. KKU Res. J. 13(7), 851–865 (2008)

4. Ochi, T., Murase, K., Fujii, T., Kawamura, M., Ikezoe, J.: Survival prediction using artificial neural networks in patients with uterine cervical cancer treated by radiation therapy alone. Int. J. Clin. Oncol. 7(5), 294–300 (2002)
5. Tomek, I.: An experiment with the edited nearest-neighbor rule. IEEE Transactions on Systems, Man, and Cybernetics SMC-6(6), 448–452 (1976)
6. Kubat, M., Matwin, S.: Addressing the curse of imbalanced training sets: one-sided selection. In: Fisher, D.H. (ed.) ICML, vol. 97, pp. 179–186. Morgan Kaufmann (1997)
7. Chawla, N.V., Bowyer, K.W., Hall, L.O., Kegelmeyer, W.P.: SMOTE: synthetic minority over-sampling technique. J. Artif. Int. Res. 16(1), 321–357 (2002)
8. Chawla, N.V., Japkowicz, N., Kotcz, A.: Editorial: special issue on learning from imbalanced data sets. ACM SIGKDD Explorations Newsletter 6(1), 1–6 (2004)
9. Soto, C.: Model for cervical cancer result prediction. (Ms.D. Thesis in Computer Science). Department of Computer Science, Khon Kaen University, Thailand (2013)
10. Parpinelli, R.S., Lopes, H.S., Freitas, A.A.: Data mining with an ant colony optimization algorithm. IEEE Transactions on Evolutionary Computation 6(4), 321–332 (2002)
11. Cohen, W.W.: Fast effective rule induction. In: Proceedings of the Twelfth International Conference on Machine Learning, pp. 115–123. Morgan Kaufmann (1995)
12. Gaines, B.R., Compton, P.: Induction of ripple-down rules applied to modeling large databases. J. Intell. Inf. Syst. 5(3), 211–228 (1995)
13. Frank, E., Witten, I.H.: Generating accurate rule sets without global optimization. In: Shavlik, J.W. (ed.) Proceedings of the Fifteenth International Conference on Machine Learning, ICML 1998, pp. 144–151. Morgan Kaufmann Publishers Inc., San Francisco (1998)
14. Freund, Y., Mason, L.: The alternating decision tree learning algorithm. In: Bratko, I., Dzeroski, S. (eds.) Proceedings of the Sixteenth International Conference on Machine Learning (ICML 1999), pp. 124–133. Morgan Kaufmann Publishers Inc., San Francisco (1999)
15. Quinlan, J.R.: C4. 5: programs for machine learning. Morgan Kaufmann Publishers Inc., San Francisco (1993)
16. Huang, G.B., Wang, D.H., Lan, Y.: Extreme learning machines: a survey. Int. J. Mach. Learn. & Cyber. 2(2), 107–122 (2011)
17. Zong, W., Huang, G.B., Chen, Y.: Weighted extreme learning machine for imbalance learning. J. Neurocomput. 101, 229–242 (2013)
18. Ganganwar, V.: An overview of classification algorithms for imbalanced datasets. International Journal of Emerging Technology and Advanced Engineering 2(4), 42–47 (2012)
19. Witten, I.H., Frank, E.: Data Mining: Practical machine learning tools and techniques, 2nd edn. Morgan Kaufmann, San Francisco (2005)

Genetic Algorithm Based Prediction of an Optimum Parametric Combination for Minimum Thrust Force in Bone Drilling

Rupesh Kumar Pandey and Sudhansu Sekhar Panda*

Department of Mechanical Engineering, Indian Institute of Technology Patna, India
rupeshiitp@gmail.com,
sspanda@iitp.ac.in

Abstract. Drilling operation on bone for screw insertion to fix the broken bones or for the fixation of implants during orthopaedic surgery is highly sensitive. It demands for minimum drilling damage of bone for proper fixation and quick recovery postoperatively. The aim of the present study is to find out an optimum combination of bone drilling parameters (feed rate and spindle speed) for minimum thrust force during bone drilling using genetic algorithm (GA). Central composite design is employed to carry out the bone drilling experiments and based on the experimental results, a response surface model was developed. This model is used as a fitness function for genetic algorithm (GA). The investigation showed that the GA technique can efficaciously estimate the optimal setting of bone drilling parameters for minimum thrust force value. The suggested approach can be very useful for orthopaedic surgeons to perform minimally invasive drilling of bone.

Keywords: Bone drilling, Thrust force, Response surface methodology, Genetic algorithm.

1 Introduction

Drilling of bone with minimum damage to the bone tissue has been increasingly gaining attention by the researchers as it helps in better fixation and quick recovery of the broken bones. Thrust force produced during bone drilling is one of the major concerns as the exposure of bone to its higher magnitudes can result in micro cracks, bursting, drill bit breakage or even the death of bone cells surrounding the drill site [1-2].

In past, many researchers have carried out bone drilling investigations to determine the influence of drilling parameters on the thrust force produced [1-4]. Drill bit breakage commonly takes place due to uncontrollable and large bone drilling forces causing considerable damage to the bone [5-7]. The removal of the drill bit becomes an additional requirement as it will produce adverse effect on the body and can also hamper the proper placement of the other fixations [8-9]. Moreover, it will increase

* Corresponding author.

S. Boonkrong et al. (eds.), *Recent Advances in Information and Communication Technology*, Advances in Intelligent Systems and Computing 265,
DOI: 10.1007/978-3-319-06538-0_11, © Springer International Publishing Switzerland 2014

the duration of the surgery. The rate of heat generation is also high with higher drilling forces [10] which can result in thermal osteonecrosis. Furthermore, larger drilling forces can initiate large number of crack formation which can result in misalignment of the fixation or can cause permanent failure [2]. Therefore, it is critical to evaluate the amount of force generated during bone drilling and to determine the optimal setting of drilling parameters for minimum force generation to facilitate better fixation and quick healing of the broken bones after surgery.

The studies on the effect of drilling parameters on bone drilling thrust force started in late 1950s [2-3]. Spindle speed and feed rate were the drilling parameters considered in most of the cases [3-4, 10-14]. It was observed that the thrust force decreases with an increase in spindle speed [13]. But, the bones drilled with higher spindle speeds were reported with increased trauma [3, 14]. The increase in feed rate increased the thrust force in bone drilling [4, 13]. Despite of the aforementioned studies, there is no clear suggestion on the optimal settings of the feed rate and spindle speed for minimum force generation.

In the present work, a statistical model for bone drilling process to predict the thrust force as a function of feed rate (mm/min) and spindle speed (rpm) is developed using response surface methodology (RSM). Next, the model is used as a fitness function in Genetic algorithm (GA) for determination of the optimal setting of feed rate and spindle speed for minimum thrust force. The adopted approach is then validated through the confirmation experiment.

RSM is an accumulation of mathematical and statistical tools which are easy, quick and effective for modeling the process in which several variables influence the response of interest [15-16]. In most of the real problems the relationship between the response and independent variable is unknown. In RSM the relationship between the response and the independent process variables is represented as (1)

$$Y = f(A, B, C) + \varepsilon \tag{1}$$

Where Y is the desired response, f is the response function and ε represents the error observed in the response. A second order model is generally employed if the response function is nonlinear or not known, shown in (2) [15-16]

$$Y = \beta_0 + \sum_{i=1}^{k} \beta_i x_i + \sum_{i=1}^{k} \beta_{ii} x_i^2 + \sum_{i} \sum_{j} \beta_{ij} x_i x_j + \varepsilon \tag{2}$$

Where β_0 is the coefficient for constant term β_i, β_{ii} and β_{ij} are the coefficients for linear, square and interaction terms respectively.

Genetic algorithm developed by Goldberg based on the Darwins theory of biological evolution is very effective for solving of both constrained and unconstrained optimization problem [17-18]. It can be applied to solve a wide variety of optimization problems including the cases having discontinuous, stochastic or highly non linear objective function which are not well handled by the standard optimization algorithms [18]. In genetic algorithm the solution is referred as chromosome which consists of the desired set of values termed as genes. The genes can be binary, numerical or symbol depending upon the type of problem to be solved. A collection of the chromosome forms population. The first step in this approach is to randomly initialize a population as a potential solution. In the next step each chromosome of the population will undergo the fitness function to determine their suitability as the solution to the problem. Further, few chromosomes from the current population will be selected as the parent and will mate through a process termed as cross over thus, producing new chromosomes as off springs having genes which are the combination of their parent genes. In a generation, few parent chromosomes also undergo random changes through mutation to produce children. The number of chromosomes undergoing crossover and mutation is controlled by using the crossover and mutation rate respectively. The chromosomes for the next generation are selected from the population based on the Darwin's theory of survival for the strongest species. The chromosomes with higher fitness value will have the greater probability of getting selected to the next generation. Over the successive generations the chromosome value will converge towards the best solution [17-18]. The flowchart of GA methodology is shown in Fig. 1.

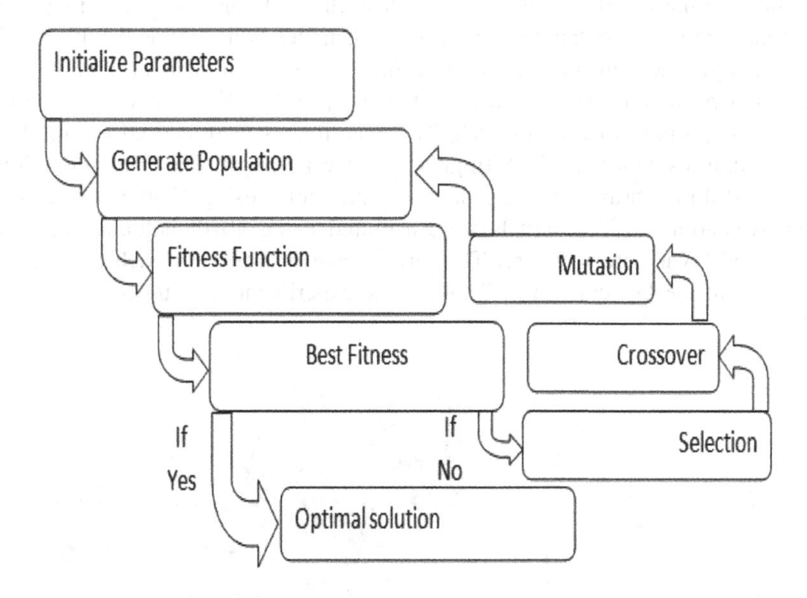

Fig. 1. The flow of GA optimization methodology

2 Experimental Procedure

2.1 Experimental Design Based on Response Surface Methodology

The bone drilling parameters considered are feed rate (mm/min) and spindle speed (rpm) (shown in Table 1) and the response taken is thrust force (N). The central composite design (CCD) of RSM was employed to design the plan of experiments for studying the relationship between the response and the bone drilling parameters. Full factorial design for factors at two levels i.e. high (+1) and low (-1) corresponding to a face centered design with thirteen runs (four factorial points, four axial points and five central points) was used as shown in Table 2. The bone drilling parameters and their levels are considered based on the wide range of experiments reported in the literature [1-4, 10-14].

Table 1. Factor and levels considered for bone drilling

	Control factor	*Low level (-1)*	*High level (+1)*
A	Feed rate (mm/min)	30	150
B	Spindle speed (rpm)	500	2500

2.2 Experimental Details

The work material used for conducting the bone drilling experiments was bovine femur, as the human bones are not easily available and it closely resembles the human bone, allowing the results to be extrapolated in the real surgical situations [19-20]. The bovine femur was obtained from a local slaughter house immediately after the slaughter and the experiments were done within few hours to maintain minimum loss in thermo-physical properties of the fresh bone [19-20]. No animal was scarified solely for the purpose of this research. The experiments were carried out on 3 axis MTAB flexmill using 4.5mm HSS (high speed steel) drill bit. The drilling depth was 6mm. The drilling thrust force signals were measured using Kistler 9257B piezo electric dynamometer. The signals were acquired using 5070 multichannel charge amplifier and Dynoware software. The thrust force obtained for each experimental run is listed in the last column of Table 2.The experimental set up is shown in the Fig. 2.

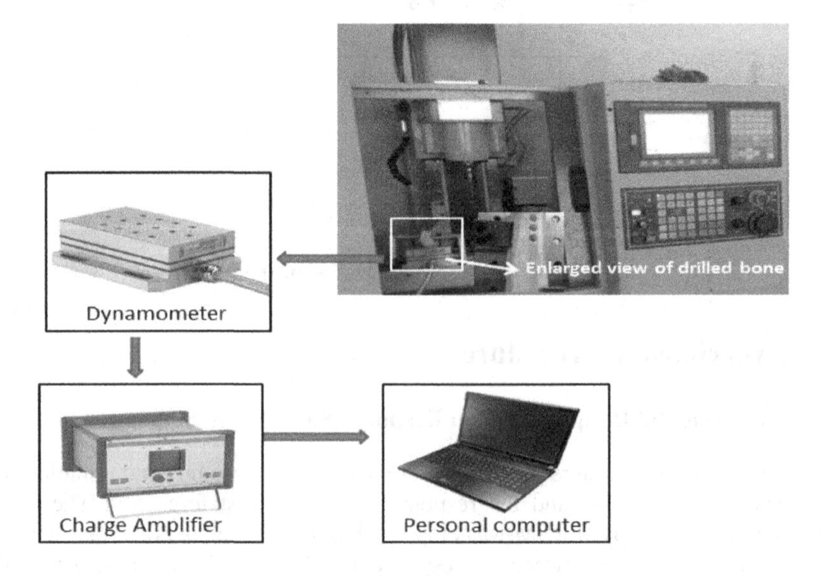

Fig. 2. Experimental set up

Table 2. Experimental condition and result

Exp No.	*Feed rate (mm/min)*	*Spindle speed (rpm)*	*Thrust Force (N)*
1	90	1500	16.93
2	150	1500	24.55
3	90	500	29.25
4	150	500	45.43
5	30	2500	4.72
6	90	2500	14.31
7	30	1500	6.155
8	90	1500	16.79
9	90	1500	17.41
10	90	1500	17.29
11	150	2500	20.33
12	90	1500	17.01
13	30	500	11.32

3 Development of Mathematical Model

A mathematical model correlating the thrust force and drilling process parameters is developed based on (2) using design expert software version 8.0.1 [21]. A quadratic model is selected based on low standard deviation and high R squared value as mentioned in Table 3 [21].

Table 3. Model summary statistics

Source	*Std. deviation*	*R-Squared*	*Adjusted R-Squared*	*Predicted R-Squared*	
Linear	4.08	0.8721	0.8466	0.6912	
2FI	3.00	0.9378	0.9171	0.7966	
Quadratic	**1.44**	**0.9888**	**0.9808**	**0.8880**	**Suggested**

The model is given by (3) as:

$$\begin{aligned} Force = 8.796 + 0.3817 \times Feed\ rate - 0.0155 \times spindle\ speed \\ - 7.7083 \times E - 5 \times Feed\ rate \times spindle\ speed \\ - 4.2713 \times E - 4 \times Feed\ rate^2 \\ + 4.8898 \times E - 6 \times spindle\ speed^2 \end{aligned} \quad (3)$$

Analysis of variance (ANOVA) carried out to find the significance of the developed model and individual model coefficients at 95% confidence interval is shown in Table 4.

Table 4. ANOVA table for the proposed model

Source	*DOF*	*SS*	*MS*	*P value*
Model	5	1287.77	257.55	<.0001
A-Feed rate	1	773.28	773.28	<.0001
B-Spindle speed	1	362.55	362.55	<.0001
AB	1	85.56	85.56	.0004
A^2	1	6.53	6.53	.1200
B^2	1	66.04	66.04	.0008
Residual	7	14.58	2.08	
Total	12	1302.35		

Where

DF= Degree of freedom

SS= Sum of squares

MS= Mean square

From the ANOVA table it can be seen that the model is significant as its p value is less than 0.0500. In this case A, B, AB, B2 are significant model terms. Values greater than 0.1000 indicate the model terms are not significant [21]. The comparison of the predicted thrust force values with the actual values is shown in the Fig. 3.

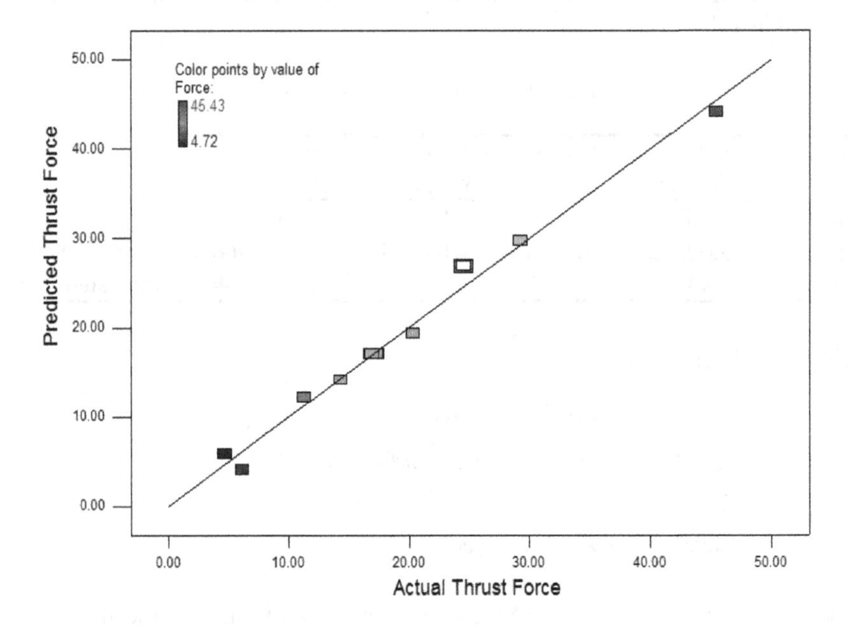

Fig. 3. Comparison between the predicted and the actual thrust force

4 Optimization of the Thrust Force with GA

The optimal setting of the spindle speed and feed rate for minimum thrust force during bone drilling is determined using GA. The developed response surface model of thrust force is taken as an objective function to be minimized for minimum thrust force value. Global optimization tool box of Matlab R2010b is used to determine the values of feed rate and spindle speed for the minimum value of the fitness function. The problem formulation is subjected to the boundaries (limitations) of the drilling parameters and is stated as follows:

$$30 \leq \text{Feed rate} \leq 150 \tag{4a}$$

$$500 \leq \text{Spindle speed} \leq 2500 \tag{4b}$$

The solution of the problem formulated depends upon various criteria such as the initial population generated type of the selection function, crossover and mutation rate selected. There is lack of guidelines in the previous literatures on the selection of the above mentioned parameters [22] for best optimal result. In this study the method of trial and error was used in Matlab global optimization tool box to find the best set of cutting conditions for minimum value of thrust force. The combination of parameters used that lead to the minimum value of the fitness function is listed in Table 5.

Table 5. Factor and levels considered for bone drilling

Parameters	*Setting Value*
Population size	20
Mutation rate	0.8
Crossover rate	0.2

The result obtained by GA using (3) as the fitness function along with the limitations imposed by (4a-4b) and the parameters mentioned in Table 5 is shown in Fig. 4 and Table 6.

Table 6. Results obtained from GA analysis

Parameters	*Value*
Minimum fitness function thrust force	3.6378 (N)
Optimal cutting conditions	
Feed rate	30
Spindle speed	1804.821

From Table 6, it can be observed that the minimum value of thrust force is 3.6378 N with the feed rate of 30 (mm/min) and spindle speed of 1804.821 (rpm). The minimum value of thrust force is obtained after 51 generations as shown in Fig. 4.The criteria used for the termination of GA is the change in the weighted average of fitness function value over stall a generation which is less than function tolerance. The Fig. 4 also shows the mean value of the fitness function as 3.6381.

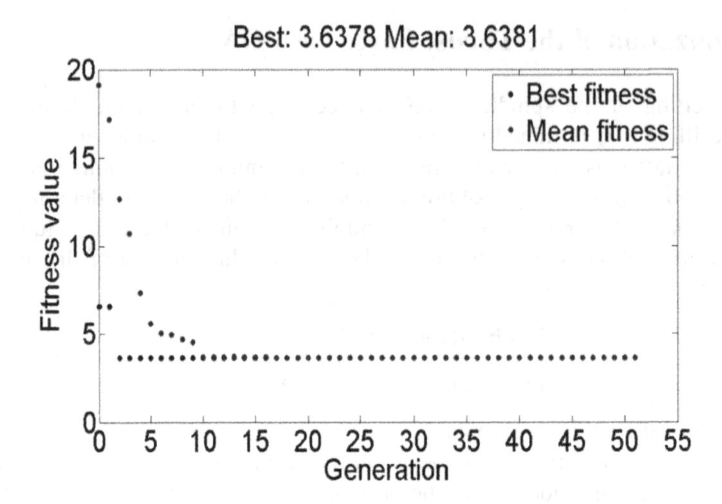

Fig. 4. Plot of the fitness value against generation

To validate the result obtained from the above analysis, confirmation experiments were carried out. The result of the confirmation experiments are shown in Table 7.

Table 7. Confirmation experiments and results

No.	*Feed rate (mm/min)*	*Spindle speed (rpm)*	*Thrust force*		*%Error*
			Exp.	*predicted*	
1	30	1000	7.10	6.94	2.25
2	30	1805	3.87	3.64	5.94
3	60	1000	15.68	14.92	4.84
4	120	500	38.63	37.30	3.44

Four experiments were carried out within the range of the parameters studied for the confirmation of the obtained results. Three settings of feed rate and spindle speed were selected randomly whereas one optimal setting predicted by GA was analyzed. From Table 7 it is clear that the predicted values and those obtained from experiment are very close hence, RSM model can effectively predict the thrust force values whereas GA can be very useful to minimize the thrust force during bone drilling.

5 Conclusions

In the present investigation, an approach involving the integration of RSM with GA is used for the optimization of thrust force in bone drilling process. From the above analysis following conclusions are drawn:

- The mathematical model developed by RSM can effectively predict the thrust force in bone drilling within the range of the parameters studied.
- GA results showed that the best combination of bone drilling parameters for minimum thrust force is 30 mm/min of feed rate and 1805 rpm of spindle speed.
- The results of the confirmation experiments validated that the combination of RSM and GA is suitable for optimizing the bone drilling process.
- The use of above approach can greatly assist the orthopaedic surgeons to decide the best level of drilling parameters for bone drilling with minimum mechanical damage.

References

1. Pandey, R.K., Panda, S.S.: Drilling of bone: A comprehensive review. Journal of Clinical Orthopaedics and Trauma 4, 15–30 (2013)
2. Lee, J., Gozen, A.B., Ozdoganlar, O.B.: Modeling and experimentation of bone drilling forces. Journal of Biomechanics 45, 1076–1083 (2012)
3. Thompson, H.C.: Effect of drilling into bone. Journal of Oral Surgery 16, 22–30 (1958)
4. Wiggins, K.L., Malkin, S.: Drilling of bone. Journal of Biomechanics 9, 553–559 (1976)
5. Brett, P.N., Baker, D.A., Taylor, R., Griffiths, M.V.: Controlling the penetration of flexible bone tissue using the stapedotomy microdrill. Proceedings of the Institution of Mechanical Engineers, Part I: Journal of Systems and Control Engineering 218, 343–351 (2004)
6. Kendoff, D., Citak, M., Gardner, M.J., Stubig, T., Krettek, C., Hufner, T.: Improved accuracy of navigated drilling using a drill alignment device. Journal of Orthopaedic Research 25, 951–957 (2007)
7. Ong, F.R., Bouazza-Marouf, K.: The detection of drill-bit break-through for the enhancement of safety in mechatronic assisted orthopaedic drilling. Mechatronics 9, 565–588 (1999)
8. Price, M., Molloy, S., Solan, M., Sutton, A., Ricketts, D.M.: The rate of instrument breakage during orthopaedic procedures. International Orthopaedics 26, 185–187 (2002)
9. Bassi, J.L., Pankaj, M., Navdeep, S.: A technique for removal of broken cannulated drill-bit: Bassi's method. Journal of Orthopaedic Trauma 22, 56–58 (2008)
10. Augustin, G., Davila, S., Mihoci, K., Udiljak, T., Vedrina, D.S., Antabak, A.: Thermal osteonecrosis and bone drilling parameters revisited. Archives of Orthopaedic and Trauma Surgery 128, 71–77 (2008)
11. Abouzgia, M.B., James, D.F.: Temperature rise during drilling through bone. International Journal of Oral and Maxillofacial Implants 12, 342–3531 (1997)
12. Hobkirk, J.A., Rusiniak, K.: Investigation of variable factors in drilling bone. Journal of Oral and Maxillofacial Surgery 35, 968–973 (1977)
13. Jacobs, C.H., Berry, J.T., Pope, M.H., Hoaglund, F.T.: A study of the bone machining process-drilling. Journal of Biomechanics 9, 343–349 (1976)
14. Albrektsson, T.: Measurements of shaft speed while drilling through bone. Journal of Oral and Maxillofacial Surgery 53, 1315–1316 (1995)
15. Myers, R.H., Montgomery, D.C.: Response surface methodology, 2nd edn. Wiley, New York (2002) ISBN 0-471-41255-4
16. Box, G.E.P., Hunter, J.S., Hunter, W.G.: Statistics for experimenters, 2nd edn. Wiley, New York (2005) ISBN 13978-0471-71813-0

17. Deb, K.: Chichester (ed.) Multi-objective Optimization using Evolutionary Algorithms. John Wiley and Sons, Ltd., England (2001)
18. Udayakumar, T., Raja, K., Afsal, T., Husain, M., Sathiya, P.: Prediction and optimization of friction welding parameters for super duplex stainless steel (UNS S32760) joints. Materials and Design 53, 226–235 (2014)
19. Karaca, F., Aksakal, B., Kom, M.: Influence of orthopaedic drilling parameters on temperature and histopathology of bovine tibia: An in vitro study. Medical Engineering & Physics 33(10), 1221–1227 (2011)
20. Lee, J., Ozdoganlar, O.B., Rabin, Y.: An experimental investigation on thermal exposure during bone drilling. Medical Engineering & Physics 34(10), 1510–1520 (2012)
21. Design Expert, `http://www.statease.com/dx8descr.html`
22. Bhushan, R.K., Kumar, S., Das, S.: GA Approach for Optimization of Surface Roughness Parameters in Machining of Al Alloy SiC Particle Composite. Journal of Materials Engineering and Performance 21, 1676–1686 (2012)

The Evolutionary Computation Video Watermarking Using Quick Response Code Based on Discrete Multiwavelet Transformation

Mahasak Ketcham[1,*] and Thittaporn Ganokratanaa[2]

[1] Department of Information Technology Management, Faculty of Information Technology, King Mongkut's University of Technology North Bangkok, Bangkok, Thailand
maoquee@hotmail.com
[2] King Mongkut's University of Technology Thonburi, Bangkok, Thailand
charisma_sbunny@hotmail.com

Abstract. Nowadays, commercial activity on Internet and media requires a protection by increasing security. The 2D Barcode with a digital watermark is a widely interest research in security field. QR Code with invisible watermark prevents information hiding text. This paper proposes QR Code (Quick Response Code) that is embedded an invisible video watermark by using Discrete Multiwavelet transformation (DMT). We have developed an optimization technique using the genetic algorithm to search for optimal quantization step in order to improve both quality of watermarked video and robustness of the watermark. This technique does not require the original image in the watermark extraction. The experimental results show that the proposed watermarking algorithm yields watermarked image with good imperceptibility and very robust.

Keywords: 2D Barcode, QR Code, Video Watermark, Evolutionary Computation (EA), Discrete Multiwavelet Transformation (DMT), Genetic Algorithm (GA).

1 Introduction

Video Watermarking is a young and rapidly evolving field in the area of multimedia. Following factors have contributed towards the triggering of interest in this field. The society is contaminated by the tremendous piracy of digital data, as copying of digital media has become comparatively easy. This is an era where need has arisen for fight against "Intellectual property rights infringements". Copyright protection must not be eroded due to malicious attacks. Tampering of the digital data needs to be concealed at some point. Barcode became widely known because of their accuracy, and superior functionality characteristics. QR Code is a kind of 2D (two dimensional) Barcode symbol which is categorized in matrix code. It contains information in both the vertical and horizontal directions, whereas a 1D (one dimensional) Barcode symbol contains data in one direction only. QR Code holds a considerably greater volume of

* Corresponding author.

S. Boonkrong et al. (eds.), *Recent Advances in Information and Communication Technology*, Advances in Intelligent Systems and Computing 265,
DOI: 10.1007/978-3-319-06538-0_12, © Springer International Publishing Switzerland 2014

information than a 1D Barcode. QR Code developed by Denso Wave [1] (a division of Denso Corporation) and released in 1994. QR Code can encode in many type of characters such as numeric, alphabetic character, Kanji, Kana, Hiragana, symbols, binary, and control codes. Approximate maximum capacity 7,089 characters can be encoded in one symbol and maximum version is 40. Features of QR Code are high capacity, and error correction. Error correction helps to restore when symbol is dirty or damaged. The highest level can be roughly 30% of code words. [2] QR Code can decode easily by upload picture on web providers. [3] So, everybody who has a file of QR Code image, they can decode it all. QR Code is seen as a weakness in security. [4] This paper proposes method for add watermark that is information hiding into QR Code. Digital watermark is a kind of information security and protection technology. Watermarking is mostly similar to steganography in a number of respects. The main idea of steganography is the embedding of secret information into data under assumption that others cannot know the secret information in data. The main idea of watermark is check the secret information embedded in data or not. Watermark is the embedding information in media for exchange the information within the group. In recent years, some multiwavelet-based image watermarking algorithms have been proposed. Kwon and Tewfik [3] proposed an adaptive image watermarking scheme in the discrete multiwavelet transform (DMT) domain using successive subband quantization and a perceptual modeling. The watermark is Gaussian random sequence with unit varianceand the original image is needed for watermark detection. In [4], Zhang et al. proposed a novel watermarking scheme for an image, in which a logo watermark is embedded into the multiwavelet domain of image using back-propagation neural network (BPN). Due to the learning and adaptive capabilities of BPN, their scheme can gain good robustness. A lot of video watermarking algorithms have been proposed in the literature employed either in spatial [5] or frequency domain [6-11]. Recently, some video watermarking algorithms have been proposed with ICA, PCA [12-14] and SVD [15]. The detailed descriptions of these schemes are as follows. Mobasseri [5] has proposed a spatial domain watermarking scheme for compressed videos. Authors have showed that it is possible to embed a watermark in raw video and still recover it from MPEG decoder, by exploiting the inherent processing gain of direct sequence spread spectrum. Tsai & Chang [6] have proposed a novel watermarking scheme for a compressed video sequence via VLC decoding and VLC code substitution. To have better imperceptibility, they used Watson's DCT-based visual model for video watermarking. Ge et al [7] have presented a novel adaptive approach to video watermarking. It takes full advantage of both intra-frame and inter-frame information of video content to guarantee the perceptual invisibility and robustness of the watermark. A major advantage of this scheme is that the watermark can be extracted without referring to the original video while embedded adaptively in accordance with the human visual system and signal characteristics. Another way to improve the performance of watermarking schemes is to make use of artificial intelligent (AI) techniques. The watermarking system can be viewed as an optimization problem. Therefore, it can be solved by genetic algorithm (GA), support vector machine (SVM) or adaptive tabu search (ATS). There has been little research in application of GA to digital image watermarking problems. In this paper, we propose QR Code embed an Intelligent Video watermarking method based on the discrete multiwavelet transform. In our algorithm, the watermark is embedded into the multiwavelet transform coefficients using quantization index modulation technique. We apply the GA to search for optimal watermarking parameters in order to achieve optimum performance. This paper is organized as follows: Section 2 describes the preliminaries. Watermark Embedding Algorithm in the DMT domain with genetic algorithm optimization in Section 3. In Section 4, the experimental results are shown. The conclusions of our study can be found in Section 5.

2 Preliminaries

2.1 Multiwavelet Transform

Discrete Multiwavelet Transformations (DMT) has received much attention in signal processing applications. It possesses diverse desirable properties including, orthogonality, symmetry and compact support with a given approximation order and offers the possibility of superior performance for image processing application. It is compared its performance with scalar wavelets which cannot possess all multiwavelet's properties at the same time. The concept of Multiwavelet Transform is based on the multiresolution analysis (MRA) which is the same as the wavelet transform, but multiwavelet transform has scaling and wavelet function more than one function. Furthermore, there are several properties, which cannot exist in the wavelet transform, can occurred simultaneously. Thus, the multiwavelet is explicitly the main motivation for constructing signal processing. [16]. Multiwavelet Transform is based on the principles of the filter bank which consists of two phases, analysis and synthesis. The multiwavelet transform coefficients is a matrix coefficient and the wavelet transform coefficient is a scalar. In the actual use of multiwavelet transform, it requires a formatted input as a discrete vector sequence due to it has the scaling functions more than one. The process before entering input signal via multiwavelet filterbank is called prefiltering or multiwavelet initialization.

2.2 Quick Response Code (QR Code)

The QR Code is prepared to be watermarked via a robust video watermarking. The robustness of watermarked image, which can avoid various image processing, attacks like noise addition, rotation and so on. Any Intruder cannot take logo because it should hide in the QR code image. The QR code image is tolerable up to 30% noise. So our watermarking method provided more security imperceptibility and certain robustness. The structures of the QR Code are shown as below.

a.) One-dimensional bar code.

b.) The QR Code

Fig. 1. Shows the structure of the QR Code

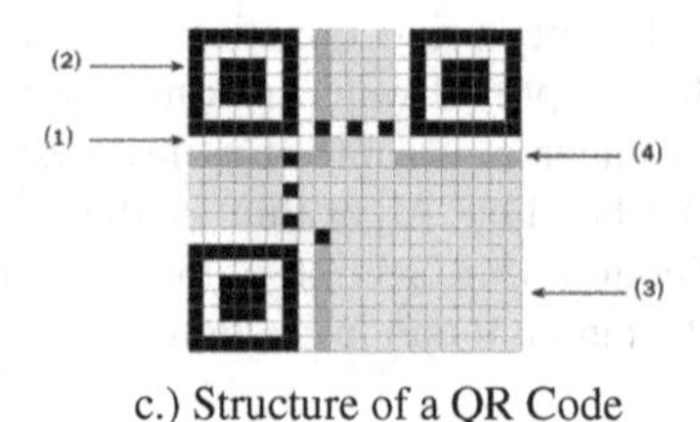

c.) Structure of a QR Code

Fig. 1. *(continued)*

Fig.1 a) One-dimensional bar code is one of the type of barcodes which has less capacity. It is used in various products such as, journal, and pill box. However, because the data is growing in nowadays, thus, the two-dimensional barcode is occurred with high capacity as shown in Fig.1 b).

- Numeric is data from 0 to 9. Using 10-bit coding per 3 digits for encryption. It can be encoded to QR code and stored information to 7089 characters.
- Alphanumeric includes numbers 0-9, lowercase a-z, uppercase A-Z and $, %, *, +, -, ., /, : characters. Using 11-bit coding per 2 characters for encryption. It can be encoded to QR code and stored information to 4296 characters.
- Data used for encoding into 8 bit. It can be encoded to QR code and stored information to 2953 characters.
- KANJI, the Japan alphabet, used for encoding into 13 bit. It can be encoded to QR code and stored information to 1817 characters.

The error correcting function of QR code for miss reading is defined in 4 level as below

- Level L (Low) 7% of codewords can be restored.
- Level M (Medium) 15% of codewords can be restored.
- Level Q (Quartile) 25% of codewords can be restored.
- Level H (High) 30% of codewords can be restored.

2.2.1 The Structure of QR Code

The character of QR code is defined by the size of barcode which consists of 40 versions including, version 1 to 40. Version 1 has 21x21 matrix size and next version has more matrix size in each 4 modules until version 40 which has 177x177 matrix. The structure of QR code is shown in Fig.1c).

1. Finder pattern is used to identify position QR code for decryption.
2. Timing pattern is used to determine a symbol's coordinate QR code for decryption.
3. Encode data is a position of encoded data.
4. Format information is used to store information of error collecting level.

2.3 Genetic Algorithm (GA)

Genetic algorithm (GA) is one of the most widely used evolutionary computations. GA is a search technique for finding the global maximum/minimum solutions for problems. Although the GA operation performs randomly, choosing candidates to avoid stranding on a local optimum solution, there is no guarantee that the global maximum/minimum will be found. In general, the possibility of obtaining the global maximum/minimum by using GA is related to the complexity of a problem. That is the more complex a problem is, the higher the difficulty of obtaining the optimum solution [23]. The GA is one of the most widely used artificial intelligent techniques for optimization. They have been successfully applied to obtain good solutions in optimal parameter of video watermark. Usually, the GA starts with some randomly selected genes as the first generation, called population. Each individual in the population corresponding to a solution in the problem domain is called chromosome. An objective, called fitness function, is used to evaluate the quality of each chromosome. The chromosomes of high quality will survive and form a new population of the next generation. By using the tree operators: selection crossover, and mutation, we recombine a new generation to find the best solution. In order to apply the GA for embedding video watermarking into the DMT the chromosomes is used to adjust position values of audio watermarking on DWT.

3 Proposed Method

In this section, we first give a brief overview of the QR Code Intelligent Video watermark embedding and watermark extracting processes in the DMT domain. We then describe the GA optimization of our proposed method.

3.1 Watermark Embedding Algorithm

In this paper, the watermark data is a QR Code. The watermark embedding algorithm is described as follows:

***Step1*:** Sequent video frames are extracted from the video V, denoted by F^i where i is the total number of frames and $i \in [1,n]$.

***Step2*:** Generate a random QR Code watermark W using the secret key, where W is a binary pseudo-random noise sequence of watermark bits, and $W = \{w_i\}$ for $i = 1, 2, \ldots, N_w$, where N_w is the length of watermark.

***Step3*:** Transform the frame image into four-level decomposition using the DMT. Then, create multiwavelet trees and rearrange them into 3072 groups.

Step4: To increase the watermarking security, we order the groups Tg_m in a pseudorandom manner. The random numbers can be generated using the secret key K. We further combine the coefficients of every three groups together to form "a triple tree: Tt_i", for $i = 1, 2, \ldots, 1024$. Each watermark bit is embedded into one triple tree.

Step 5: For watermark embedding, we select the first N_w triple trees, which have the largest mean values. Then, the watermark sequence $\{w_i\}$ is embedded into the selected triple trees by quantization index modulation technique.

Step 6: In order to improve both quality of watermarked Video and robustness of the intelligent video watermark, this work employs the genetic algorithm to search for the quantization steps. The details of GA optimization process will be described in details in next Section.

Step 7: Pass the modified DMT coefficients through the inverse DMT to obtain the watermarked Video.

3.2 Watermark Extracting Algorithm

The watermark extracting algorithm is outlined as follows:

Step1: Sequent video frames are extracted from the video V_w, denoted by $(F_w)^i$ where i is the total number of frames and $i \in [1,n]$.

Step2: Transform the frame image into four-level decomposition using the DMT. Then, create multiwavelet trees and rearrange them into 3072 groups.

Step3: We order the groups in a pseudorandom manner by a similar secret key which was used in the embedding process. Then, combine every 3 groups to form a triple tree Tt_n, for $n = 2,1,...,1024$.

Step4: Let Tt_i denote the first N_w triple trees, which have the largest mean values. The embedded watermark can be extracted from Tt_i.

Step5: After extracting the watermark, we used normalized correlation coefficients (NC) to quantify the correlation between the original watermark and the extracted one.

3.3 Improving Performance Using Genetic Algorithm

In the design of Intelligent Video watermarking system, there are two goals that are always conflicted. These goals are imperceptibility and robustness. In order to minimize such conflicts, this work employs the GA to search for optimal watermarking parameters. This allows the system to achieve optimal performance for Intelligent Video watermarking. For the optimization process, GA is applied in the watermark embedding and the watermark extracting processes to search for quantization step 6. The objective function of searching process is computed by using factors that relate to both robustness and imperceptibility of a watermark. A high quality output video and robust watermark can then be achieved. The diagram of our proposed algorithm of applying GA and details of GA are described as follows: The most critical step in the GA optimization process is the definition of a reliable objective function. In this paper, the objective function of GA uses both normalized correlation (NC) and peak signal-to-noise ratio (PSNR), PSNR is an objective quality assessment related with perceptual distortion, while NC is a robustness measure.

Chromosomes in GA represent desired parameters to be searched [17]. Number of chromosomes used in this work is 30. The encoding scheme is binary string with 32 bit resolutions for each chromosome. The parameter is then represented by chromosome with length of 96 bits. The objective function uses both a peak signal-to-noise ratio (PSNR) and normalized correlation as performance indices. An objective value W can be calculated from (1):

$$W = \delta_{PSNR} \times PSNR + \delta_{NC} \times NC \tag{1}$$

Where δ_{PSNR} and δ_{NC} are weighting factors of *PSNR* and *NC*, respectively. These weighting factors represent the significance of each index used in GA searching process. In this work, a ranking selection is chosen for selection mechanism. The crossover and mutation probability is fixed at 0.7 and 0.05.

4 Experimental Results and Analysis

In this section, we are embedding and extracting a video watermark through a standard QR Code method. We use a QR watermark sizes 64x64 pixels. The original binary QR Code watermark image is shown in Fig. 2

Computer

Fig. 2. Original text and watermark

To evaluate the proposed algorithm, several video sequences with format AVI are used. The Fig. 3 shows some video sequences used in the evaluation. The proposed algorithm is evaluated from embedded watermark imperceptibility and robustness points of view.

Fig. 3. Some video sequences used for evaluation of the proposed algorithm

The content is information about data to encode and decode QR Code and PSNR value is the quantity of efficiency in embedding that defined as (2).

$$PSNR = 10\log_{10}(\frac{255^2}{MSE}) \quad dB \tag{2}$$

For MSE can define as (3):

$$MSE = \frac{\sum (f_w(x,y) - f(x,y))^2}{n} \tag{3}$$

$f_W(x, y)$ is a data of frame image that is embedded watermark already. $f(x, y)$ is a data of original frame image and n is size of pixel. NC value is the quantity of efficiency in extracting that defined as (4).

$$NC = \frac{\sum_i \sum_j W_{(i,j)} - W'_{(i,j)}}{\sum_i \sum_j \left[W_{(i,j)}\right]^2} \tag{4}$$

$W_{(i,j)}$ and $W'_{(i,j)}$ are represent an intensity of original watermark at position (i, j) and watermark from extracting respectively. We can calculate all PSNR as follow (in vertical is value of PSNR that has unit in dB and in horizontal is order of QR Code image that is encoded).

The results of watermarked video quality are also shown in Table 1. The results obtained from our proposed method which is called GA. The watermarked video quality is measured using PSNR (Peak Signal to Noise Ratio). For a video, PSNR is calculated by taking average of PSNR values of all frames and called Average PSNR. Watermarked videos are having average PSNR values of 48.7937 dB and NC values is1. This means that the presented method results in a good quality.

Table 1. The experiment results of PSNR and NC

Content	watermarking	Watermarked frame
Computer		
PSNR	50.2418	
NC	1	
Engineer		

Table 1. *(continued)*

PSNR	48.2526	
NC	1	
QR Code		
PSNR	49.3112	
NC	1	
Watermark		
PSNR	47.3692	
NC	1	

5 Conclusion

Present research on intelligent video watermarking using QR Code Based on Discrete Multiwavelet Transformation. This paper proposed a video watermarking algorithm in the multiwavelet domain. The embedding technique is based on the quantization index modulation. In our optimization process, we use genetic algorithms to search for optimal parameters. These parameters are optimally varied to achieve the most suitable watermarked video. The watermark insertion and watermark extraction are based on the quantization index modulation technique and the watermark extraction algorithm does not need the original frame image in the extraction process. The testing results of the watermarked video quality show that our proposed method can improve the performance of the watermarking process such that the better watermarked video quality and watermark robustness are achieved.

References

1. Denso wave incorporated, `http://www.denso-wave.com/qrcode/index-e.html`
2. ISO/IEC 18004:2000(E), Information technology Automatic identification and data capture techniques Bar code symbology QR Code (2000)
3. Kwon, K.-R., Tewfik, A.H.: Adaptive watermarking using successive subband quantization and perceptual model based on multiwavelet transform. In: Proc. SPIE-Security and Watermarking of Multimedia Contents IV, vol. 4675, pp. 334–348 (2002)

4. Zhang, J., Wang, N., Xiong, F.: Hiding a logo watermark into the multiwavelet domain using neural networks. In: Proc. IEEE ICTAI 2002, pp. 477–482 (2002)
5. Mobasseri, B.G.: A spatial digital video watermark that survives MPEG. In: Proc. Int. Conf. Information Technology: Coding and Computing, Las Vegas, USA, pp. 68–73 (2000)
6. Tsai, H.M., Chang, L.W.: Highly imperceptible video watermarking with the Watson's DCT-based visual model. In: Proc. IEEE Int. Conf. on Multimedia and Expo, Taipei, Taiwan, pp. 1927–1930 (2004)
7. Ge, Q., Lu, Z., Niu, X.: Oblivious video watermarking scheme with adaptive embedding mechanism. In: Proc. Int. Conf. Machine Learning and Cybernetics, Xian, China, pp. 2876–2881 (2003)
8. Hsu, C.T., Wu, J.L.: A DCT-based watermarking for videos. IEEE Transactions on Consumer Electronics 44(1), 206–216 (1998)
9. Hong, I., Kim, I., Han, S.S.: A blind watermarking technique using wavelet transform. In: Proc. IEEE Int. Sym. Industrial Electronics, Pusan, Korea, pp. 1946–1950 (2001)
10. Liu, H., Chen, N., Huang, J., Huang, X., Shi, Y.Q.: A robust DWT-based video watermarking algorithm. In: Proc. IEEE Int. Sym. Circuits and Systems, Scottsdale, Arizona, pp. 631–634 (2002)
11. Niu, X., Sun, S., Xiang, W.: Multiresolution watermarking for video based on gray-level digital watermark. IEEE Transactions on Consumer Electronics 46(2), 375–384 (2000)
12. Joumaa, H., Davoine, F.: Performance of an ICA video watermarking scheme using informed techniques. In: Proc. IEEE Int. Conf. Image Processing, Genoa, Italy, pp. 261–264 (2005)
13. Sun, J., Liu, J.: Data hiding with video independent components. IEEE Electronics Letters 40(14), 858–859 (2004)
14. Mirza, H.H., Thai, H.D., Nagata, Y., Nakao, Z.: Digital video watermarking based on principal component analysis. In: Proc. Int. Conf. Innovative Computing, Information and Control, Kumamoto, Japan, pp. 290–294 (2007)
15. Kong, W., Yang, B., Wu, D., Niu, X.: SVD based blind video watermarking algorithm. In: Proc. Int. Conf. Innovative Computing, Information and Control, Beijing, China, pp. 265–268 (2006)
16. Attakitmongcol, K., Hardin, D.P., Wilkes, D.M.: Multiwavelet Prefilters II: Optimal Orthogonal Prefilters. IEEE Trans. on Image Processing 10, 1476–1487 (2001)
17. Kumsawat, P., Pasitwilitham, K., Attakitmongcol, K., Srikaew, A.: An Artificial Intelligent Technique for Robust Digital Watermarking in Multiwavelet Domain. World Academy of Science, Engineering and Technology 60 (2011)
18. Wang, X.Y., Yang, H.Y., Cui, C.Y.: An SVM-Based Robust Digital Image Watermarking Against Desynchronization Attacks. Signal Processing 88(9), 2193–2205 (2008)
19. Hirakawa, M., Iijima, J.: A Study on Data Management Using Mobile Computing With Digital Watermark Technology. In: International Conference on Service Systems and Service Management, June 8-10, pp. 186–191 (2009)
20. Vongpradhip, S., Rungraungsilp, S.: QR code using invisible watermarking in frequency domain. In: ICT and Knowledge Engineering (ICT & Knowledge Engineering), pp. 47–52 (2012)
21. Pholsomboon, S., Vongpradhip, S.: Rotation, scale, and translation resilient digital watermarking based on complex exponential function. In: TENCON 2004 (2004 IEEE Region 10 Conference), vol. 1, pp. 307–310 (2004)

22. Ketcham, M., Vongprahip, S.: An Algorithm for Intelligent AudioWatermarking Using Genetic Algorithm. In: IEEE Congress on Evolutionary Computation (CEC), Swissôtel The Stamford, Singapore, September 25-28 (2007)
23. Ketcham, M., Vongprahip, S.: Genetic Algorithm Audio Watermarking using Multiple Image-based Watermarks. In: ISCIT (October 2007)
24. Ketcham, M., Vongprahip, S.: Intelligent Audio Watermarking using Genetic Algorithm in DWT Domain. International Journal of Intelligent Technology 2(2) (2007) ISSN 1305-6417

Mining N-most Interesting Multi-level Frequent Itemsets without Support Threshold

Sorapol Chompaisal[1], Komate Amphawan[2], and Athasit Surarerks[1]

[1] ELITE Laboratory, Chulalongkorn University, Bangkok, Thailand
capukampan22@gmail.com, Athasit.S@chula.ac.th
[2] Computational Innovation Laboratory, Burapha Univerisity, Thailand
komate@gmail.com

Abstract. Mining multi-level frequent itemsets from transactional database is one of the most important tasks in data mining community. It aims to discover correlation among items with their hierarchical categories under support-confidence values and thresholds. However, it is well-known that the task of providing an appropriate support threshold to mine the most interesting patterns without prior knowledge in advance is very difficult and it is more reasonable to ask the users to specify the number of desired patterns. Therefore, in this paper, we propose an alternative approach to mine the most interesting multi-level frequent patterns without the setting of support threshold, called *N-most interesting multi-level frequent pattern mining*, where N is the number of desired patterns with the highest support values per each category level. To mine such patterns, an efficient adaptive *FP-growth* algorithm, called *NMLFP*, is proposed. Extensive performance studies show that *NMLFP* has high performance and linearly scalable on the number of desired results.

Keywords: Association Rules, N-most interesting patterns, Multi-level frequent itemsets.

1 Introduction

Association rule mining (*ARM*) [1, 2] under the support-confidence framework is the task of discovering relationship or correlation among items appearing together in large database. It is applied in a wide range of applications such as marketing, medical diagnostics, web analysis, decision making, etc. The process of *ARM* can be divided two steps: (*i*) discovering frequent itemsets from the given database that meet the support threshold, and (*ii*) generation rules from the frequent itemsets found in the first step that satisfy the confidence threshold.

In general, *ARM* focused on investigation of interesting rules at a single concept level. However, there are some applications which need to discover relations at multiple concept level. For example, besides of the need of finding 75% of customers that purchase milk may also purchase cookies, it could be informative to also illustrate that 65% of people buy almond cookies if they buy chocolate milk. The latter information expresses the lower concept level with more specific and concrete

S. Boonkrong et al. (eds.), *Recent Advances in Information and Communication Technology*, Advances in Intelligent Systems and Computing 265,
DOI: 10.1007/978-3-319-06538-0_13,

information than that of the former. Therefore, Han et al. [3] introduced an approach to mine multiple-level association rules under user-given concept hierarchy of items and support-confidence thresholds to discover more general/specific knowledge from database in real-world applications.

To effectively and efficiently explore multiple-level concept level of association rules, there are two simple frameworks of defining the support threshold to measure interestingness of itemsets and association rules. Firstly, the framework of setting only one support threshold for all level items (also called *uniform minimum support*) is proposed. Under this approach, it is quite convenience to users to determine a support threshold. However, it may give some uninteresting rules at higher level if the support threshold is small. Then, the setting different minimum support threshold on each level by reducing the thresholds at lower levels of hierarchy, called *reduced minimum support* [3, 4], is introduced to alleviate a drawback of the first framework. This approach cannot only be found interesting rules at different level, but may also have high potential to find nontrivial, informative association rules.

However, it is well-known that the setting of a suitable support threshold without prior knowledge in advance is a difficult task. If the support threshold is too high, then there may be only a small or even no result. In that case, the user may have to guest a smaller threshold and do the mining again. If the threshold is too low, then there may be too many results for the users. Thus, asking the number of desired outputs is considered easier and mining top-*k* frequent patterns/*N*-most interesting patterns [8-11] has become a very popular task.

Thus, in this paper, we aim to alleviate difficulties of setting support threshold and then introduce an alternative approach to mine the most interesting multi-level frequent pattern, called *N-most interesting multi-level frequent itemset mining*. This approach allows the users to specify a simple threshold that is the number of desired result (*N*). Then, a set of *N* most frequent itemsets at each level is discovered. To quickly discover such itemsets, an efficient algorithm namely *NMLFP* is proposed. *NMLFP* is a tree-based algorithm under FP-tree and FP-growth technique. It also applies top-down traversal to quickly cut-down the search space.

The rest of this paper is organized as follows. In Section 2, the concepts related to the multi-level frequent patterns and *N*-most interesting multiple-level frequent patterns are introduced. An efficient method to discover *N*-most interesting multiple-level frequent itemsets are described in Section 3. Section 4 reports the performance analysis under several experiments. Lastly, the conclusion is in Section 5.

2 Problem Definitions

In this section, basic definitions of the concept hierarchy for identifying multi-level frequent patterns and notations for *N*-most interesting multi-level frequent patterns are described.

2.1 Multi-level Frequent Patterns

Let $I = \{i_1, i_2, \ldots, i_m\}$ be a set of products, objects or events, called *items*, where each item is associated with multiple meaning. For example, given the product hierarchy of

Fig. 1, "*foremost chocolate milk*" may be viewed on three different perspectives. At the lowest level (level 1), we can consider "*foremost chocolate milk*" as a specific *product*. At the higher level (level 2), it simply represents the *kind of milk*, meanwhile the highest level (level 3) refers to *milk* which is a kind of food. In this context, the terms of lower and higher level is applied to express the levels of items' meaning.

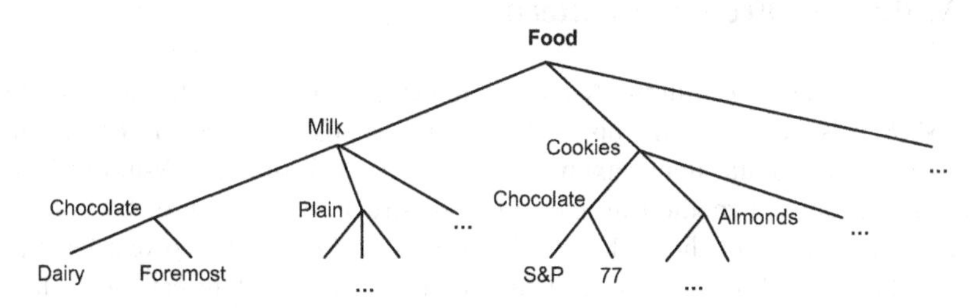

Fig. 1. A concept hierarchy for the relevant data items

A transactional database $TDB= \{t_1, t_2, \ldots, t_n\}$ is a set of transactions where a transaction t_j is composed of (*i*) a unique transaction identifier (*tid*) and (*ii*) a set of items *Y*, in which each item is encoded into a concise numeric form (also with the sequences of the symbol '*' according to their positions in the hierarchy concept) that can identify not only the given products, but its hierarchical level as well. For example, node of "*Milk*" in Fig.1 would be represented by "1**", the node of "*Chocolate*" after "*Milk*" by "11*" and the last level node of "*Foremost*" by "112", respectively. Then, the encoded items "1**", "11*" and "112" can be regarded as a set of distinguish items occurring on different level.

A set $X \subseteq I$ is called a *k*-itemset, if it contains *k* items at the same level. If $X \subseteq Y$, it is said that *X* occurs in t_j or t_j contain *X*. The support vale of *X*, denoted as s^X, is the ratio of *X*'s occurrence to the total number of transactions in database. Then, the support of X is used to define the concept of multi-level frequent itemsets as follows.

Definition 1: An itemset *X* is called *a multi-level frequent itemset* if *it contains encoded items at the same level of the concept hierarchy and its support is no less than the user-given support threshold.*

From above definition, the problem of mining multi-level frequent patterns is the task of discovering a complete set of frequent patterns on all the levels of concept hierarchy under user-given support thresholds. However, the setting of support thresholds is a difficult task. Then, we proposed to mine the *N*-most interesting frequent itemsets without any support threshold which can be defined as follows.

Definition 2: An itemset *X* is called *a N-most interesting multi-level frequent itemset* if *it contains at least two encoded items at the same level of the concept hierarchy and there is no more than N – 1 itemsets at the same level of X having support more than that of X.*

Thus, the mining of *N*-most interesting multi-level frequent patterns is to mine *N* itemsets at each level of concept hierarchy that have highest value of support under user-given the number of desired pattern *N*.

3 *NMLFP*: An Efficient Method for Mining *N*-most Interesting Multi-level Frequent Patterns

An efficient method for mining *N*-most interesting multi-level frequent patterns, called *NMLFP*, is described in this section. The proposed method is based on the concept of relative data item taxonomic together with mining *N*-most frequent itemsets. The proposed method applies *FP-tree* structure to capture the content of database and it consists of three phases: (*i*) construct a *FP-tree* for lowest-level items with two scanning of transactional database, (*ii*) mine *N* itemsets (with highest support) from the *FP-Tree* and then store all of them into *N*-most multi-level table and (*iii*) create a *FP-tree* for higher-level *FP-tree* from the current lower-level *FP-tree*, respectively. The details of *NMLFP* are described below:

STEP 1. Create header table for all of lowest-level items and then scan each transaction of *TDB* in order to collect occurrence frequency/support of each item into the header table. Lastly, sort the header table by support descending order.

STEP 2. Build *FP-tree* at lowest-level in the same manner as [2] in which each transaction is scanned, trimmed and sorted in the same order as the header table. Then, the sorted items in the scanned transaction are sequentially inserted into *FP-tree* with frequency to be 1 (if there exists a node of any regarded item in the *FP-tree*, its frequency is updated by one.)

For illustration, we use an example with the encoded transactions shown in Table 1. Let the number of required results *N* for each level be 5 and the number of concept hierarchy level be 3. Our task is to discover the 5 itemsets with highest support from the three level of concept hierarchy level.

Table 1. The encoded transactional database

TID	Items
T01	111, 121, 212, 221, 311
T02	111, 211, 221, 312
T03	111, 122, 211, 212, 311
T04	112, 122, 212, 311, 322, 412
T05	112, 121, 221, 222, 312
T06	111, 112, 122, 311, 321
T07	111, 211, 221, 312, 422
T08	122, 212, 222, 312, 322

With the first and second scan of the encoded database on step 1 and 2, we get the header table and FP-tree for all of the lowest-level items as shown in Fig. 2.

STEP 3. Generate *N*-most frequent itemsets from the constructed *FP-tree* by (i) creating a table for storing a set of *N*-most interesting itemsets during mining process namely *N-most table*, (ii) applying the top-down traversal technique to consider items in the header table (consider from the second to the last item in the header table), (*iii*) traversing all the paths of the considered item *X* in the *FP-tree* in bottom-up manner in order to collect support of items appearing with *X* (consider only the items having higher position than that of *X* in the header table). If the support of an item *Y* occurring with *X* is greater than that of the Nth itemset in the *N-most table*, the Nth itemset is removed and the itemset '*XY*' is inserted into the *N-most table*. If there are itemsets '*XY*' (more than one), where *Y* is an item occurring with *X*, inserted into the *N-most table*, a *small FP-tree* for *X* with occurring with all items *Y* is created, (*iv*) traversing the *small FP-tree* of *X* by using *COFI-tree* traversal technique[8] to generate high-support itemsets with larger size.

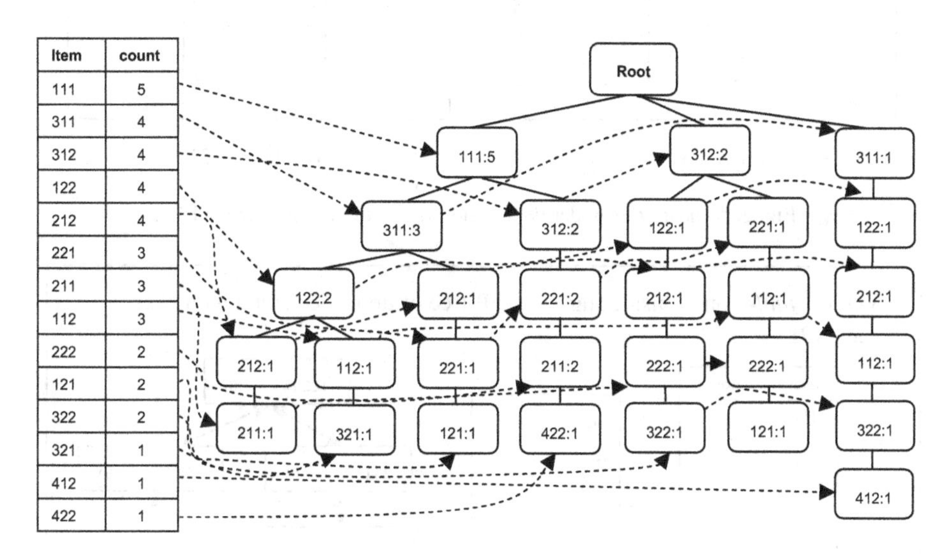

Fig. 2. *FP-tree* for lowest-level items (level 3)

On step 3, *NMLFP* starts to consider the second item in the header table that is item '311', then all the paths of item '311' is traversed with the considering bottom items of the paths. We get {('111, 311':3), ('311':1)} and then we know that item '111' appear together with '311' three times. The support of itemset '111,311' is compared with the Nth itemset in the *N-most table*. Fortunately, there is none of itemset in the *N-most table*. Then, the itemset '111, 311' is inserted into the *N-most table* (as shown in Fig. 3(a)). Next, item '312' (the 3rd item in the header table) is considered. With the traversal of all paths we got {(111,312:2), (312:2)} and know that '111' appears together with the considered item twice. The support of '111, 312' is compared with the Nth itemset in the *N-most table*. Since there is only one itemset in the *N-most table*, the itemset '111, 312' is inserted in to the *N-most table* by support descending order (as shown in Fig. 3(b)).

With the considering and traversing of item '122', we gain {(111, 311, 122:2), (312, 122:1), (311, 122:1)}. Then, the itemsets '111, 122', '311, 122' and '312, 122' with support 2, 3 and 1 are inserted into the *N-most table*. In this case, there is more than one itemsets of '122' inserted into *N-most table*, and then a *small FP-tree* of '122' with its header table is created as shown in Fig. 4.

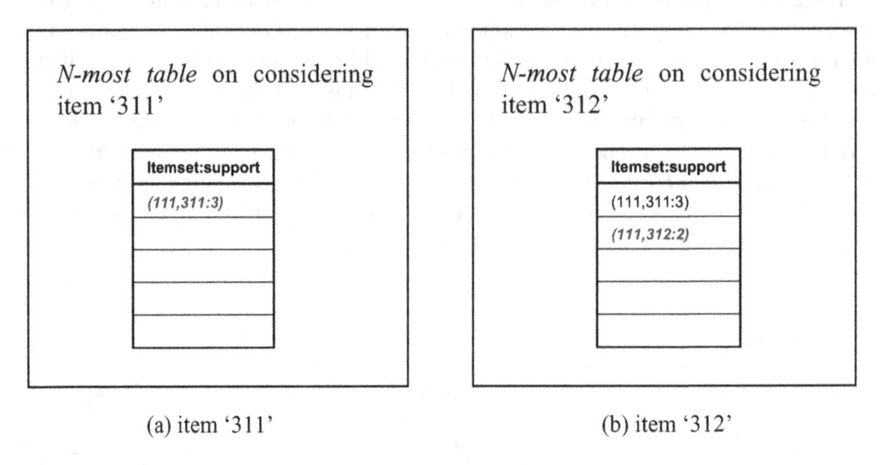

Fig. 3. *N-most table* under the considering of items '311' and '312'

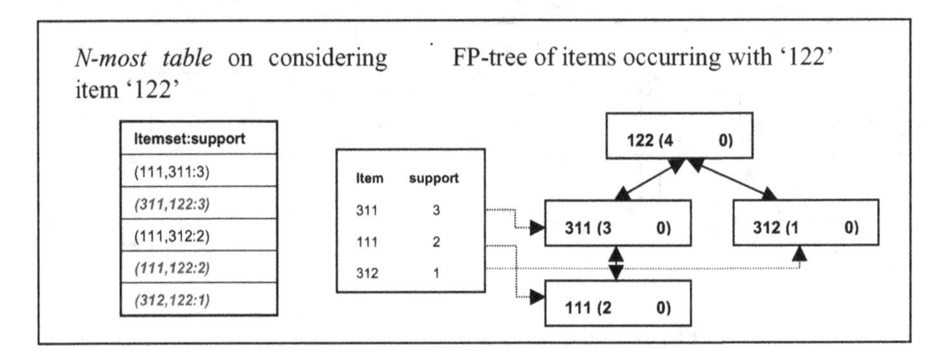

Fig. 4. *N-most table* and *a small tree* during the considering of item '122'

Next, the small tree of '122' is traversed to get support of long itemsets. From Fig. 4, we can observe that itemset '111, 311, 122' appear together twice. Then, its support is compared with the Nth itemset in the *N-most table* (*i.e.* itemset '312, 122' with support 1). Since support of itemset '111, 311, 122' is greater than that of '312,122', the itemset '312, 122' is eliminated and then the itemset '111, 311, 122' is inserted into the *N-most table* by support descending order.

By repeating these processes on all items in the header table, we gain *N* itemsets with highest support at lowest level contained in the *N-most table* (as shown in the leftmost table in Fig. 6).

STEP 4. Construct a new *FP-tree* for higher-level items from the current *FP-tree* (low-level *FP-tree*) by (*i*) creating header table for all higher-level items, (*ii*) traversing each branch in the low-level *FP-tree* by encoding the last digit of item to be '*' (for example, item '111' is encoded to be '11*', or item '12*' is encoded to be '1**') in order to collect support of all higher-level item into the header table (if there is a high-level item appearing more than once in the considered branch, we will count only the highest support value), (*iii*) sorting the header table by support descending order, (*iv*) traversing each branch in the low-level *FP-tree* again to collect high-level items appearing in the considered branch (collect only one with highest support, if there is an item having multiple appearances in the considered path) and then sorting the items by the order of header table, (*v*) creating a path for the collected higher-level items into the higher-level *FP-tree*, and (*vi*) removing the regarded branch from the low-level *FP-tree*.

On step 4, the leftmost path of the lowest-level FP-tree (of Fig.2) is traversed and encoded to be {11*:5, 31*:3, 12*:2, 21*:1, 21*:1} (in this case the item '21*' appears twice, we only consider one with highest support). Then, the supports of all items in the path are used to update into the header table. Then, all of paths in the *FP-tree* are scanned, encoded and used to update support values of the header table. After considering all branches, the header table is ordered by support descending value (as shown in the left of Fig. 5). Next, all branches is scanned again. For the leftmost path, {111:5, 311:3, 122:2, 212:1, 211:1} is converted into {31*:3, 11*:3, 21*:1, 12*:1} and then sequentially inserted into the higher-level *FP-tree*. After traverse all branches of the lowest-level *FP-tree*, we gain the higher-level *FP-tree* (*FP-tree* of level 2) and its corresponding header table as shown in Fig. 5.

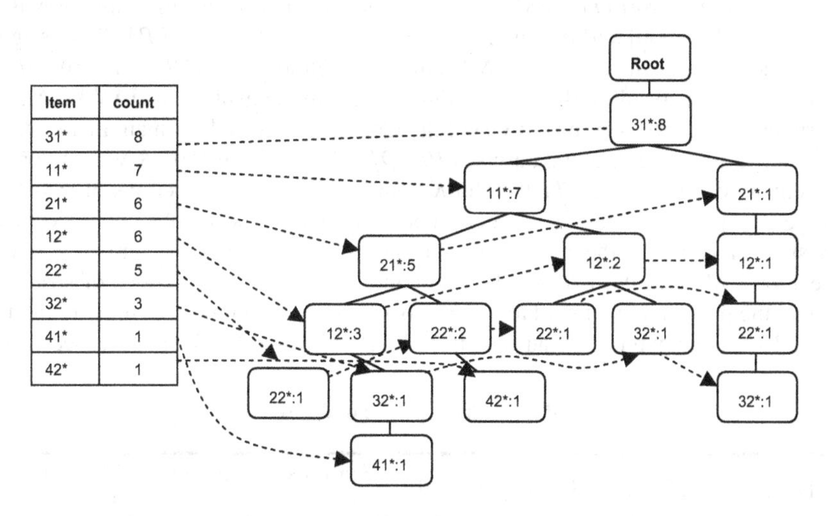

Fig. 5. *FP-tree* for higher-level items (level 2)

STEP 5. Repeat step 3 and 4 on the new generated higher-level *FP-tree* until the highest level.

In our example, after creating and mining *N*-most interesting itemset from the *FP-tree* of all levels, the results of 5-most itemsets for all levels are shown as Fig. 6.

N-most interesting itemsets for all levels

Level 3
Itemset:support
(111,311:3)
(122,311:3)
(211,111:3)
(212,311:3)
(212,311:3)

Level 2
Itemset:support
(31*,11*:7)
(31*,21*:6)
(31*,12*:6)
(31*,11*,21*:5)
(31*,11*,12*:5)

Level 1
Itemset:support
(1**,3**:8)
(1**,2**:7)
(3**,2**:7)
(1**,3**,2**:7)
(1**,4**:2)

Fig. 6. *N-most table* for all levels of concept hierarchy

4 Performance Evaluation

We here study the performance of our proposed method by performing several experiments and then comparing with a modification of *mining multiple-level frequent itemsets algorithm, FPM-T* [6]. To conduct experiments, the value of *N* is set in range of [50, 2,000] and the support threshold at each level for *FPM-T* is assigned corresponding to the support of the Nth itemset at each level of *NMLFP* in order to get the same set of results. Both algorithms were implemented in JAVA. All the experiments are performed on 2.5 GHz Intel Core i5 and with 4GB main memory.

As shown in Table 2, two datasets, *T10I4D100K* and *T20I6D100K* with 4 concept hierarchical levels are used. *T10I4D100K* contains 100,000 transactions of size 10 in average, the potential maximal large itemsets of size 4, the number of node in the highest level is 8 and the number of fan-outs from level to level are 5, 5 and 5, respectively. Meanwhile, *T20I6100K* also contains 100,000 transactions of size 20 in average, the potential maximal large itemsets of size 6, the number of node in the first level is 15 and the number of fan-outs from level to level are 6, 3 and 4, respectively.

Table 2. Characteristic of datasets

Database	|*T*|	|*I*|	#node at level 1	Fan-outs 1 to 2	Fan-outs 2 to 3	Fan-outs 3 to 4
DB1	10	4	8	5	5	5
DB2	20	6	15	6	3	4

4.1 Execution Time

We consider the execution time of the both algorithms with all processes for mining the results. As shown in Fig. 7, the runtime of *NMLFP* is mainly efficient than that of *FPM-T* since *NMLFP* can take advantages from top-down traversal to mine results (*i.e.* firstly consider itemsets with highest support). This can help us to quickly gain *N* itemsets with highest support and raise the support of the Nth-itemset which is used as the criterion to cut-down the search space. Consequently, it slightly increases as the value of *N* increases. Whenever the value of *N* increases, we have to consider and discover more itemsets to be the results. Then, we have to take more computational time as well.

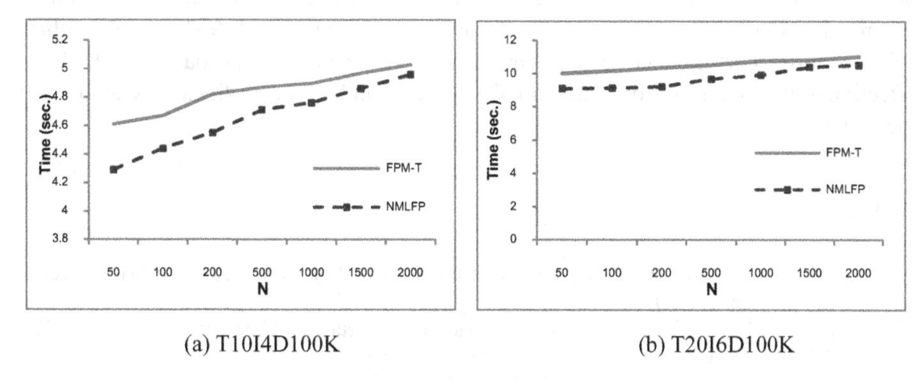

(a) T10I4D100K (b) T20I6D100K

Fig. 7. Execute time on (a) T10I4D100K and (b) T20I6D100K

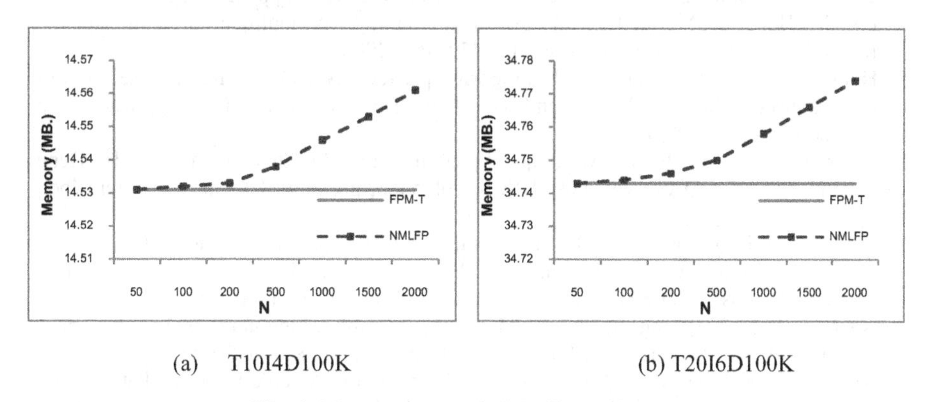

(a) T10I4D100K (b) T20I6D100K

Fig. 8. Memory usage of NMLFP and FPM-T

4.2 Memory Consumption

To evaluate memory consumption of the both algorithms, we consider all components for mining the results including header table, and FP-tree. As indicated in Fig. 8, we can observe that *NMLFP* consumes a bit more memory than *FPM-T* due to *NMLFP* have to maintain a set of *N* most interesting itemsets for each level during mining process. Then, *NMLFP* uses a bit more memory than *FPM-T* for all cases. However, the gap of different is not significant. In all experiments, the gap of memory consumption is less than 1MB.

5 Conclusion

In this paper, we study the problem of multi-level frequent itemsets mining and then try to avoid difficulties from the setting of suitable support thresholds. Therefore, we propose the alternative approach to mine *N*-most interesting multi-level frequent itemsets without support threshold. This can discover the most *N* frequent itemsets for each level. To mine such itemsets, we introduce an efficient *NMLFP* algorithm based on the concept of FP-tree with single scan and top-down traversal to quickly generate the results. Our performance studies show that the proposed *NMLFP* algorithm achieves high performance in comparing with the previous approach. *NMLFP* is effective and efficient on both the small and large and it is also linearly scalable on the value of *N*.

References

1. Agrawal, R., Srikant, R.: Fast Algorithms for Mining Association Rules. IBM Almaden Research Center (1994)
2. Han, J., Pei, J., Yin, Y.: Mining Frequent patterns without candidate generation. SIGMOD Rec. 29(2) (2000)
3. Han, J., Fu, Y.: Discovery of Multi-Level Association Rules from Large Databases. In: 21st VLDB Conference on Very Large Data Base, Switzerland, pp. 420–431 (1995)
4. Lee, Y., Hong, T., Wang, T.: Multi-level fuzzy mining with multiple minimum supports. In: Expert Systems with Applications, pp. 459–468 (2008)
5. Hong, T., Huang, T., Chang, C.: Mining Multiple-level Association Rules Based on Pre-large Concepts. In: Data Mining and Knowledge Discover in Real Life Application, pp. 187–200. In Tech (2009)
6. Eavis, T., Zheng, X.: Multi-level frequent pattern mining. In: Zhou, X., Yokota, H., Deng, K., Liu, Q. (eds.) DASFAA 2009. LNCS, vol. 5463, pp. 369–383. Springer, Heidelberg (2009)
7. Mohammad, E., Osmar, R.: COFI-tree Mining: A New Approach to Pattern Growth with Reduced Candidacy Generation. In: Workshop on Frequent Itemset Mining Implementations (FIMI 2003) in Conjunction with IEEE-ICDM (2003)
8. Ngan, S., Lam, T., Wong, R., Fu, A.: Mining N-most interesting itemsets without support threshold by the COFI-tree. Int. J. Bus. Intell. Data Mining, 88–106 (2005)
9. Amphawan, K., Lenca, P., Surarerks, A.: Mining top-*k* Periodic-Frequent Pattern from Transactional Databases without Support Threshold. In: Papasratorn, B., Chutimaskul, W., Porkaew, K., Vanijja, V. (eds.) IAIT 2009. CCIS, vol. 55, pp. 18–29. Springer, Heidelberg (2009)
10. Amphawan, K., Lenca, P., Surarerks, A.: Mining top-*k* regular-frequent itemsets using database partitioning and support extimation. Expert Systems with Applications 39, 1924–1936 (2012)
11. Amphawan, K., Lenca, P.: Mining top-*k* frequent-regular patterns based on user-given trade-off between frequency and regularity. In: Papasratorn, B., Charoenkitkarn, N., Vanijja, V., Chongsuphajaisiddhi, V. (eds.) IAIT 2013. CCIS, vol. 409, pp. 1–12. Springer, Heidelberg (2013)

Dominant Color-Based Indexing Method for Fast Content-Based Image Retrieval

Ahmed Talib[1,2], Massudi Mahmuddin[1], Husniza Husni[1], and Loay E. George[3]

[1] Computer Science Dept., School of Computing,
University Utara Malaysia, 06010 Sintok, Kedah, Malaysia
s91707@student.uum.edu.my, {ady,husniza}@uum.edu.my
[2] IT Dept., Technical College of Management,
Foundation of Technical Education, 10047 Bab Al-Muadham, Baghdad, Iraq
[3] Computer Science Dept., College of Science,
Baghdad University, 10071 Al-Jadriya, Baghdad, Iraq
loayedwar57@yahoo.com

Abstract. Content-based image retrieval is an active research area in image processing and computer vision. Color represents an important feature in CBIR applications, thus many color descriptors were proposed. Sequential search is one of the common drawbacks of most color descriptors especially in large databases. In this paper, dominant colors of an image are indexed to avoid sequential search in the database. Dominant colors in query image are used independently to find images that containing similar colors to create reduced search space instead of the whole database search space. This will speed up the retrieval process in addition to improve the accuracy of color descriptors. Experimental results show effectiveness of the proposed color indexing method in reducing the search space to less than 25% without degradation the accuracy.

Keywords: Color indexing, Dominant colors, MPEG-7 descriptors, RGB color space, Database search space.

1 Introduction

Image retrieval become one of the most famous research directions nowadays because it uses to search an image in archive, domain-specific, personal and web image databases. For retrieving color images from multimedia database, low level features and especially color feature have been widely used in this regard. This is because color represents the most distinguishable feature compared with other visual features, such as texture and shape [1-2]. In this respect, MPEG-7 Committee proposed many color, texture and shape descriptors to be used in image and video retrieval [3]. Authors in [4-5] maintain that human visual system first identifies prominent colors in the image and second it processes any other details. The whole process resembles the way humans recognize image from its dominant colors without paying any attention to their distribution. MPEG-7's Dominant Color Descriptor (DCD) provides compact and effective representations for colors in the image [3]. Recently, compactness

S. Boonkrong et al. (eds.), *Recent Advances in Information and Communication Technology*, Advances in Intelligent Systems and Computing 265,
DOI: 10.1007/978-3-319-06538-0_14, © Springer International Publishing Switzerland 2014

property of dominant colors representation becomes more attractive for many researchers to reduce size of color descriptors from several hundred bins (histogram-based methods) into few colors (8 colors in MPEG-7 DCD) such as the works that have been achieved in [6-8]. Additionally, this compactness property becomes mandatory in specific applications such as web-based image retrieval [1].

Searching a large images database imposes many challenges because the time required to retrieve the results of query image is high, consequently this will degrade performance of CBIR. Researchers address some issues that are related with image retrieval performance. These issues can be divided into two categories [9]. First category concerns with retrieval robustness (accuracy of image retrieval), it ignored the retrieval efficiency where most works in this category rely on sequential search [2],[10-11]. Second category concerns with retrieval efficiency, where the time of retrieval (retrieval speed) get more focus. In this paper, the second category is focused.

Color is a salient feature among the low level visual features [2],[7]. As a representative of color feature, color histogram and dominant color descriptors are the widely used features in content based image retrieval (CBIR) [5]. The high dimensionality of the histogram feature vector make it highly computational cost in similarity measure and inefficient in searching/indexing process. It is called "curse of dimensionality" problem that represent the first problem of color-based indexing methods. To solve this problem, many dimensions reduction approaches are used. Color quantization techniques are used in this context but fixed quantization techniques lead to accuracy degradation. Therefore, dynamic color quantization techniques were proposed, which are Dominant Colors (DCs) extraction methods. DCs consider as the most effective solutions in this context where few colors are extracted to represent the image. To index images' DCs, vector quantization methods are used but "color approximation" problem is emerged. Color approximation problem is occurred due to using clustering algorithms to find the representative (centroid) of the cluster. This color approximation degrades the retrieval accuracy because the matching process is performed with cluster centroid (which is obtained by averaging all cluster's colors) instead of actual color values. Therefore, color indexing method is proposed. It belongs to space partitioning methods, where RGB color space is divided into small partitions using uniform Octree color quantization method [12]. It combined with B+-tree method, which is used for representing the colors percentages, to filter irrelevant images out in early stage.

The paper is organized in the following way. Section 2 is concerned with explicating the general CBIR indexing methods and specifically color-based indexing methods because the color feature represent the scope of this paper. Section 3 is mainly concerned with the proposed RGB-based indexing method and the newly proposed color percentage filtering scheme that helps improve and speed up the retrieval process. Section 4 illustrates the extensive experiments that contain quantitative results. Finally comes the conclusion in Section 5.

2 Related Works

In large image database, indexing is urgent matter to reduce the search space of the retrieval process and in turn to speed up the process. Most color-based methods perform sequential search in their retrieval process; this will impose delay in the time of image retrieval process. Color histogram is one of the common color descriptors. Despite of the color histogram is characterized by simplicity in its implementation but it results large feature vector that is difficult to index, *"curse of dimensionality"* problem. Therefore, DCs can be used to avoid this problem. The benefits of DCDs over histogram-like methods is the former find image's representative colors from image itself instead of making it fixed in the color space as the latter. This will make accurate and compact descriptor. Therefore, DCD is selected to be the base of this paper to perform the image indexing.

For indexing the image features, there are two main approaches in general: *multi-dimensional indexing* and *vector quantization techniques*. Multi-dimensional indexing techniques are divided into two categories, Space Partitioning (SP) and Data-Partitioning (DP) methods. Both of them divide the space or data into small partitions but the difference lies in how the partitioning process is achieved. SP methods such as kd-tree divides the whole space into disjoint partitions without consideration of the data whereas in DP methods such as R-tree, feature space is divided depending on features (data) distribution in the database [13]. The advantage of SP method is it performs complete and disjoint partitions of the whole space that means there is no overlapping between these partitions.

In vector quantization techniques, there are many techniques that have been proposed, such as hierarchical K-means clustering. In these methods, there is no partitioning of space or data into small parts, instead grouping the data into clusters or groups is achieved. Each cluster (group) is represented by a single value called cluster's centroid. Cluster's centroid is computed by averaging all cluster members, thus the query point is compared with cluster's centroid instead of original value of the members. Disadvantages of these methods is comparing with clusters' centroids instead of original values lead to inaccurate results because some cluster's members are far from the query points in spite of having suitable distance from cluster's centroid. The latter problem is called "color approximation problem".

For 3-dimensional color indexing, several methods have been proposed in CBIR field. High dimensional histogram indexing method, that used by Deng et al. [14] for comparison, is considered as the simplest and most expensive indexing method. This method suffers from high dimensional problem, 1024-D of color histogram bins. Babu et al. [15] combine color clustering and spatial indexing method (R-Tree) for indexing colors of flags and trademarks databases. Sudhamani and Venugopal [16] also proposed a method for color clustering and indexing. They used mean shift algorithm for color clustering, R*-Tree for spatial indexing and perceptually uniform LUV color space instead of RGB. The above two methods depend on clustering that are suffered from aforementioned problems of clustering (one of VQ techniques). In general, color-based indexing methods depend on fixed range of colors in similarity measure. Therefore, spatial indexing methods such as R-Tree and R*-Tree are not necessary and fixed indexing structure is more efficient [14]. Accordingly, Lattice structure was proposed in [14] that is characterized by efficient finding the nearest neighbors of given point (color) in 3-dimensional LUV color space. Nevertheless, this efficiency depends on careful selection of radius in hexagonal lattice cell and this is not a straight forward process, hence there is no comparison (in the literature) has been made with this method. Additionally, it suffers from same problem of SP and clustering methods, which is the query point may locate at the border of lattice cell.

Recent research is proposed by Yildizer et al. [13] to solve this problem. The significant contribution of that research is introducing two threshold values C_G and Cs that can be considered as search space parameters. C_G represents the distance from query point that can be searched around it (in the closest cluster) to find similar images instead of considering all cluster members. Cs represents the distance from query point that can be considered to add other clusters to the similarity searching process.

As conclusion, all vector quantization indexing methods, which most color-based indexing methods are depended on, suffered from color approximation problem. Approximation process is carried out during vector quantization process, such as K-means clustering method, to produce color centroids. Therefore, this problem is addressed in this paper to discuss and solution is proposed.

3 Proposed Indexing Method

In this section, DC-based indexing method is proposed to reduce search space of color-based methods (i.e.: it dedicates to all methods, not only DC-based methods) to speed up retrieval process as well as preserve retrieval accuracy.

The proposed method motivated by the question, "where are we need to search exactly, to reduce search space rather than doing whole database search space?" The key answer is searching depends on fixed range queries. In other words, searching on images that only have colors of distance less than or equal the maximum distance to the query image colors. As mentioned before, using tree-like indexing in the fixed range queries is ineffective. Therefore, fixed space partitioning method is used in this research. Before building database index structure, similarity between two colors must be considered and maximum distance between these colors also needs to be determined because the index structure will depend upon them; this will be explained in the next section.

3.1 Maximum Distance between Similar Colors

The key of the proposed method is a similarity among colors within fixed range. The searching can be done only in a specific range within distance, which is the maximum distance between two colors to consider them as similar colors. The Euclidian distance between two 3-D colors to assume them similar was 10, 20 or 25 [17]. Therefore, the maximum difference value (MxDV) for each color channels (Red, Green and Blue in RGB color space) is set to 25.

3.2 Indexing Structure

In DC-based methods (e.g. MPEG-7 DCD), DCs are extracted using dynamic quantization method (GLA) and most likely the image is quantized to most significant 5-bits of color channels (bit3 to bit7). Since the maximum difference between two channels is 25, the changing of the two bits (bit3 and bit4) is within this range. This because the weights of these bits are 8 and 16 respectively; their summation is 24 that approximately equals to MxDV (25). In this regard, bit7, bit6 and bit5 are out of tolerance range of colors to be similar. Thus, these three bits are the first level of color similarity; they are used to differentiate among not similar colors. The other two bits (bit3 and bit4) are used separately to be second and third levels respectively of color similarity. Hence, first indexing dimension contains 512 cell. Second and third indexing dimensions contains 8 cells, as presented in Fig. 1.

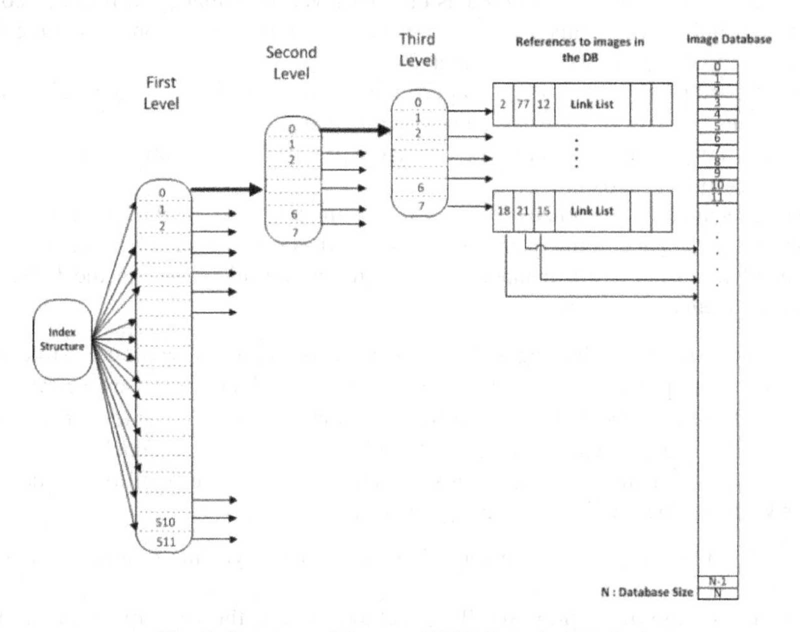

Fig. 1. Structure of Proposed RGB Indexing Method

Color percentage plays an important role in similarity measure of certain color with its corresponding color in other images where the similar colors consider as dissimilar if their percentages have large difference such as similarity measure of different color descriptors such as MPEG-7 DCD, Correlogram, and most color descriptors. Therefore, filtering images (that have large difference in percentage) out in early stage helps in reducing search space and speeding up the retrieval process. In Deng [14], filtering process is performed online during query processing. Filtering out the dissimilar images in terms of colors is achieved firstly; then percentages of colors are matched through pass by all images sequentially to perform second level of filtering. This online process is time consuming process in large database; thus, achieving it offline is mandatory in large size database. Therefore, the proposed indexing structure is extended to include partitions of color percentages. Single level B+-tree is used to represent color percentages. The unique node of B+-tree contains four pointers each pointer refers to the certain list as <0.25, 0.25-0.5, 0.5-0.75 and >0.75 respectively.

3.3 Searching Process

In this process, a query is required to find its similar images in the database. Searching process includes the following steps:

1. For each DC in the query image, find database images that are similar in both colors values and colors percentages; this is by reaching to suitable node(s) in index structure and in turn to the database images that associated with this node(s).

A. Reaching to the nodes in the third level of index structure, which are similar to the query color, is considered as the first level of similarity (it is called color-based similarity). In this level, all false matched images that don't contain colors similar to the query will be eliminated.

B. Second level of similarity is a percentage-based similarity. Each reached node in point (A) has B+-tree structure of color percentages. Two paths of B+-tree that nearest to the query color percentage will be selected to obtain the candidate images for comparison.

2. Merging images references that resulted from each DC of the query image to produce search space of the query, which is called as reduced search space (RSS).

3. Calculate dissimilarity distance between query and all images in the RSS, and then rank them accordingly.

In step 1, most false match images are removed according to different color tolerance value and color percentages. Three color tolerance values are used according to maximum distance value that extracted from Section 3.1. These color tolerance values (CTV) are 0, 8 and 24 regarding to 2-bits (bit3 and bit4) of color channels.

The difference of the proposed indexing scheme over space partitioning methods such as kd-tree structure lies in different aspects:

1. The fixed-size representation (array) is faster than dynamic-size representation (tree) of kd-tree.
2. The new design of the proposed RGB method exploits the range query in building the index structure (3-levels structure), instead of changing query parameters to perform range query as in kd-tree that have 8-levels structure to represent RGB color space. This new structure speeds up the search mechanism.
3. Embedding B+-tree representation in the last level of the proposed representation makes the search result more accurate and fast. This is because, the color percentage is used as fourth level of filtering to exclude the images that have different color percentage than query's colors percentages.

4 Experimental Evaluation

Performance of the proposed technique is evaluated based on the following:

1. **Number of Indexed Colors:** Indexing of images database can be performed with different numbers of colors (8, 5, 3 or 1) to measure the effect of each one on the retrieval performance. It is noteworthy to mention here that DCs is sorted in descending order according to their color percentage before indexing process.
2. **Evaluation Metrics:** Two types of metrics are used in this research:
   - *Efficiency Metrics:* The main goal of indexing is to reduce search time compared to sequential search by reducing database images that can matched the query image. The reduced time needed for searching in the RSS can be computed by the percentage RSS/WSS, WSS denoted to the Whole Search

Space. The percentage RSS/WSS% can be called as Search Space Ratio (SSR) that represents ratio of images that are actually searched to the all images in the database. An Overhead Ratio (OHR) of the proposed indexing is also necessary to be computed. This ratio represents the time needed to create RSS.
- *Accuracy Metrics:* Three quantitative performance metrics are utilized to measure the accuracy of different color descriptors that are used in the proposed indexing method. These metrics are ARR, ANMRR and P(10).

3. **Evaluation Datasets:** Evaluating the proposed indexing techniques will be conducted on two datasets, newly introduced Cartoon-11K (11,120 images) and well-known Corel-10K (10,800 images). These datasets are different in terms of image content (color and variety) as well as their sizes are large enough to fit the objective of designing the indexing methods. The main dataset in this research is cartoon dataset that is used to evaluate color descriptors. This is because the characteristic of the most cartoon characters is appearing with the same colors in all or most images [18].
4. **Competing Indexing Methods:** Indexing methods that are selected to compete with the proposed method are sequential search, k-Means (KM), and recent k-Means with B+-tree methods (KMB) [13]. Sequential search is a conventional method in CBIR for searching in the database. The accuracy resulted from sequential search is considered as optimal accuracy because searching in this method include whole database. Therefore, all competing indexing methods accuracies are compared with it to check the degradation that can be obtained from these methods due to the reduction of search space.
5. **Evaluation Color Descriptors:** the color descriptors that can be used to test the proposed indexing method are MPEG-7 DCD and Color Correlogram (ColGrm). MPEG-7 DCD is used because it contains dominant colors whereas the Correlogram is complicated color descriptor and it is general color descriptor (it does not have dominant colors). This is to prove that the proposed indexing method can be generalized for all color descriptors not just for DCDs.

Table 1. Accuracy and Efficiency metrics for Color Correlogram Descriptor using sequential search and competing indexing methods applied on Cartoon-11K Dataset

Color ColGrm	Indexed color=8		Indexed color=5		Indexed color=3		Indexed color=1	
	P(10)/ ARR/ ANMRR	SSR+ OHR	ARR/ ANMRR/ P(10)	SSR+ OHR	ARR/ ANMRR/ P(10)	SSR+ OHR	ARR/ ANMRR/ P(10)	SSR+ OHR
Sequential Search	0.35/ 0.118/ 0.852				100%			
K-Means Clustering	0.31/ 0.100/ 0.874	45.8% + 0.7%	0.32/ 0.102/ 0.872	40.8%+ 0.4%	0.27/ 0.089/ 0.889	24.1%+0 .1%	0.22/ 0.076/ 0.905	14.5%+0 .06%
K-Means with B+Tree	**0.35**/ 0.115/ 0.856	76.6% + 1.3%	**0.35**/ 0.116/ 0.856	71%+1. 1%	0.31/ 0.104/ 0.870	39.8%+0 .6%	0.23/ 0.080/ 0.899	26.7%+0 .4%

Table 1. (*continued*)

Proposed Octree CTV =24	**0.36/ 0.122/ 0.848**	**57.3 %+ 0.9%**	**0.36/ 0.122/ 0.848**	**49.8%+ 0.08%**	**0.36/ 0.120/ 0.850**	**40.5%+. 007**	0.32/ 0.097/ 0.879	24.5%+ .0008%
Proposed Octree CTV=8	**0.35/** 0.115/ 0.857	**27%+ .075 %**	**0.35/** 0.113/ 0.859	**25%+ 0.02%**	**0.35/** 0.108/ 0.866	**22.1%+ .004%**	0.31/ 0.087/ 0.890	16.3%+ .0005%
Proposed Octree CTV=0	0.34/ 0.100/ 0.875	14%+ 0.003 5%	0.34/ 0.099/ 0.877	13.6%+. 0005	0.32/ 0.093/ 0.884	12.8%+ .0002%	0.27/ 0.081/ 0.898	10.8%+ .00001%
Proposed Octree+CPF CTV =24	**0.36/ 0.122/ 0.848**	**47.2 %+ 0.8%**	**0.35/ 0.122/ 0.848**	**39.2%+ .07%**	**0.35/ 0.121/ 0.850**	**29.7%+ 0.005**	0.31/ 0.093/ 0.883	15.1%+ .0006%
Proposed Octree+CPF CTV=8	**0.36/** 0.115/ 0.857	**18.9 %+ .065 %**	**0.35/** 0.113/ 0.859	**17.2%+ .015%**	**0.35/** 0.108/ 0.865	**14.8%+ .003%**	0.31/ 0.084/ 0.892	10.5%+ .0004%
Proposed Octree+CPF CTV=0	0.34/ 0.100/ 0.874	9.3%+ .0003 %	0.34/ 0.099/ 0.876	9%+ 0.004	0.32/ 0.092/ 0.884	8.5%+ .00015%	0.27/ 0.079/ 0.900	7.4%+ 0.0003%

Analysis the results of all competing indexing methods that shown in Table 1 can be summarized in the following points:

- KM reduces search space (that mean, the time) to the half (and more) according to number of indexed colors; but the accuracy is degraded. This is because; the comparison of query DCs is performed with the cluster centroids instead of actual colors inside the clusters. Cluster centroid represents an approximation to all cluster members, thus comparison with it will produce some errors.
- KMB outperforms the KM in enhancing the retrieval accuracy but with increasing the space that have been searched. This is because; some missing nearest images to the query that located in the other clusters are reached in this method. Additionally, the colors in the suitable range inside one cluster are selected only instead of all cluster members; this helps to avoid searching in the whole space. Therefore, this method succeeds in obtaining good accuracy (compared with K-means) with reasonable search space.
- The Proposed Octree indexing and Octree with Color Percentage Filter (CPF) methods have different settings involve four different number of indexed colors (8, 5, 3, and 1) and three color tolerance values (24, 8 and 0). The accuracy value of the proposed indexing scheme is increased in some settings (that presented in the bold font) than sequential search method. This is due to the following reasons; first one, the query's DC is reached to the exact corresponding color value in the index structure and some colors around it (according to tolerance value). This will lead to the query will be compared with images that have similar colors only. In this case, no need to approximate color value (as in KM that have some errors and leads to compare with some images of not similar colors). Second reason for increasing the accuracy is narrowing the search space to include images of similar colors only and in turn the rank of some relevant images will be enhanced. As depicted in Table 1, search space is significantly reduced into 22% without degradation to the accuracy

that measured by P(10), which represent the accuracy of the first page of retrieval results of CBIRs. P(10) is very important in web-based application [1]. On the other hand, KM could not maintain the retrieval accuracy whereas KMB could, in the 8 and 5 indexed colors but with significantly increasing to the search space. Moreover, the accuracy of the proposed Octree method (in most settings) is better than the accuracy of KM and KMB with outperforming in reducing search space, the Octree has SSR lower than KM and KMB.

– Color Percentage-based Filtering (CPF) method using B+-tree is proposed to speed up the retrieval process by considering only the images that have similar colors as well as similar color percentage. The result showed that this filtering process succeed in reducing SSR of the proposed indexing method by 10% (in average) without degradation to the accuracy.

Experiments on MPEG-7 DCD also show similar accuracy to the ColGrm descriptor but limitation of space in this paper led to exclude it here. Additionally, the results of Corel-10K dataset is also excluded due to the same reason.

5 Conclusions

In this paper, indexing methods of CBIR are presented. Specifically, the problems of color-based indexing methods such as high-dimensional problem for histogram-based methods and color approximation problem of DC-based methods are addressed. Both problems are tackled in this paper by proposing DC-based indexing method, where uniform RGB color space is used. The superiority features of the proposed RGB indexing method over the existing methods lies in utilization of dominant color for indexing images using Octree quantization method where this indexing method can be used for different color descriptors. Moreover, the proposed method characterized by using static representation for index structure that allow speed retrieval as well as it has color percentage filtering scheme (using B+-tree) to filter irrelevant images out in early stage.

References

1. Penatti, O.A.B., Valle, E., Torres, R.D.S.: Comparative Study of Global Color and Texture Descriptors for Web Image Retrieval. Journal of Visual Communication and Image Representation (2012)
2. Talib, A., Mahmuddin, M., Husni, H., George, L.E.: A weighted dominant color descriptor for content-based image retrieval. Journal of Visual Communication and Image Representation 24, 345–360 (2013)
3. Yamada, A., Pickering, M., Jeannin, S., Jens, L.C.: MPEG-7 Visual Part of Experimentation Model Version 9.0-Part 3 Dominant Color. ISO/IEC JTC1/SC29/WG11/N3914, Pisa (2001)
4. Yang, N.-C., Chang, W.-H., Kuo, C.-M., Li, T.-H.: A fast MPEG-7 dominant color extraction with new similarity measure for image retrieval. Journal of Visual Communication and Image Representation 19(2008), 92–105 (2008)

5. Kiranyaz, S., Birinci, M., Gabbouj, M.: Perceptual color descriptor based on spatial distribution: A top-down approach. Journal of Image and Vision Computing 28(2010), 1309–1326 (2010)
6. Kiranyaz, S., Birinci, M., Gabbouj, M.: Perceptual Color Descriptors. Foveon, Inc., Sigma Corp., San Jose, California, USA (2012)
7. Talib, A., Mahmuddin, M., Husni, H., George, L.E.: Efficient, Compact, and Dominant Color Correlogram Descriptors for Content-based Image Retrieval. Presented at the MMEDIA 2013: Fifth International Conference on Advances in Multimedia, Venice, Italy, April 22-26 (2013)
8. Wong, K.M., Po, L.M., Cheung, K.W.: A compact and efficient color descriptor for image retrieval. In: Proceedings of IEEE International Conference on Multimedia and Expo (ICME 2007), Beijing, China, pp. 611–614 (2007)
9. Jouili, S., Tabbone, S.: Hypergraph-based image retrieval for graph-based representation. Pattern Recognition 45, 4054–4068 (2012)
10. Kunttu, I., Lepistö, L., Rauhamaa, J., Visa, A.: Image correlogram in image database indexing and retrieval. In: Proceedings of 4th European Workshop on Image Analysis for Multimedia Interactive Services, London, UK, April 9-11, pp. 88–91 (2003)
11. Hou, A.L., Zhao, L.-Q., Shi, D.-C.: Garment image retrieval based on multi-features. In: International Conference on Computer, Mechatronics, Control and Electronic Engineering (CMCE), pp. 194–197 (2010)
12. Gervautz, M., Purgathofer, W.: A simple method for color quantization: Octree quantization. In: New Trends in Computer Graphics, pp. 219–231. Springer (1988)
13. Yildizer, E., Balci, A.M., Jarada, T.N., Alhajj, R.: Integrating wavelets with clustering and indexing for effective content-based image retrieval. Knowledge-Based Systems 31, 55–66 (2012)
14. Deng, Y., Manjunath, B.S., Kenney, C., Moore, M.S., Shin, H.: An efficient color representation for image retrieval. IEEE Trans. Image Process 10(1), 140–147 (2001)
15. Babu, G.P., Mehtre, B.M., Kankanhalli, M.S.: Color indexing for efficient image retrieval. Multimedia Tools and Applications 1, 327–348 (1995)
16. Sudhamani, M., Venugopal, C.: Grouping and indexing color features for efficient image retrieval. International Journal of Applied Mathematics and Computer Sciences i3, 150–155 (2007)
17. Stehling, R.D.O., Nascimento, M.A., Falcão, A.X.: Cell histograms versus color histograms for image representation and retrieval. Knowledge and Information Systems 5(3), 315–336 (2003)
18. Khan, F.S., Rao, M.A., van de Weijer, J., Bagdanov, A.D., Vanrell, M., Lopez, A.: Color Attributes for Object Detection. In: Twenty-Fifth IEEE Conference on Computer Vision and Pattern Recognition, CVPR 2012 (2012)

A Pilot Study on the Effects of Personality Traits on the Usage of Mobile Applications: A Case Study on Office Workers and Tertiary Students in the Bangkok Area

Charnsak Srisawatsakul[1,*], Gerald Quirchmayr[2], and Borworn Papasratorn[1]

[1] Requirement Engineering Laboratory, School of Information Technology
King Mongkut's University of Technology Thonburi, Bangkok, Thailand
charnsak.sri@st.sit.kmutt.ac.th,
borworn@sit.kmutt.ac.th

[2] Faculty of Computer Science, University of Vienna, Vienna, Austria
gerald.quirchmayr@univie.ac.at

Abstract. Recent research suggests that the "big five personality traits" influence the purchasing and usage preferences of mobile application. However, the impact of monetizing of applications and personality traits has so far been largely unattended. We have therefore extended our research to cover monetizing models of mobile applications. In this paper, we aim to enhance the understanding of the relationship between the "big five personality traits" and the usages and purchase intention of mobile applications in difference categories. Our initial data for the pilot study consists of 173 individuals, collected from smart device consumers who live in Bangkok, Thailand. Pearson's correlation and multiple linear regressions were used to analyze the data. The initial results indicate that some personality traits are associated with the usages and intention to purchase mobile applications. It is highly possible to conclude from the data that conscientious persons placed more intention to use productive applications. Specifically, this personality trait has a positive relationship with utilities, education, business and maps and navigation. Neuroticism reported only significant relation with in-app purchase in utilities applications. Agreeableness showed no significance during our regressions analysis. The most widely used paid application among all traits is entertainment. The findings of this pilot study will serve as indicators for the direction of our planned future research in this field.

Keywords: Mobile Applications, Personality Traits, Monetizing, Purchase Intention, In-App Purchase, Application Usages.

1 Introduction

In recent years, mobile technologies and apps (short for applications) have become an important part in our personal and professional live. This is mainly due to the constant

* Corresponding author.

S. Boonkrong et al. (eds.), *Recent Advances in Information and Communication Technology*, Advances in Intelligent Systems and Computing 265,
DOI: 10.1007/978-3-319-06538-0_15,

improvements of the mobile devices technology over the past decade. For instance, devices are today equipped with better processing power, mobile broadband communication, Assisted Global Positioning System (AGPS), Accelerometer, and Gyroscope etc. Moreover, the prices of those high-performing devices have become relatively cheap. Consequently, this makes them affordable for the average consumer. According to the latest research report from ITU (International Telecommunication Union) [1] in 2013 the number of world wide Mobile-cellular subscribers has now reached 6.8 billion (2 billion are active). In Thailand, There are 84 million subscribers [1].

Predictably, the mobile apps market has rapidly grown. ABI Research suggest that the market value is expected to reach $27 billion by the end of 2013 [2]. As a consequence, this will create new opportunities for developers. So many researchers pay attention to mobile apps development. However, most of the existing academic research focuses only on technological aspects [3–6]. Few studies have been carried out in the past to investigate the impact of individual behavior on the usage of mobile apps [7, 8]. In this study, we now aim to analyze the relationship of individual differences with the usage of mobile apps on mobile devices such as smartphones and tablets. According to the literatures, individual behavior is typically determined by abstract personality traits [9, 10]. Hence, the objective of our study is to explore the relationship between personality traits and categories of mobile apps through the application of well-established methods. Therefore, we used Pearson's correlation and multiple linear regressions methods to test our model. One of the early papers in the field that we can build our study on, is by Lane [7]. His paper already focused on analyzing the influence of personality traits in the context of mobile app usages. We need to expand it with more specifics on mobile app monetizing. Specifically, free, paid, in-app purchase and we also focus on consumers' intentions to purchase mobile apps. In addition, we cover the usage of mobile apps on tablets as a complement to smart phones. The big five personality traits framework [10, 11] has been use to categorized the personality of the participants. We also use the Mini International Personality Item Pool (Mini-IPIP) [12] to measure the big five factors of personality. For reasons of practical relevance, we have grouped the categories of apps in a way very similar to Apple App Store and Android Play Store into 8 apps types. These are Entertainment, Utilities, Education, Business, Social Networking, Photography, Maps and Navigation, and Healthcare and Fitness.

The major results of our pilot study analysis will hopefully be a good first indication of interesting patterns and will therefore be used for our further empirical research. It should also help developers, marketers, and other related professionals to create more effective mobile apps and have various app preferences for each individual behavior.

2 Literature Review

2.1 Personality Traits and Measurement

Personality traits are enduring characteristics of the individual that summarize trans-situational consistencies in characteristic styles of responding to the environment

[9, 11]. There are many studies on personality emerging since 1990s. Allport [9] proposed framework that organized personality into groups. Prior research implies that the Big Five traits [11] (Neuroticism, Extraversion, Openness, Conscientiousness, and Agreeableness) are the basic dimensions, which explain much of the shared variance of human personality.

Goldberg et al [13] proposed the International Personality Item Pool (IPIP). It is a free inventory of questionnaire construct items for measuring personality traits. Therefore, There are numerous of free and commercial instruments for measuring the Big Five personality traits. For example, Costa et al [11] proposed the revised version of his NEO Five-Factor Inventory (60 items). 44-item Big Five Inventory by John et al [14], The 40 items Mini Markers of Big Five [15], 75 items Traits Personality Questionnaire 5 (TPQue5) [16] Donnellan [12] proposed 20 questions called Mini-IPIP and etc. Goldberg et al [13] lists three reasons why the commercial inventory publishers are lack of progress in part to the policies and practices. First, the publisher rarely permits researchers to make change of an inventory. For Instance, they cannot change the item order, and rewording. That is to say, a researcher has to purchase and use an inventory "as is". Second, most publishers disallow to use their inventory on the World Wide Web to prevent their copyright. Third, publishers mostly have not released their entire scoring key. Finally, publishers design to the market rather than to improve their tests. Donnellan et al [12] provides in-depth analysis of the IPIP. They suggest that one potential problem with the IPIP are the large number of questionnaires. As a consequence, this might end up producing transient measurement errors. Participants may respond carelessly due to annoyance with the length of assessment. For this reasons, they have developed and validated a shorter form of IPIP that is called Mini-IPIP. It consisted of 20 questionnaire items. The result suggested that the Mini-IPIP is psychometrically acceptable and practically useful short measure of the Big Five personality traits. Furthermore, Cooper [17] conducted a research for confirmatory factor analysis of the Mini-IPIP. The results highlight that the measure form Mini-IPIP has acceptable reliability and clearly interpretable factor structure which support the original result from Donnellan [12]. However, if we raise the importance on the size of questions, Mini-IPIP stills not the shortest Big Five measurement tools. There are some shorter Big Five instruments, the Ten-Item Personality Inventory (TIPI) [18]. Moreover, some instruments offer five-item inventories [19–21]. Thus, many studies have revealed that the shorter number of questions is not the better. To illustrate, Baldasaro et al [22] identify that the score of TIPI might not be adequately reliable. The scales are moderately correlated together. Credé et al [23] suggested that the use of very short measures of personality traits may significantly affecting the validity and reliability. The results have shown that even slightly longer measures can substantially increase the validity of research findings without significant inconvenience to the researchers or research participants. Moreover, scores on the six-item questions by Shafer [24] demonstrated average validities that were lower or equal to those observed for scores on the Mini-IPIP. In addition, empirical findings suggested that Mini-IPIP is nearly have the validity and reliability as good as the longer 50-item IPIP[12, 17, 22, 23]. Thus, in this study, we use 20 items Mini-IPIP as our Big Five instrument.

2.2 Theories Connecting Personality Traits and Mobile Applications

There are two major mobile apps stores in the market. First, Apple app store is the apps store for iOS users. It offers more than 850,000 apps to iPhone, iPad and iPod touch users in 155 countries around the world [25]. Also, they announced that customers have downloaded over 50 billion apps from Apple app store on May 2013. Second, Google play store for android users, which have more than 700,000 apps. Interactive Advertising Bureau reported that mobile advertising revenues reach 8.852 billion in 2012 [26]. One of the most important mobile advertising media is the mobile app. Srisawatsakul et al [27] conducted a research on the acceptance of mobile advertising among Thai people. The result shows that intention to accept mobile advertising of Thai people is mostly related to individual prospective. The growth of both mobile apps usage and mobile advertising show a great potential. Therefore, we need a better understanding of how consumers perceived the apps in a human context. Personality traits have been explained to be associated with technology in various ways. Svendsenet al [28] implies that researchers need to be concerned with the personality of the participants when doing research in technology adoption. Previous studies of mobile apps usages have not dealt with the monetizing of mobile app types, which have 4 main options for monetizing from a developer perspective [29]. First, the most straightforward model, developers sell the app as one-time payment. Consumers can easily re-install it anytime they want by the provider's ecosystem. Second, they adopt the freemium strategy by offering their app for free with some limitations. For example, allow users to use it fully functional for 15 days or allow reduced function versions of their apps for free. They can pay for removing the limitations. Third, developers can also make money from premium content through apps. For instance, in the racing games, players can purchase new cars, upgrade or unlock for higher levels. In the e-books store, readers can subscribe to magazines, newspaper or buy new books. All of this can be done via in-app purchase. Fourth, developers can make some money from the free space in their app for advertising purposes.

3 Approach

3.1 Model

The conceptual framework of this study is illustrated in Figure 1. For the apps type, there are 24, 26 and 18 genres of apps on the Apple App Store, Google Play Store and Windows Phone Store, respectively. Therefore, some apps share the same characteristics. Moreover, each user's usage converges to small group of apps. Thus, we carefully defined applications categories for analysis by grouping applications in the App store genres to 8 categories, as shown in Table 1. We survey the frequency of the mobile apps usages used based on monetizing (free, paid, and in-app purchase). Moreover, we also focus on the intention to purchase mobile apps in the future. The current purchasing behavior of a user can for example serve as indicator for future purchases in the same category of apps.

3.2 Application of the Model (Data)

In this study, data were collected using a survey instrument. We got a total of 173 participants (N=173). Our participants are office workers and tertiary students who acquainted with mobile apps and live in Bangkok, Thailand. The design of the personality traits questionnaires was based on Mini-IPIP [12]. All of the participants were aged between 18 and 44. Just over half the sample 56.6% was female, Of these, the participants are using Android, iOS, Windows Phone, Black Berry and other operating system as 43.7%, 46.5%, 4.2%, 3.7% and 1.9%, respectively.

Table 1. Mobile Application Categories Based on Mobile Apps Store

Application Type	Application type from mobile apps store	Acronym
Entertainment	Games, Entertainment, Media Files Player, Music	ENT
Utilities	Utilities And Productivity	UTI
Education	News, Books, Reference, Sports, Travel, Weather, Magazine and Learning.	EDU
Business	Business, Finance, Lifestyle (shopping)	BUS
Social Networking	Social Networking, Social Media	SOC
Photography	Camera app, Effect, Photo Filter, Photo editor	PHO
Maps and Navigation	GPS based navigation, Maps	MAP
Healthcare and Fitness	Health, Medical, Fitness	HEA

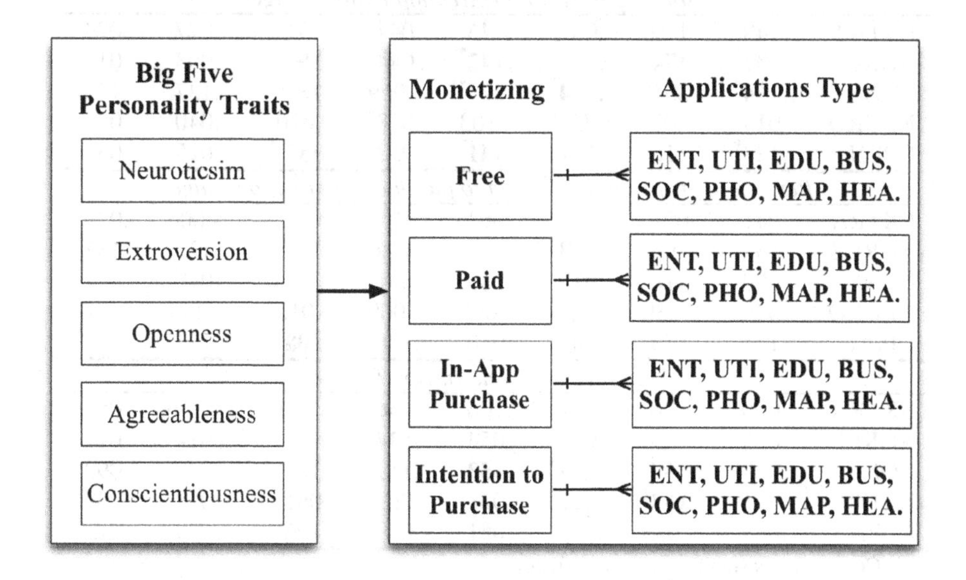

Fig. 1. Proposed model. Each monetizing option has 8 categories of mobile apps type.

4 Results

The descriptive statistics shows that skewness values were in the range of ≥ 1 or ≤ -1. Thus, it suggested that the items are univariate normal. Statistical significance was analyzed using Pearson's correlation and Multiple Linear Regressions as appropriate. The 8 categories of app type (ENT, UTI, EDU, BUS, SOC, PHO, MAP and HEA) are dependent variables and the personality trait variables (EXTRA, AGREE, CONS, NEURO, and OPEN). The personality traits have total of 20 questions Cronbach's alpha were well above .60 across all five traits. Cronbach's alphas for the free, paid, in-app and intention to purchase items, which have 8 questions each, were .80, .85, .94 and .90, respectively. The standardized regression coefficient (β) and adjusted R^2 were computed. These results are tabulated in Tables 2 and 3 respectively (EXTRA=Extraversion, AGREE=Agreeableness, CONS=Consciousness, NEURO= Neuroticism, OPEN=Openness to Experience).

Table 2. Pearson Correlations between dependent and independent variables

Variables	ENT	UTI	EDU	BUS	SOC	PHO	MAP	HEA
Personality traits and FREE application usage								
EXTRA	.026	.061	.065	.078	.007	.034	**.181****	**.137***
AGREE	.125	**.146***	**.158***	.077	.090	**.163***	.103	.112
CONS	.068	**.195****	**.229****	**.194****	.092	.099	**.154***	.097
NEURO	.024	.028	.003	-.032	-.074	.022	-.050	-.041
OPEN	.123	.107	**.146***	.044	.164*	**.244****	**.212****	.116
Personality traits and PAID application usage								
EXTRA	**.149***	.119	.064	**.146***	.047	.091	.077	.073
AGREE	**.187****	**.174***	.111	**.142***	.044	**.198****	.003	.015
CONS	**.144***	**.231****	**.174***	**.224****	-.069	**.157***	**.137***	.121
NEURO	.014	.007	-.024	-.103	.068	-.020	.040	.004
OPEN	**.134***	.116	**.191****	**.141***	.052	**.185****	.025	.053
Personality traits and Intention to purchase application usage								
EXTRA	.041	.004	.012	.063	.115	.028	-.004	.034
AGREE	**.168***	**.190****	**.148***	.082	**.140***	**.172***	.122	**.132***
CONS	.090	**.216****	.075	**.146***	.031	.041	.099	.047
NEURO	.099	-.039	**-.136***	-.067	-.009	-.012	-.111	-.099
OPEN	**.146***	**.160***	**.166***	.040	.108	**.208****	.125	**.131***
Personality traits and In-App Purchase								
EXTRA	.000	.020	.002	.106	.056	-.053	-.066	.051
AGREE	.085	.029	.097	.054	**.134***	.104	.014	.012
CONS	.069	.087	**.138***	**.153***	.119	.125	.079	.092
NEURO	.101	**.153***	.022	.018	.069	.061	.048	.038
OPEN	.041	-.033	.114	.051	.019	.106	.032	-.028

**. Correlation is significant at the 0.01 level (1-tailed).

*. Correlation is significant at the 0.05 level (1-tailed).

4.1 Observation from Pearson's Correlation Analysis

Extraversion shows significant correlated to the use of free and paid apps. For the free apps, they have positive correlation with map and navigation ($r = 0.181$, $p < 0.01$) and healthcare and fitness ($r = 0.137$, $p < 0.05$). Also, the paid entertainment and business have positive correlated ($r = 0.149$, $p < 0.05$ and $r = 0.146$, $p < 0.05$). Agreeableness have the highest number of positive correlation with 6 categories of apps in intention to purchase applications which an R-value from 0.132 to 0.190. Also, it correlated with paid entertainment ($r = 0.187$, $p < 0.01$), utilities ($r = 0.174$, $p < 0.05$), business ($r = 0.142$, $p < 0.05$) and photography ($r = 0.198$, $p < 0.05$).

Table 3. Standardized Regressions Coefficients (β) of dependent and independent variables

	ENT	UTI	EDU	BUS	SOC	PHO	MAP	HEA
			Personality traits and FREE application usage					
EXTRA	**-.049****	-.036	-.043	.032	-.054	-.082	.132	.102
AGREE	.097	.069	.048	-.018	.002	.077	-.081	.026
CONS	.017	**.165***	**.198****	**.195****	.065	.031	.101	.037
NEURO	.062	.061	.043	-.030	-.035	.089	-.034	-.034
OPEN	.098	.052	.092	-.014	**.154***	**.240****	**.179****	.058
Adj-R^2	*-0.004*	*0.190*	*0.360*	*0.011*	*0.005*	*0.046*	*0.043*	*-0.001*
			Personality traits and PAID application usage					
EXTRA	.071	.024	-.022	.088	.035	-.011	.055	.051
AGREE	.115	.072	-.028	-.010	.069	.113	-.090	-.082
CONS	.061	**.185****	**.147***	**.183****	-.121	.080	**.158***	.131
NEURO	.036	.031	.016	-.093	.078	.025	.034	.003
OPEN	.050	.032	**.175****	.054	.055	.118	.017	.043
Adj-R^2	*0.019*	*0.034*	*0.026*	*0.042*	*-0.008*	*0.027*	*-0.002*	*-0.009*
			Personality traits and Intention to purchase application usage					
EXTRA	-.069	-.120	-.050	.031	.075	-.065	-.064	-.013
AGREE	.148	.110	.093	.014	.116	.134	.070	.096
CONS	.023	**.179****	.014	.134	-.056	-.045	.063	-.017
NEURO	.151	.017	-.096	-.066	.006	.046	-.077	-.070
OPEN	.118	.098	.113	-.024	.047	**.184****	.078	.080
Adj-R^2	*0.028*	*0.046*	*0.017*	*-0.003*	*-0.001*	*0.029*	*0.004*	*0.000*
			Personality traits and In-App Purchase					
EXTRA	-.070	-.027	-.084	.071	-.013	-.158	-.120	.039
AGREE	.094	.035	.032	-.042	.139	.074	-.006	-.020
CONS	.048	.098	.126	**.147***	.078	.117	.107	.104
NEURO	.129	**.159****	.062	.016	.084	.114	.075	.027
OPEN	.029	-.037	.102	.015	-.048	.107	.055	-.052
Adj-R^2	*-0.004*	*0.006*	*0.004*	*0.000*	*0.003*	*0.020*	*-0.008*	*-0.016*

**. *P* is significant at the 0.05 level.

*. *P* is significant at the 0.10 level.

For free apps the significant relationship show with utilities ($r = 0.146$, $p < 0.05$), education ($r = 0.158$, $p < 0.05$), and photography ($r = 0.163$, $p < 0.05$). Conscientiousness found to have the most significant correlation with all generous of apps. The most positive correlations are in the form of paid apps with R-value from 0.137 to 0.231. For the free app type they also show correlation with utility ($r = 0.195$, $p < 0.01$), education ($r = 0.229$, $p < 0.01$), business ($r = 0.194$, $p < 0.01$) and map and navigations ($r = 0.154$, $p < 0.05$). Neuroticism shows significant correlation on intention to purchase education apps and in-app purchase of utilities. Openness to experience draw some positive correlation with free, paid and intention to purchase.

4.2 Observations from Multiple Linear Regressions Analysis

Participants with low extraversion score showed minor regression coefficient with the use of free entertainment categories only ($\beta = -0.049$, $p < 0.05$). Agreeableness did not draw any significant regression coefficient with any app categories. Inspection of the indices suggested significant positive regressions between Conscientiousness and every monetizing type with β-value range from 0.147 to 0.198. Neuroticism have positive regression coefficient with in app purchase of utilities apps ($\beta = 0.153$, $p < 0.05$). Openness to experience showed positive regressions coefficient with free social network ($\beta = 0.154$, $p < 0.10$), photography ($\beta = 0.240$, $p < 0.05$), maps and navigation ($\beta = 0.179$, $p < 0.05$) along with intention to purchase photography apps ($\beta = 0.184$, $p < 0.05$).

5 Interpretation and Discussion of Results

This research shows that personality traits can influence the usages and purchase decision of mobile apps in different monetizing context. There are several possible explanations that could be draw from the result. Paid entertainment is the most acceptable category for every trait, which have showed high level of both Pearson's correlation coefficient and standardized regressions coefficients. Participants who have low scores in extraversion did not have any significant in any categories except entertainment. The reasons that should explain this is because extraversion described as active, outgoing, talkative and etc. [10]. That is to say they choose to spend time with people more than mobile devices. However, the findings do not support the previous research from Lane [7], which identifies that extraversion have much more interest on entertainment apps. Agreeableness does not report any significant in the multiple linear regressions analysis. However, It reported some positive related with entertainment (paid), utilities (free and intention), education (free), business (free and paid), social network (intention), photography (free), and healthcare (intention). This supported the description of agreeableness as generous, and considerate [10]. Conscientiousness represents the tendency to be efficient, organized, planning, reliable, and responsible [10]. This can assumed that conscientiousness is more likely to use mobile devices for their productivity. Surprisingly, our results support those assumptions. They have the positive relationship with utilities, education, business and maps and navigation in free, paid and intention to purchase, which all of them are productive apps. This finding confirms the results described in several previous studies. For example, Lane [7] implied that conscientious people draw significantly more interest on communications, productivity, and utilities apps. Chittaranjan [8] explained the strong relation of conscientiousness and the use of work related apps.

Neuroticism reported a few relationships with mobile apps categories. The finding suggested that they only interest on the in-app purchase of utilities apps. Openness to experience draws some positive relationship on the free social network, photography, and maps and navigation. Similarly, they also indicate the importance of paid app in categories of education and intention to purchase photography. Nonetheless, this results are opposite of the descriptive of openness to experience traits. They should be wide interests in different of mobile app categories.

6 Limitations of the Study and Future Research

In this section, we acknowledge the limitations of our study and highlight areas of future research. First, participant personality traits were self-reported during a single session. Accordingly, common method bias is a possible weakness. Second, since this is a pilot study. The potential limitations of this study involve the sample itself. The samples size was relatively small, with the small number of samples, the adjusted R2-values are relatively low that is to say the results of the 5 predictors could explained very low percentage of the variance. Therefore, the results of this study are good indications for future researches.

In the future work, we also aim to compare the results from our hypotheses with different control group. As the respective data has already been collected for office workers and students, which can be differentiated by age, income, and lifestyle the same data set can be used for a pilot. The in-app purchase has turned out to be a special case, which seems to have some different characteristics. A separate analysis will therefore be carried out in the future to analyze them in more detail.

In conclusion, we have presented good indications between big five personality traits and mobile app usage and purchase decision. However, additional work is needed to experiment by using a larger samples size. Moreover, from those indications it also suggested that we could test in more deeply detail in each apps genre.

Acknowledgements. The authors would at this point like to express their appreciation of the financial support provided by Austrian Federal Ministry for Science and Research (Ernst Mach-ASEA UNINET scholarship) and the infrastructure support of the Faculty of Computer Science at the University of Vienna. Also, the authors would like to thank the reviewers for their valuable suggestions, which have either been incorporated in this paper or will be the subject of current or future research.

References

1. ITU: ITU World Telecommunication/ICT Indicators database, http://www.itu.int/en/ITU-D/Statistics/Documents/facts/ICTFactsFigures2013.pdf

2. Research, A.B.I.: The Mobile App Market will be Worth $27 Billion in, as Tablet Revenue Grows (2013), https://www.abiresearch.com/press/the-mobile-app-market-will-be-worth-27-billion-in-
3. Gu, X., Xu, Z., Wang, T., Fang, Y.: Trusted Service Application Framework on Mobile Network. In: Gu, X., Xu, Z., Wang, T., Fang, Y. (eds.) 2012 9th Int. Conf. Ubiquitous Intell. Comput. 9th Int. Conf. Auton. Trust. Comput., pp. 979–984 (2012)
4. Charland, A., LeRoux, B.: Mobile Application Development: Web vs. Native. Queue 9, 20 (2011)
5. Holzer, A., Ondrus, J.: Mobile application market: A developer's perspective. Telemat. Informatics 28, 22–31 (2011)
6. Liu, C., Zhu, Q., Holroyd, K.A., Seng, E.K.: Status and trends of mobile-health applications for iOS devices: A developer's perspective. J. Syst. Softw. 84, 2022–2033 (2011)
7. Lane, W., Manner, C.: The Influence of Personality Traits on Mobile Phone Application Preferences. J. Econ. Behav. Stud. 4, 252–260 (2012)
8. Chittaranjan, G., Blom, J., Gatica-Perez, D.: Who's Who with Big-Five: Analyzing and Classifying Personality Traits with Smartphones. In: 2011 15th Annu. Int. Symp. Wearable Comput., pp. 29–36 (2011)
9. Valentine, C.W.: Personality—A Psychological Interpretation by Gordon W. Allport, pp. xiv + 588. Constable, London (1943); Br. J. Educ. Psychol. 13, 48–50 (1943)
10. McCrae, R.R., John, O.P.: An introduction to the five-factor model and its applications. J. Pers. 60, 175–215 (1992)
11. Costa, P.T., MacCrae, R.R.: Psychological Assessment Resources, I.: Revised NEO Personality Inventory (NEO PI-R) and NEO Five-Factor Inventory (NEO FFI): Professional Manual. Psychological Assessment Resources (1992)
12. Donnellan, M.B., Oswald, F.L., Baird, B.M., Lucas, R.E.: The mini-IPIP scales: tiny-yet-effective measures of the Big Five factors of personality. Psychol. Assess. 18, 192–203 (2006)
13. Goldberg, L.R., Johnson, J.A., Eber, H.W., Hogan, R., Ashton, M.C., Cloninger, C.R., Gough, H.G.: The international personality item pool and the future of public-domain personality measures. J. Res. Pers. 40, 84–96 (2006)
14. John, O.P., Srivastava, S.: The Big Five trait taxonomy: History, measurement, and theoretical perspectives. In: Pervin, L.A., John, O.P. (eds.) Handbook of Personality: Theory and Research, pp. 102–138. Guilford Press, New York (1999)
15. Saucier, G.: Mini-Markers: A Brief Version of Goldberg's Unipolar Big-Five Markers. J. Pers. Assess. 63, 506–516 (1994)
16. Tsaousis, I., Kerpelis, P.: The Traits Personality Questionnaire 5 (TPQue5). Eur. J. Psychol. Assess. 20, 180–191 (2004)
17. Cooper, A.J., Smillie, L.D., Corr, P.J.: A confirmatory factor analysis of the Mini-IPIP five-factor model personality scale. Pers. Individ. Dif. 48, 688–691 (2010)
18. Gosling, S.D., Rentfrow, P.J., Swann, W.B.: A very brief measure of the Big-Five personality domains. J. Res. Pers. 37, 504–528 (2003)
19. Bernard, L., Walsh, R., Mills, M.: Ask once, may tell: Comparative validity of single and multiple item measurement of the Big-Five personality factors. Couns. Clin. Psychol. J. 2, 40–57 (2005)

20. Aronson, Z.H., Reilly, R.R., Lynn, G.S.: The impact of leader personality on new product development teamwork and performance: The moderating role of uncertainty. J. Eng. Technol. Manag. 23, 221–247 (2006)
21. Woods, S., Hampson, S.: Measuring the Big Five with single items using a bipolar response scale. Eur. J. Pers. 390, 373–390 (2005)
22. Baldasaro, R.E., Shanahan, M.J., Bauer, D.J.: Psychometric properties of the mini-IPIP in a large, nationally representative sample of young adults. J. Pers. Assess. 95, 74–84 (2013)
23. Credé, M., Harms, P., Niehorster, S., Gaye-Valentine, A.: An evaluation of the consequences of using short measures of the Big Five personality traits. J. Pers. Soc. Psychol. 102, 874–888 (2012)
24. Shafer, A.B.: Brief Bipolar Markers for The Five Factor Model of Personality. Psychol. Rep. 84, 1173–1179 (1999)
25. Apple Press: Apple - Press Info - Apple's App Store Marks Historic 50 Billionth Download, http://www.apple.com/pr/library/2013/05/16Apples-App-Store-Marks-Historic-50-Billionth-Download.html
26. (IAB):IABinternetadvertisingrevenuereport, http://www.iab.net/media/file/IAB_Internet_Advertising_Revenue_Report_HY_2013.pdf
27. Srisawatsakul, C., Papasratorn, B.: Factors Affecting Consumer Acceptance Mobile Broadband Services with Add-on Advertising: Thailand Case Study. Wirel. Pers. Commun. 69, 1055–1065 (2013)
28. Svendsen, G.B., Johnsen, J.-A.K., Almås-Sørensen, L., Vittersø, J.: Personality and technology acceptance: the influence of personality factors on the core constructs of the Technology Acceptance Model. Behav. Inf. Technol. 32, 323–334 (2013)
29. Cortimiglia, M., Ghezzi, A., Renga, F.: Mobile Applications and Their Delivery Platforms. IT Prof. 13, 51–56 (2011)

Intelligent Echocardiographic Video Analyzer Using Parallel Algorithms

S. Nandagopalan[1], T.S.B. Sudarshan[2], N. Deepak[3], and N. Pradeep[3]

[1] Department of Computer Science and Engineering,
Bangalore Institute of Technology, Bangalore, India
snandagopalan@gmail.com
[2] Department of Computer Science & Engineering,
Amrita Vishwa Vidyapeetham, Amrita School of Engineering,
Bangalore, India
tsb_sudarshan@blr.amrita.edu
[3] Network Division Eshamount Technologies, Bangalore, India
pradeepp12@gmail.com

Abstract. This paper proposes an intelligent framework to accurately analyze these echo images in order to discover disease category and assess the severity automatically. Typically, each video consists of 90-100 frames of 2D echo, color Doppler image video, and several .jpg images. Our framework consists of parallel algorithms developed under OpenMP environment. The major tasks are cardiac boundary tracing, quantifying the heart chambers and extracting 2D features, and other features required for computing statistical features and build a classifier model for categorization. Segmentation of an echo image is done using parallel implementation of K-Means algorithm and they are boundary extracted using active contour method. Bayesian model is used to classify a given patient into normal or abnormal. The experiment involves videos taken from 60 normal and abnormal patients from a local cardiology Hospital. The results obtained with our algorithms outperformed with respect to the results that have already been reported.

Keywords: Echocardiography, Parallel K-Means, OpenMP, Bayesian.

1 Introduction

It is estimated that every year, 0.5 million people in the Unites States die because of coronary heart disease. India will have 62 million patients with heart disease by 2015, compared to 16 million in the US. Speed of diagnosis is therefore vital. One of the popular techniques is echocardiographic image analysis for early detection of heart diseases. For the study of echocardiographic images, segmenting and identifying the objects such as LV, RV, LA, and RA from successive frames are important. Segmentation subdivides an image into its constituent regions or objects [14], [16], [17]. Automatic LV segmentation is a difficult task due to the relatively poor quality (speckle noise) of echocardiographic images.

S. Boonkrong et al. (eds.), *Recent Advances in Information and Communication Technology*, Advances in Intelligent Systems and Computing 265,
DOI: 10.1007/978-3-319-06538-0_16, © Springer International Publishing Switzerland 2014

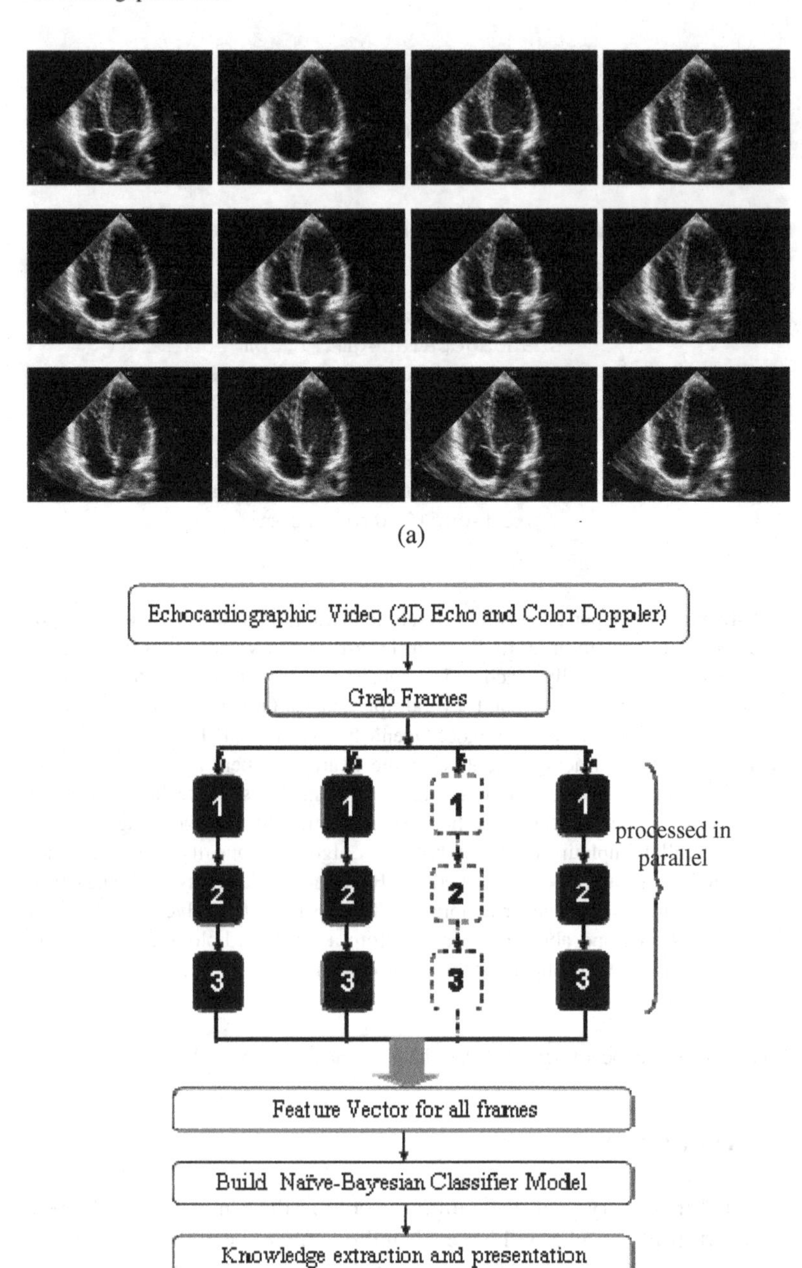

1 - Preprocessing and Parallel K-Means.

2 - Automatic Cavity Boundary Extraction/Color Doppler segmentation.

3 - Quantification/Statistical Feature Extraction

(b)

Fig. 1. (a) First 12 frames of a typical Echo video (b) Block Diagram of the Intelligent Echocardiographic Analyzer

Our main contributions in this paper are as follows: efficient parallel K-Means algorithm for segmentation, cardiac cavity tracking using active contour, naive-Bayesian model for classifying the patient into normal or abnormal based on a strong feature vector built based on actual patients. In addition to these, the color Doppler image segmentation and feature extraction are also added for a better feature vector. All these tasks are designed using parallel algorithms in the OpenMP environment which is the de facto standard for parallel programming [5].

Unfortunately, the sequential K-Means algorithm is quite time consuming especially when the number of data points is large. It is estimated that the theoretical time complexity of K-Means is $O(nkdt)$, where n is the number of data points, k is the number of clusters, d is the number of dimensions, and t is the number of iterations [11], [12]. A parallel version of K-Means is proposed to speedup the execution. The block diagram shown in Figure 1 depicts the various modules of the proposed Echocardiographic analyzer.

The patient echo video and images (2D and color Doppler) are acquired by experts using an ultrasound machine and stored in the image database. Each video consists of approximately 30-90 frames and are grabbed using an algorithm and stored as still jpg images. These images are preprocessed before applying the segmentation process.

Parallel K-Means algorithm is run on each frame concurrently so that we achieve a higher speedup. The clustered image frames are then boundary extracted using the active contour method. In the case of color Doppler images the segmentation process is slightly different. That is, we do not apply parallel K-Means algorithm. From the segmented objects, it is straight forward to compute the quantification parameters for each frame such as chamber area, volume, height, width, ejection fraction, etc. For the color Doppler images we compute statistical data such as histogram, kurtosis, skewness, texture properties, etc. These features are integrated to formulate as a feature vector which will be used by Bayesian model for training the analyzer. This model gives us accurate details of the patient disease level, if abnormal. The feature vector can also be used for various other tasks in the diagnosis process.

2 Previous Work

Sequential K-Means algorithm was used for biomedical image segmentation using adaptive techniques and morphological operations by Chang Wen Chen, *et al* [11]. Three algorithms were proposed by Carlos Ordonez using DBMS SQL and C++ and demonstrated how K-Means can be of practical importance for clustering large data sets [10]. Another approach to speed up the K-Means was suggested by Khaled Alsbti, *et al*, based on *k-d* tree structure [12]. The bottleneck of this algorithm is that of distance computation and more number of iterations.

To overcome this parallel algorithms are suggested by Piotr Kraj, et. al and many other authors [1], [3]. The software described in this paper is a high performance multithreaded application that implements a parallelized version of the K-Means Clustering algorithm. In another paper written by Snapawat, et. al, used a network of homogeneous workstations with Ethernet network and use message-passing for communication between processors for parallelizing the K-Means algorithm [2]. However, the results obtained by these authors are poor. Manasi N. Joshi proposed an algorithm for Parallel K-Means Algorithm on Distributed Memory Multiprocessors [4].

For LV contour tracing, Santos [13] has adopted windows adaptive technique, Chunming Li used active contour method [14], and others have shown results based on template matching. A semi-automatic procedure for segmentation of echocardiogram images based on histogram and edge detection techniques was proposed by Boonchieng, *et al* [15]. Due to image noise and poor image contrast these methods give either inaccurate boundary detection or take more running time. Most of the past works in the area of image segmentation use some preprocessing steps for noise elimination [11], [13].

3 Proposed Parallel K-Means Algorithm

Large datasets, such as echo image data, pose new challenges for clustering algorithms. Algorithms with linear complexity, like K-Means clustering, need to be scaled-up and implemented in a more efficient way. Making the algorithm parallel instead of serial is one potential solution when a sequential clustering algorithm cannot be further optimized. With a parallel algorithm, the computational workload is divided among multiple CPUs and the main memory of all participating computers. For this we use a quad-core machine under Windows operating system. C#.NET language is used for this purpose with the help of OpenMP environment [5].

The input for K-Means is a data set D containing n points with d dimensions, $D = \{i_1, i_2, i_3, .., i_n\}$. For our case, we assume $k = 3$, because any echocardiographic image is segmented into three regions, i.e. cardiac cavity (black region), near endocardium (white region), and the rest (gray region) [10]. The dataset is the pixel values of the given image $f(x, y)$ of size $M \times N$, where $f(x, y)$ is the gray scale value of a pixel at location (x, y). The initial seed values for the K-Means algorithm are selected at random. The proposed parallel K-Means algorithm distributes the dataset to all the available processors equally (in our case it is 4.) However, the centroid vector is kept globally which will be updated and shared in every iteration [1]. The parallel K-Means algorithm is shown in Figure 2.

Algorithm Parallel_KMeans(D, k, m)
// Input: D: The pixel dataset of frame-i, $k = 3$, m - # of iterations
// Output: Three clusters with respective pixel data
1. Select k centroids from D using random number generator
2. p = omp_get_num_proc(); // p - number of processors
3. Divide the dataset D into p subsets, $D_1, D_2, \ldots D_p$
4. **for** $u = 1$ to m do
    5. Execute the **for** loop in **parallel** for each subset on each p
        #pragma omp parallel shared <*variables*>
        #pragma omp for private <*variables*>
            foreach pixel j **do**
                Find the nearest cluster center using Euclidean distance metrics
                Assign/Reassign the cluster id for pixel j
        Update the cluster centroids

Fig. 2. Parallel K-Means algorithm

The algorithm executes m iterations, where m is given as an input decided by the domain experts. In the case of echo image clustering it is experimentally found that m is set to 4 to 8 which gives satisfactory results. This algorithm is suitable for multidimensional dataset, including synthetic data.

4 Echo Image Frame Segmentation and Quantification

Active contours (snakes) are often used in computer vision and image analysis to detect and locate objects, and to describe their shape. Conventionally these snakes are allowed to trace the object on a raw image. But in this paper we follow a different technique. To segment the echo images we first apply the parallel K-Means algorithm as explained in Section 3 and then select the initial contour for boundary tracing of any heart chamber. This approach would speedup the tracing as the clustered image has uniform intensity on which the active contour moves quickly.

Here, the geometric active contour with level set approach is used due its advantages over parametric approach. We use a modified geometric active contour algorithm as presented in [6]. The next step in the image analysis is to extract clinically relevant and useful features like area, volume, ejection fraction, etc. A series of image processing and geometric methods are applied to compute these parameters automatically and is shown in Figure 3(a) to (e).

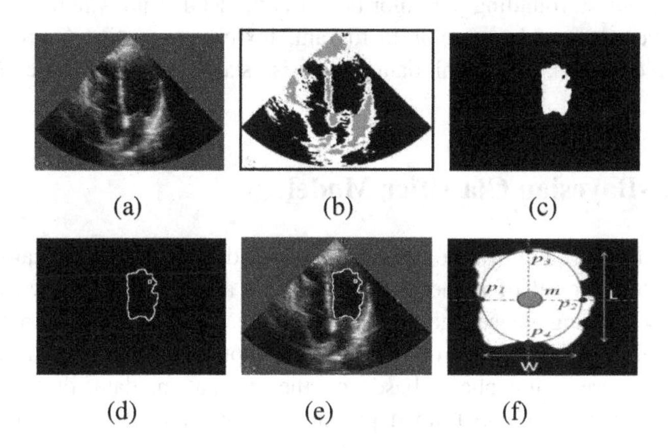

Fig. 3. Image Quantification (a) Original image (b) K-Means Clustered (c) LV boundary marked (d) Sobel edge detected (e) Final image (f) Quantification

Figure 3(e) is to display the automated traced LV object and it depicts the precision in locating the boundary which is difficult work for a human operator. The LV is approximated as an ellipse and hence its major-axis, minor-axis, area, volume, etc. are computed using the mid-point method [6], [7]. This technique can be extended to other image views as well.

5 Active Contour Based Color Segment Boundary Tracing

For color images, the contour must trace the outer boundary of the ROI. This is achieved by setting the constant value, c_0 to positive which is used to define binary level set function as initial contour. On the other hand, if c_0 is negative, the contour will expand [7].

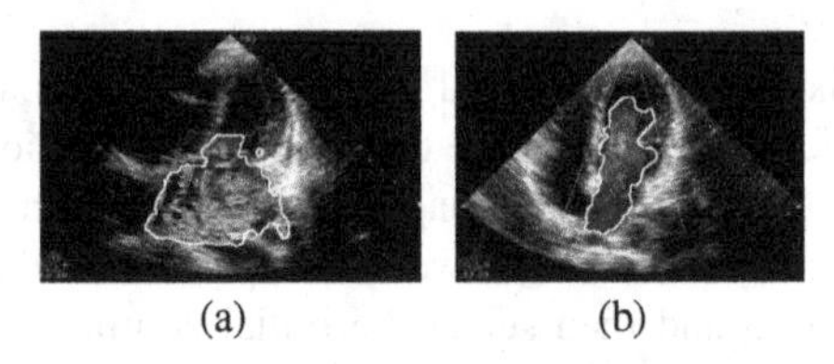

(a) (b)

Fig. 4. Color Doppler Image Boundary tracing

In the color Doppler images, the color object in the entire echo image is important as it provides diagnostic information. For instance, if the object contains uniform blue or red pixels it signifies that the patient is normal. However, if the object contains mosaic color pixels it is due to the turbulent flow of the blood. This type of flow indicates that the patient has stenosis or regurgitation abnormality. The working of this algorithm is shown in Figure 4(a), an abnormal case, and in Figure 4(b), a normal case. The white contour surrounding the color object is the ROI from which the knowledge is obtained. Features such as color histogram, texture properties (entropy, energy, homogeneity, contrast), statistical data (kurtosis, skewness), and edge density are extracted.

6 Naïve-Bayesian Classifier Model

In the proposed model all attribute values are assumed as continuous because the echo image parameters are all of numeric values. There are two phases involved in this design namely training phase and prediction phase. During the training phase the posterior probabilities and class probabilities are computed and stored in a suitable data structure. In the prediction phase, based on the test patient data, the probability is calculated by referring to the trained probability table and the final class label is obtained as described in [8].

We have implemented the classifier in three ways: (i) SQL based approach with a tightly coupled database model that uses categorical values. (ii) OLAP-SQL based method to simplify the queries that uses categorical values (iii) direct approach in C# with continuous values. These methods are elaborately explained in [8-9].

7 Results and Discussions

The echocardiographic videos and still images of normal and abnormal patients were obtained from Sri Jayadeva Institute of Cardiovascular Sciences and Research

Hospital, Bangalore, India. From these videos the frames are grabbed using our software and resized to 680×512. For the purpose of various experiments, a total of 69 patient echo videos and images have been used. The computing environment includes a quad-core DELL Optiplex 780 desktop computer running Windows7 operating system. We have developed the entire framework using C#.NET 2010 and MATLAB. The results obtained through our algorithms (automatic) and manual are closely matching. This section presents the results of the proposed algorithms considering appropriate datasets.

7.1 Performance of Parallel K-Means

The computational performance of parallel K-Means is tested by considering varying data sizes keeping the dimensions to 3 for synthetic data. It is then compared with its sequential performance and is shown in Table 1.

Table 1. Computational Performance of Parallel K-Means versus Sequential K-Means. p - number of processors, $d = 3$, number of dimensions

Data Size (n)	Sequential K-Means (Secs)	Parallel K-Means p=2 (Secs)	Parallel K-Means p=4 (Secs)	Parallel ([2]) (Secs)	Speedup $S_p = \frac{T_s}{T_p, 4}$
1K	0.031	0.075	0.058	-	-
10K	0.393	0.732	0.521	-	-
100K	4.406	3.53	2.19	268	2.0
500K	21.107	16.67	8.76	1044	2.4
1000K	48.062	28.04	16.65	8216	2.8
10000K	484.187	305.4	160.94	-	3.0

It can be observed that the computation time of parallel K-Means is much less compared to the sequential K-Means algorithm (speedup with $p = 4$ is shown in the last column of Table 1.) Also the proposed parallel algorithm is computationally efficient compared to the algorithm of [2].

7.2 Performance of Heart Chamber Boundary Tracing

Figure 5(a) to (d) shows the images after applying parallel K-Means algorithm followed by image filtering and finally the ventricles and atriums traced automatically.

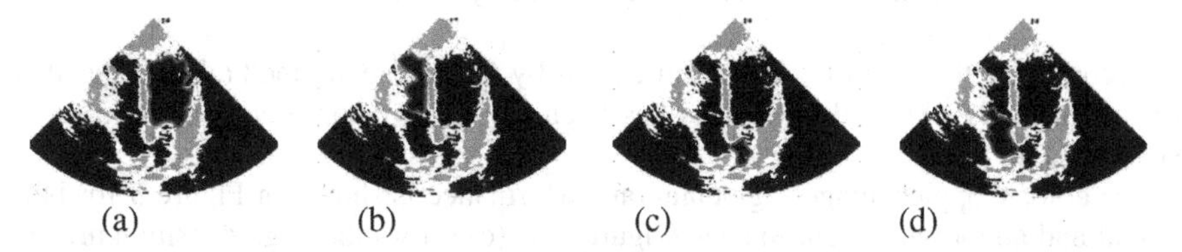

(a) (b) (c) (d)

Fig. 5. Contour of ventricle and atrium after 400 iterations (a) LV (b) RV (c) LA (d) RA

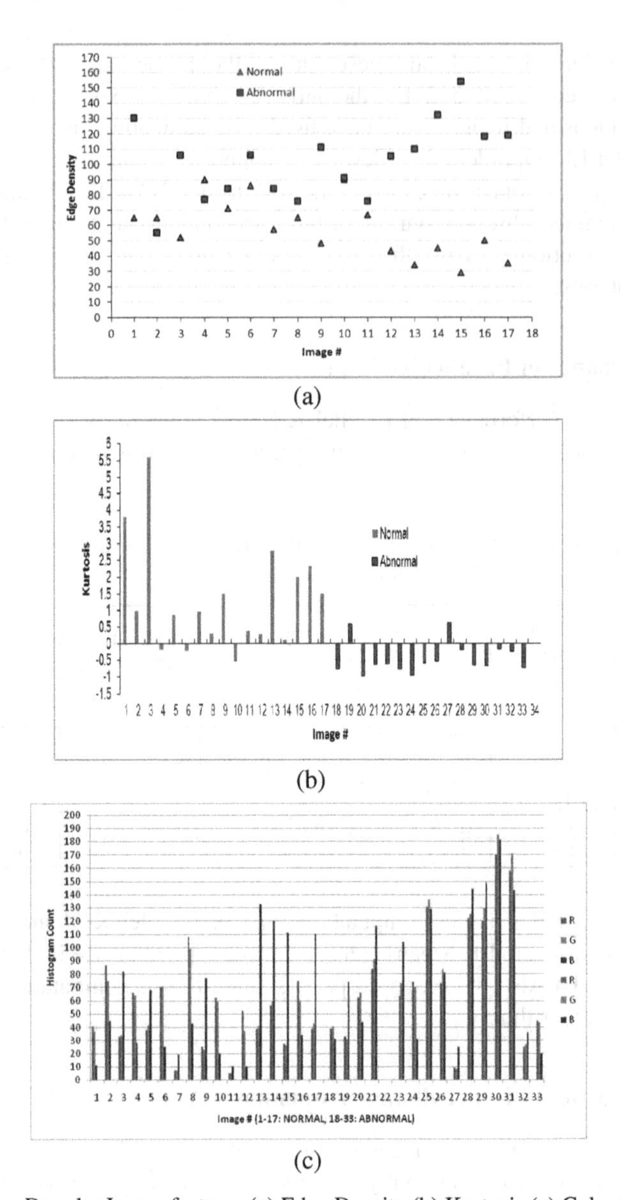

Fig. 6. Color Doppler Image features (a) Edge Density (b) Kurtosis (c) Color Histogram

Accuracy of the quantifications carried out by the automatic method is better than that of manually obtained data. This has been verified by considering a large set of patient images.

The color Doppler image segmentation performance is shown in Figure 5 for both normal and abnormal patient images. Figure 6(a)-(c) shows the edge density, kurtosis, and color histogram features for the normal and abnormal images respectively. With these features it is obvious that we can easily distinguish between normal and abnormal patients automatically.

7.3 Performance of Naïve-Bayesian Classifier Model

The proposed classifier has been trained with 60 patient data. Each patient data consists of 5 videos (each video has 40-90 frames) and around 20 -30 still images. High classification accuracy was achieved, both for training and test sets. Initial classification accuracy was 83.78%, error rate was 16.21%, sensitivity was 100.00%, and specificity was 66.66%.

8 Conclusion

Our aim was to construct a decision support system for echocardiographic video and image analysis combining high performance accuracy with simplicity of use which will allow doctors and technicians for better patient care. It has been noticed that most of the ultrasound machines provide echo images but not intelligence and therefore the decision making process is more subjective. With the proposed system extensive analysis and automatic interpretation of echo images can be obtained. Results achieved are very encouraging and justify further development of this system as valuable tool for supporting everyday clinical decisions.

References

1. Kraj, P., Sharma, A., Garge, N., Podolsky, R.: ParaKMeans: Implementation of a parallelized K-means algorithm suitable for general laboratory. BMC Bioinformatics 2008, 1–13 (2008), doi:10.1186/1471-2105-9-200
2. Kantabutra, S., Couch, A.L.: Parallel K-means Clustering Algorithm on NOWs. Technical Journal 1(6), 243–248 (2000)
3. Kim, W.: Parallel Clustering Algorithms: Survey (2009), `http://www.cs.gsu.edu/~wkim/index_files/SurveyParallelClustering.pdf`
4. Joshi, M.N.: Parallel K - Means Algorithm on Distributed Memory Multiprocessors (2003)
5. `http://openmp.org/wp/`
6. Nandagopalan, S., et al.: Automatic Segmentation and Ventricular Border Detection of 2D Echocardiographic Images Combining K-Means Clustering and Active Contour Model. In: Proc. International Conf. IEEE Computer Society and ACM Digital Library, ICSEM 2010, Bangkok, Thailand, April 23-25, pp. 447–451 (2010)
7. Nandagopalan, S., et al.: A Novel Approach to Medical Image Segmentation. J. Computer Science 7(5), 657–663 (2011)
8. Nandagopalan, S., Adiga, B.S., Sudarshan, T.S.B., Dhanalakshmi, C., Manjunath, C.N.: A Naïve-Bayesian Methodology to Classify Echocardiographic Images through SQL. In: Theeramunkong, T., Kunifuji, S., Sornlertlamvanich, V., Nattee, C. (eds.) KICSS 2010. LNCS(LNAI), vol. 6746, pp. 155–165. Springer, Heidelberg (2011)
9. Nandagopalan, S., et al.: SQL Based Cardiovascular Ultrasound Image Classification. International J. Data Mining and Bioinformatics (2011)
10. Ordonez, C.: Integrating K-Means clustering with a Relational DBMS using SQL. IEEE Transactions on Knowledge and Data Engineering 18(2), 188–201 (2006)

11. Chang, W.C., Jiebo, L., Kevin, J.P.: Image segmentation via adaptive K-Mean clustering and knowledge-based morphological operations with bio-medical applications. IEEE Transactions on Image Processing 7(12), 1673–1683 (1998)
12. Alsabti, K., Ranka, S., Singh, V.: An efficient K-Means clustering algorithm. In: Proceedings of IPPS/SPDP Workshop on High Performance Data Mining (1998)
13. Santos, J.B., Celorico, D., Varandas, J., Dias, J.: Automatic segmentation of echocardiographic Left Ventricular images by windows adaptive Thresholds. In: International Congress on Ultrasonics, Vienna (April 2007)
14. Chunming, L., Chenyang, X., Changfeng, G., Martin, D.F.: Level set evolution without re-initialization: A new variational formulation. In: Proc. of International Conf. on Computer Vision, pp. 259–268 (1987)
15. Donna, J.: Williams, Mubarak Shah: A fast algorithm for active contour and curvature estimation. IEEE Transactions on Image Understanding CVGIP 55(1), 14–26 (1992)
16. Kass, M., Witkin, A.: Snakes: Active Contour Models. In: Proc. of IEEE Computer Society Conference on Computer Vision and Pattern Recognition, CVPR 2005 (2005)
17. Lacerda, S.G., et al.: Left Ventricle segmentation in echocardiography using a radial-search-based image processing algorithm. In: IEEE EMBS Conference, Vancouver, Canada, pp. 222–225 (2008)

Durian Ripeness Striking Sound Recognition Using N-gram Models with N-best Lists and Majority Voting

Rong Phoophuangpairoj

Department of Computer Engineering, College of Engineering, Rangsit University, Thailand
rong.p@rsu.ac.th

Abstract. Durians are green spiky fruits, which are considered as a delicacy throughout Southeast Asia. They are valued for their unique flavor and powerful taste. It is desirable to be able to determine the quality of durians without cutting them because it is difficult to quantify the ripeness from the external appearance and they are expensive to purchase. In Southeast Asia and China, consumers have found that after buying and cutting durians, they are not ripe or ready to eat. Therefore, studying striking signal characteristics and developing an automated method of recognizing durian ripeness levels without cutting or destroying them could benefit consumers of the fruit. The following method of recognizing durian ripeness by studying striking signals using N-gram models with N-best lists and majority voting is proposed. The recognition process is composed of three stages: 1) extract the acoustic features from the striking signals, 2) recognize unripe and ripe durian striking signals using the N-gram models and 3) find the ripeness from the N-best lists using majority voting. The results indicate that using the 3-best lists and majority voting method it was possible to recognize durian ripeness efficiently. Average ripeness recognition rates of 95.8%, 90.4% and 93.1% were obtained from the untrained, unknown and both test sets, respectively. The results demonstrate that the method is accurate enough to be used by consumers to help them select a ripe durian.

Keywords: N-gram, HMM, MFCC, durian ripeness, durian, striking signals, striking sound recognition, N-best lists, majority voting.

1 Introduction

In Asia, Thailand, Malaysia, Indonesia, India, Vietnam and the Philippines are the main growers of durians. Thailand exports durians to countries such as the U.S.A., China, Hong Kong and Singapore. The skin of a durian is thick and consists of hundreds of hard spikes, as shown in Fig. 1. It is difficult to determine durian ripeness by observing the skin. Buyers depend on vendors' recommendations and sometimes they may not get value for money. Consumers have felt disappointed because cutting durians too early results in missing out on the delicious taste and texture. Therefore, it will impact people worldwide if we can develop a practical automated method that can determine the ripeness of durians in advance without the need to damage them.

S. Boonkrong et al. (eds.), *Recent Advances in Information and Communication Technology*, Advances in Intelligent Systems and Computing 265,
DOI: 10.1007/978-3-319-06538-0_17,

Fig. 1. A durian

Previously, signal-processing methods have been studied and applied to various fields. There has been limited research that has applied signal processing to animal sounds [1,2,3]. In gender recognition, fundamental frequencies, pitch contours and MFCCs extracted from speech signals have also been widely used to distinguish females from males [4,5,6,7]. The duration of human speech segments has also been investigated for gender classification tasks [8]. In tone recognition, MFCC-based acoustic features and the pitch contours extracted from human voices have been applied to recognize tones in languages such as Thai and Cantonese [9,10,11]. In emotion recognition, acoustic features including fundamental frequencies, spectral features, energy features and their augmentations have been studied to classify emotional states of the human voice [12]. In speech recognition, digitized speech signals have been converted to words or text using signal processing, pronunciation dictionaries and language models such as grammar [13,14] and N-gram [15,16,17]. The Hidden Markov Model (HMM) is one of the most proficient techniques used in speech recognition to create acoustic models of speech units. HMMs of syllable units have been created and applied to speech recognition systems [9]. Mel Frequency Cepstral Coefficients (MFCCs) are also acoustic features that have been widely used to recognize speech signals [18,19] and determine guava freshness from flicking signals [20]. Although N-gram is a language model used in speech recognition, it has not been applied to the recognition of durian striking signals. Therefore, this paper proposes a durian ripeness striking sound recognition method that uses N-gram models with N-best lists and majority voting.

The remainder of this paper is divided into the following parts: Section 2 discusses durian striking and striking signals. Section 3 describes N-gram models and durian ripeness striking signal recognition. Section 4 explains the proposed method. Section 5 shows the experimental results and section 6 gives conclusions.

2 Durian Striking and Its Striking Signals

Striking a durian is hitting it with a tapping stick, as shown in Fig. 2. Experienced vendors judge durian ripeness by listening attentively to the striking sounds made when the durian is hit with the tapping stick that can be bought from shops around a durian market.

Fig. 2. Striking a durian

However, buyers find it difficult to purchase durians because the duration of the striking-part of the sound is very short and human perception varies. Fig. 3 illustrates signals resulting from striking a durian five times.

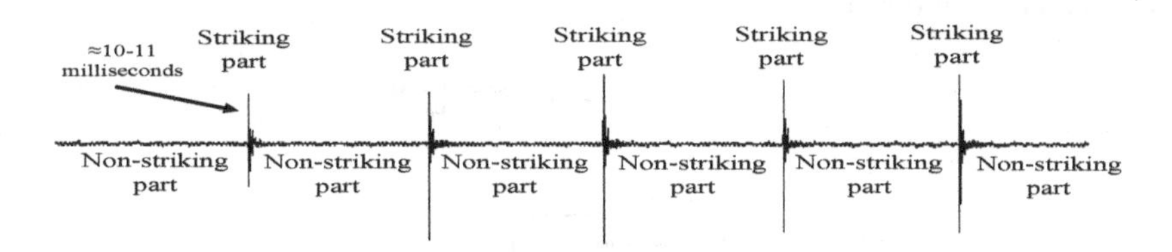

Fig. 3. Durian five-striking signals

Fig. 3 shows that the duration of a striking part can be only around 10-11 milliseconds. When striking, a non-striking part is usually generated at the beginning, middle and end of striking signals. The striking part occurs when the stick hits a durian while the non-striking part occurs when the stick does not hit or contact it.

3 N-gram Models and Durian Ripeness Striking Signal Recognition

To use an N-gram model to recognize durian ripeness striking signals, a recognizer finds the most likely striking sequence S = (s_1, s_2,... , s_n) containing n strikes of "UNRIPE" and "RIPE" that gives the highest probability. To do so, Equation 1 is applied to compute P(S), which is the probability that a striking sequence that can occur.

$$\begin{aligned} P(S) &= P(s_1, s_2, \dots, s_n) \\ &= P(s_1)P(s_2|s_1)P(s_3|s_1s_2)\dots P(s_n|s_1\dots s_{n-1}) \end{aligned} \quad (1)$$

Computing this probability is difficult in practice, therefore, it is common to represent strike history using an N-gram model. This method estimates the probability of a strike by taking into account n the preceding strikes. When n=2, the N-gram model is referred to as a bigram and when n=3, it is referred as a trigram model. The probability of a striking sequence using a bigram and a trigram model is expressed using Equation 2 and Equation 3, respectively.

$$P(S) = P(s_1, s_2, \dots, s_n) \\ = P(s_1)P(s_2|s_1)P(s_3|s_2) \dots P(s_n|s_{n-1}) \quad (2)$$

$$P(S) = P(s_1, s_2, \dots, s_n) \\ = P(s_1)P(s_2|s_1)P(s_3|s_1 s_2) \dots P(s_n|s_{n-2} s_{n-1}) \quad (3)$$

4 The Proposed Method

The proposed method consists of 3 stages: 1) extract acoustic features from the striking signals, 2) recognize unripe and ripe durian striking signals using the N-gram models, and 3) find the final recognition results from N-best lists using majority voting, as shown in Fig. 4.

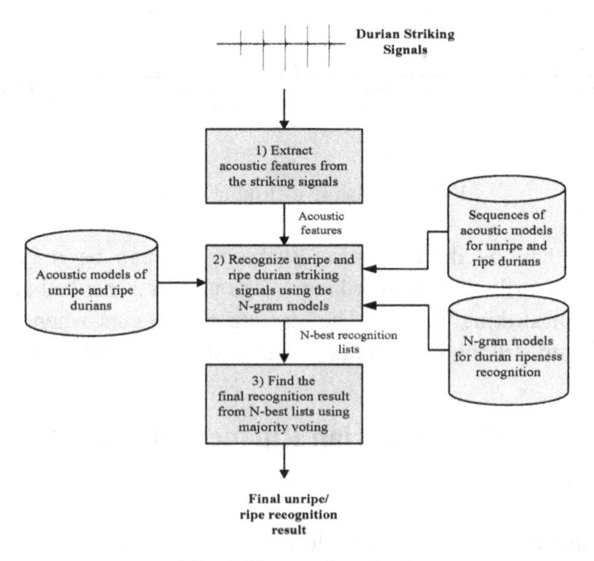

Fig. 4. Proposed method

At the first stage of the process, acoustic features are extracted from the striking signals. At the second stage, acoustic models of unripe and ripe durians, sequences of acoustic models for unripe and ripe durians and N-gram models for durian ripeness recognition are used to recognize unripe and ripe durian striking signals to obtain N-best recognition results (or N-best recognition lists). N-best results or lists are used in favor of a single result because a single result may consist of equal numbers of both unripe and ripe strikes. For example, when striking a ripe durian twice, the first strike may be recognized as ripe and the second one may be differently recognized as unripe. Consequently, the final recognition result cannot be determined using only the most probable result. At the last stage, the N-best lists are searched for the final unripe and ripe recognition result using the majority voting method.

4.1 Extract Acoustic Features of the Striking Signals

39-dimension acoustic features, consisting of 12 MFCCs with energy as well as their 1st- and 2nd-order derivatives, are extracted and used to distinguish unripe and ripe durians. To recognize durian ripeness from the extracted acoustic features, acoustic models of unripe and ripe durians, sequences of acoustic models for unripe and ripe durians, and N-gram models for durian ripeness recognition are prepared, as explained in the following sections.

4.2 Acoustic Models of Unripe and Ripe Durians

The HMM acoustic models of unripe and ripe durians are created using striking signals and prepared transcriptions without the matched positions between the "UNRIPE" and "RIPE" signals and labels. For example, if five striking sounds derived from an unripe durian are prepared for the acoustic model creation, they are transcribed as "sil UNRIPE sil UNRIPE sil UNRIPE sil UNRIPE sil UNRIPE sil". When five-striking sounds are obtained from a ripe durian, they are transcribed as "sil RIPE sil RIPE sil RIPE sil RIPE sil RIPE sil". For the acoustic model creation, the acoustic features are extracted from the striking signals. Then, the obtained acoustic features with their transcriptions are used to train the unripe, ripe and non-striking part of the acoustic model (sil: silent model). After obtaining the acoustic models, they are further applied to define sequences of acoustic models for unripe and ripe durians.

4.3 Sequences of Acoustic Models for Unripe and Ripe Durians

The sequences of acoustic models for unripe and ripe durians are created based on the characteristics of each striking sound and adjusted to obtain high recognition rates. silB and silE denote the beginning and end of striking sounds that are represented with a non-striking part (sil). Since a non-striking part has a much longer duration than a striking part, the "UNRIPE" and "RIPE" sounds are represented with sil sil sil UNRIPE sil sil sil and sil sil sil RIPE sil sil sil, respectively. The sequences for unripe and ripe durians are as follows:

```
silB    sil
silE    sil
UNRIPE    sil sil sil UNRIPE sil sil sil
RIPE    sil sil sil RIPE sil sil sil
```

4.4 N-gram Models for Durian Ripeness Recognition

The characteristics of the durian striking sounds and the permitted number of strikes are considered when creating the N-gram models. To determine durian ripeness, buyers may strike a durian more than once. Therefore, the data below represents N-gram training for the characteristics of 1- to 5-striking sounds resulting from hitting unripe and ripe durians.

```
silB UNRIPE silE
silB RIPE silE
silB UNRIPE UNRIPE silE
silB RIPE RIPE silE
silB UNRIPE UNRIPE UNRIPE silE
silB RIPE RIPE RIPE silE
silB UNRIPE UNRIPE UNRIPE UNRIPE silE
silB RIPE RIPE RIPE RIPE silE
silB UNRIPE UNRIPE UNRIPE UNRIPE UNRIPE silE
silB RIPE RIPE RIPE RIPE RIPE silE
```

The Carnegie Mellon University Statistical Language Modeling toolkit (CMU-SLM) is an efficient tool that can be used to train the N-gram models (2-gram and 3-gram language models) for durian ripeness recognition. In this work, 100 times of the data is used to create the N-gram models, which is used to recognize unripe and ripe durian striking signals.

4.5 Recognize Unripe and Ripe Durian Striking Signals Using the N-gram Models

To recognize speech signals using the N-gram language model, first, the acoustic features are extracted. After that, the acoustic models and sequences of acoustic models for unripe and ripe durians and a combination of 2-gram and 3-gram language models are used by the Open-Source large vocabulary continuous speech recognition decoder to recognize the acoustic features and obtain N-best unripe/ripe recognition lists. The final recognition result is then determined from the N-best lists.

4.6 Find the Final Recognition Result from N-best Lists Using Majority Voting

Given that a majority result of either UNRIPE or RIPE is found in the N-best lists, the final result can be determined from the N-best lists using the following algorithm.

```
for i=1 to N step by 1                // for each recognition result in the N-best lists
    Find the majority of voting between UNRIPE and RIPE
    if(a majority result is found in the ith result in the N-best list)
        q* = the majority of voting between UNRIPE and RIPE
        break;                        // Exit loop
    endif
endfor
where q* is the final recognition result
        N is the number of possible results (N-best lists)
```

For each recognition result in the N-best lists, the majority of voting between UNRIPE and RIPE (q^*) is determined using Equation 4.

Given a possible result R (in N-best lists), consisting of s strikes, R_1, R_2,..., R_s, respectively.

$$q^* = arg\max_{q} \sum_{j=1}^{s} \delta(q, R_j) \tag{4}$$

where $\delta(a,b) = 1$ if $a = b$, and $\delta(a,b) = 0$ otherwise.

q is the label of the recognition result, 1 = UNRIPE and 2 = RIPE

q^* is the majority of voting result, UNRIPE, RIPE or no majority result

If there is a majority result found, it will be used as the final recognition result. Otherwise, repeat the step to find the majority result from the next result in the N-best lists.

5 Experimental Results

Experiments were conducted to investigate the performance of the proposed method. Durian striking sounds were recorded using the 16-bit PCM format at 11,025 Hz. Durian striking sounds for training and testing were collected from 100 unripe and 100 ripe durians. The training set was composed of 50 unripe and 50 ripe durians. Each durian in the training set was struck five times. There were two sets for testing: untrained and unknown. The durian striking signals for the untrained set were recorded, by striking them at different times, from the 50 unripe and 50 ripe durians used in training. The unknown set was recorded from the 50 unripe and 50 ripe durians that were not included in the training set. Since vendors or experts can usually ascertain the ripeness of a durian by using less than five strikes, in both the untrained and the unknown test sets, each durian was struck from one to five times, respectively. The Hidden Markov Toolkit (HTK) [21] was used to create the HMM acoustic models. The Open-Source Large Vocabulary continuous speech recognition decoder, namely Julius [22], was applied to recognize the ripeness features and obtain N-best lists. Each HMM acoustic model was composed of 3 emitting states and each state contained 4 Gaussian mixtures. The results are divided into 2 parts: 1) durian ripeness recognition rates, and 2) percentages that can determine durian ripeness from N-best lists.

5.1 Durian Ripeness Recognition Rates

The accuracy of using a different number of strikes on unripe and ripe durian recognition was evaluated and the results are shown in Table 1.

Table 1. Durian ripeness recognition rates

Number of Strikes	Durian Ripeness Recognition Rate (%)		
	Untrained Set	Unknown Set	Average
1	96.0	88.0	**92.0**
2	96.0	91.0	**93.5**
3	94.0	91.0	**92.5**
4	96.0	92.0	**94.0**
5	97.0	90.0	**93.5**
Average	**95.8**	**90.4**	**93.1**

Average recognition rates of 95.8%, 90.4% and 93.1% were obtained from the untrained, unknown and both test sets, respectively. Average recognition rates were 92.0%, 93.5%, 92.5%, 94.0% and 93.5% using 1 through 5 strikes, respectively. For the untrained set, recognition rates of 96.0%, 96.0%, 94.0%, 96.0% and 97.0% were obtained using 1 through 5 strikes, respectively. For the unknown set, recognition rates of 88.0%, 91.0%, 91.0%, 92.0% and 90.0% were obtained using 1 through 5 strikes, respectively. The results show that the proposed method resulted in high durian ripeness recognition rates when an arbitrary number of strikes from 1 through 5 was used.

5.2 Percentage That Can Determine Durian Ripeness from N-best Lists

Sometimes, the recognition results contain an equal number of UNRIPE and RIPE results. When this happens, the system is unable to determine the ripeness from the 1-best result or the list. Therefore, N-best lists were used instead. The percentages that can determine the durian ripeness from N-best lists are reported in Table 2.

Table 2. Percentages that can determine durian ripeness from N-best lists

Number of Strikes	Percentages that Can Determine Durian Ripeness (%)								
	Untrained Set			Unknown Set			Average		
	1-best	2-best	3-best	1-best	2-best	3-best	1-best	2-best	3-best
1	99.0	99.0	100.0	98.0	100.0	100.0	98.5	99.5	100.0
2	98.0	100.0	100.0	97.0	100.0	100.0	97.5	100.0	100.0
3	100.0	100.0	100.0	99.0	100.0	100.0	99.5	100.0	100.0
4	99.0	100.0	100.0	99.0	100.0	100.0	99.0	100.0	100.0
5	100.0	100.0	100.0	99.0	100.0	100.0	99.5	100.0	100.0
Average	**99.2**	**99.8**	**100.0**	**98.4**	**100.0**	**100.0**	**98.8**	**99.9**	**100.0**

When recognizing striking sounds using 1 through 5 strikes, the average percentages that can determine durian ripeness from the 1-, 2-, and 3-best lists were 98.8%, 99.9% and 100.0%, respectively. For untrained test set, the average percentages from the 1-, 2-, and 3-best lists were 99.2%, 99.8% and 100.0%, respectively. For the unknown test set, the average percentages from the 1-, 2-, and 3-best lists were 98.4%, 100.0% and 100.0%, respectively. The results show that the 3-best lists could be efficiently used for durian ripeness signal recognition.

6 Conclusions

It is difficult for buyers to classify durian ripeness by ear because the duration of the durian striking parts is audible for only a short period. Therefore, a durian ripeness recognition method that uses striking sounds is useful because the fruit is undamaged during the process and expert ability is not required to ascertain the ripeness of the fruit. The durian ripeness recognition method presented in this paper used N-gram models with N-best lists and majority voting. The results show that the durian ripeness classification method is quite efficient. Average recognition rates of more than 90% were obtained from both the untrained and unknown test set. The 3-best lists (including a few more possible recognition results such as 4- and 5-best lists) can be used in durian ripeness striking signal recognition. For further works, a greater quantity of the durian samples should be used. Other methods should be attempted and compared with this method. Then, a durian-striking machine with a more efficient recognition method could be created to help consumers and the fruit industry worldwide to classify durians.

References

1. Yeo, C.Y., Al-Haddad, S.A.R., Ng, C.K.: Animal Voice Recognition for Identification (ID) Detection System. In: Proceedings of the IEEE 7th International Colloquium on Signal Processing and Its Applications, pp. 198–201 (2011)
2. Mitrovic, D., Zeppelzauer, M., Breiteneder, C.: Discrimination and Retrieval of Animal Sounds. In: Proceedings of the 12th International Multi-Media Modelling Conference, pp. 339–343 (2006)
3. Guo, G., Li, Z.: Content-based Classification and Retrieval by Support Vector Machines. IEEE Transactions on Neural Networks 14, 209–215 (2003)
4. Phoophuangpairoj, R., Phongsuphap, S., Tangwongsan, S.: Gender Identification from Thai Speech Signal Using a Neural Network. In: Leung, C.S., Lee, M., Chan, J.H. (eds.) ICONIP 2009, Part I. LNCS, vol. 5863, pp. 676–684. Springer, Heidelberg (2009)
5. Ting, H., Yingchun, Y., Zhaohui, W.: Combining MFCC and Pitch to Enhance the Performance of the Gender Recognition. In: Proceedings of the 8th International Conference on Signal Processing (2006)
6. Azghadi, S.M.R., Bonyadi, M.R., Sliahhosseini, H.: Gender Classification Based on Feedforward Backpropagation Neural Network. In: Boukis, C., Pnevmatikakis, L., Polymenakos, L. (eds.) Artificial Intelligence and Innovations 2007: From Theory to Applications. IFIP, vol. 247, pp. 299–304. Springer, Boston (2007)

7. James, M.H., Michael, J.C.: The Role of F0 and Formant Frequencies in Distinguishing the Voices of Men and Women. Attention, Perception, & Psychophysics 71(5), 1150–1166 (2009)
8. Sigmund, M.: Gender Distinction Using Short Segments of Speech Signal. International Journal of Computer Science and Network Security 8(10), 159–162 (2008)
9. Tangwongsan, S., Po-Aramsri, P., Phoophuangpairoj, R.: Highly Efficient and Effective Techniques for Thai Syllable Speech Recognition. In: Maher, M.J. (ed.) ASIAN 2004. LNCS, vol. 3321, pp. 259–270. Springer, Heidelberg (2004)
10. Thubthong, N., Kijsirikul, B.: Tone Recognition of Continuous Thai Speech Under Tonal Assimilation and Declination Effects Using Half-tone Model, International Journal of Uncertainty. Fuzziness and Knowledge-Based Systems 9(6), 815–825 (2001)
11. Lee, T., Lau, W., Wong, Y.W., Ching, P.C.: Using Tone Information in Cantonese Continuous Speech Recognition. ACM Transactions on Asian Language Information Processing (TALIP) 1(1), 83–102 (2002)
12. Ververidis, D., Kotropoulos, C.: Automatic Speech Classification to Five Emotional States Based on Gender Information. In: Proceedings of the European Signal Processing Conference, vol. 1, pp. 341–344 (2004)
13. Tangwongsan, S., Phoophuangpairoj, R.: Boosting Thai Syllable Speech Recognition Using Acoustic Models Combination. In: Proceedings of the International Conference on Computer and Electrical Engineering, pp. 568–572 (2008)
14. Phoophuangpairoj, R.: Using Multiple HMM Recognizers and the Maximum Method to Improve Voice-controlled Robots. In: Proceedings of the International Conference on Intelligent Signal Processing and Communication Systems (2011)
15. Pohl, A., Ziółko, B.: Using Part of Speech N-Grams for Improving Automatic Speech Recognition of Polish. In: Perner, P. (ed.) MLDM 2013. LNCS (LNAI), vol. 7988, pp. 492–504. Springer, Heidelberg (2013)
16. Lee, A., Kawahara, T., Shikano, K.: Julius — An Open Source Real-time Large Vocabulary Recognition Engine. In: Proceedings of European Conference on Speech Communication and Technology, EUROSPEECH, pp. 1691–1694 (2001)
17. Lee, A., Kawahara, T.: Recent Development of Open-source Recognition Engine Julius. In: Proceedings of PSIPA Annual Summit and Conference (2009)
18. Deemagarn, A., Kawtrakul, A.: Thai Connected Digit Speech Recognition Using Hidden Markov Models. In: Proceedings of the 9th International Conference on Speech and Computer (2004)
19. Li, F., Ma, J., Huang, D.: MFCC and SVM Based Recognition of Chinese Vowels. In: Hao, Y., Liu, J., Wang, Y.-P., Cheung, Y.-m., Yin, H., Jiao, L., Ma, J., Jiao, Y.-C. (eds.) CIS 2005, Part II. LNCS (LNAI), vol. 3802, pp. 812–819. Springer, Heidelberg (2005)
20. Phoophuangpairoj, R.: Determining Guava Freshness by Flicking Signal Recognition Using HMM Acoustic Models. International Journal of Computer Theory and Engineering 5(6), 877–884 (2013)
21. The Hidden Markov Model Toolkit (HTK), `http://htk.eng.cam.ac.uk/`
22. The Open-Source Large Vocabulary CSR Engine Julius, `http://julius.sourceforge.jp/en_index.php`

An Exploratory Study on Managing Agile Transition and Adoption

Taghi Javdani Gandomani, Hazura Zulzalil, Azim Abd Ghani, Abu Bakar Md. Sultan, and Khaironi Yatim Sharif

University Putra Malaysia, Malaysia
tjavdani@yahoo.com, {hazura,azim,abakar,khaironi}@upm.edu.my

Abstract. Software companies are replacing traditional software development methods with Agile methods due to coping with inherent problems of traditional methods. Due to the different nature of traditional and Agile methods, adaptation to Agile methods is not a simple process and needs to be managed in a sustainable way. In recent years, several studies have conducted on investigation of Agile migration journey, but less effort on identifying the serious managerial attentions in Agile transition process. Conducting a Grounded Theory in context of Agile software development, showed various aspects of the transition to be considered in order to having a successful change management process. This paper shows the important role of the emergent managerial attentions on success of Agile transition and adoption process.

Keywords: Agile software development, Agile transition, Agile adoption, Agile transformation, Managerial concerns, Grounded Theory.

1 Introduction

Agile software development was introduced to software market by creating Agile manifesto that focused on the new values [1]. The new values came from new development style which in turn focused on new activities and practices which in turn were underpinning of Agile approach [2]. The process of changing software development approach to Agile approach, called Agile Transformation/Transition/ Adoption, is a serious organizational mutation in which all levels of company and organization need to be involved. Agile approach promised values such as higher quality, customer satisfaction, early and frequently releases, light-weight documentation, embracing changes, etc. motivate software companies to employ Agile methods. At the same time, in the recent competitive economy, using Agile methods would be an advantage. Lots of reputed software companies are using Agile methods, at least in some of their projects [3, 4], and many others are tempted to move to Agile [5].

Several studies discussed about the challenges, barriers, and problems that software companies and teams are faced with during Agile transformation [6-8]. Since Agile transition and adoption process requires a huge change in the organizational culture

S. Boonkrong et al. (eds.), *Recent Advances in Information and Communication Technology*, Advances in Intelligent Systems and Computing 265,
DOI: 10.1007/978-3-319-06538-0_18,

and people roles and responsibilities, this process most often will be faced with a large variety of challenges. In general, involved team members, senior managers, customers, tools, and technologies are the origins of the transition challenges [8, 9].

Agile transition needs to be managed in a sustainable way mainly due to the huge amount of changes, the potential challenges and barriers, and involving various people in the transition. This denotes that companies need to be familiar with various aspects of Agile transition and try to minimize the required effort and cost by focusing on specific managerial attentions to this organizational mutation.

Conducting a Grounded Theory in context of Agile software development showed various necessary managerial attentions to be considered before the transformation.

With respect to the above discuss, this paper has been organized as follows: Section 2 presents a concise literature about Agile transition issues, followed by Section 3 which presents the employed research methodology. Section 4 provides the results and continued by Section 5 which presents a discussion on the results. Section 6 explains the limitation of the research study and finally Section 7 presents a conclusion along with recommendation for future work.

2 Agile Transition and Adoption

Transitioning to Agile approach and methods are not an easy and ordinary task. Especially, for those teams that have been using the traditional methods for a long time and are fully adapted to them. Indeed, such a process requires a significant shift in the mindset of everybody involved, enough investment and many other stimuli [7, 10].

Agile transition as an infrastructural project needs a comprehensive action plan comprising all the transition related aspects and issues. So for, several research studies have been done on almost all aspects of Agile transition. While most of them have focused on describing the journey of Agile transformation [4, 11, 12], some of them put more emphasis on the potential challenges and barriers [7, 8, 13]. Also, some other studies have focused on Agile principles and Agile promised values [14, 15]. At the same time a few studies have proposed some theoretical transformation frameworks or guidelines [16, 17]. However, they have not been successful enough in practice to be employed in real environments [13, 18].

Beside the above studies, in some other research studies human aspects as critical factors of the transition have been studied in detail [9, 19, 20]. In these research studies, cultural and people-related issues have been emphasized. Due to the people-intensive nature of Agile methods, paying enough attention to the people-related issues is so crucial. Moreover, in some other studies, facilitators, prerequisites, and drivers of Agile transition have been investigated [20-22]. In general, effective training, customer collaboration, people commitment to change, Agile champions, appropriate pilot project selection, suitable method or practice selection are some of the recommended drivers and facilitators. Lesson learned and weaknesses and strengths of the transition are discussed in other studies [9, 13, 23]. Clearly, the results of the previous studies give other software companies better vision to prevent colliding with the problems and challenges within migration to Agile.

With respect to the literature, it seems that Agile coaches or those who are responsible for managing or handling Agile transition should be aware of the general conditions covering the whole process. Furthermore, those who are involved in the transition need to be familiar with the change process and the factors that affect the transition. Undoubtedly, such knowledge increases the chance of success when attempting to migrate to Agile and decreases the demanded time and cost significantly.

In sum up, a lot of the factors should be considered together for a successful Agile transition, but it is impossible to cover all of them in a single article. Following the selected research methodology, only a minor literature review was conducted before inception of this research study [24]. However, after theory development, the findings were investigated in light of a major literature review [25]. In the next sections, this study presents its findings focusing on the managerial attentions in real Agile transition process that have been explored from the real data.

3 Research Methodology

Grounded Theory (GT) was employed as the most suitable research methodology for conducting this research study. The founders of GT, Glaser and Strauss, defined it as a systematic approach which tries to discover main concern of people involved the context under study [25]. GT, in general, is helpful to answering questions like, "what is going on in an area?" by generating formal or substantive theory [26]. Although this method usually have been employed in social research studies, it is also useful for wide range of research studies in context of software engineering in general and in Agile context in particular [27, 28].

3.1 Procedure of GT

GT defines some particular steps for conducting a qualitative research, as depicted in Figure 1.The following steps can be defined generally [25]:

- Data Collection: Contrary to the other research methodologies, GT requires minimum literature review and starts with data collection to cover general questions and concerns of people involved in area under study. There is no limitation about the type of data. However, most often data collection is carried out via conducting several interviews.
- Data Analysis - Open coding/Substantive coding: Once some data is collected, data analysis will be started. Collected data should be reviewed line by line/ sentence by sentence/ paragraph by paragraph... and *key points* should be extracted. Then, each key point will be assigned with an *open code.* Using *constant comparison* technique, the newly assigned code should be compared with the previous open codes in the same and previous interviews. This technique checks whether the newly assigned code is created previously or not. Also, it helps the researchers to

identify concepts [25]. Iterative using of constant comparison leads to emergence of *categories,* higher abstraction level, that each of them encompass several related concepts [25].

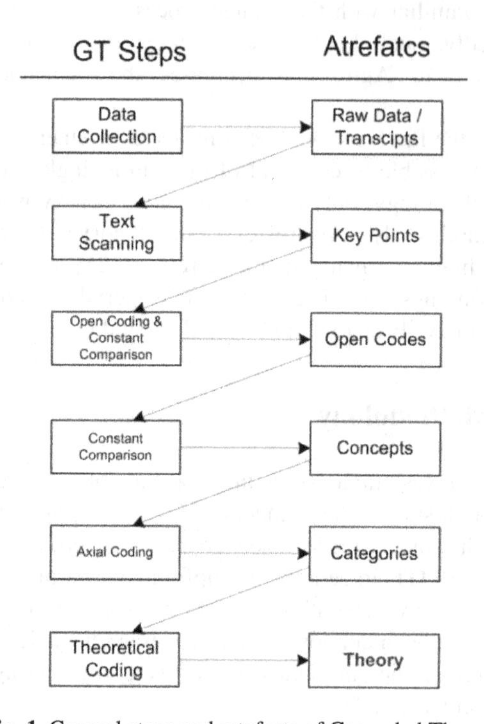

Fig. 1. General steps and artefacts of Grounded Theory

- Theoretical Memoing: After each interview, for clarifying viewpoints of respondents, supplementary data can be collected in form of memos [29]. This process helps the researchers to enrich the relationships between categories.
- Data Analysis- Theoretical coding: Theoretical coding or theory building is the last step of data analysis in which the researchers look for finding relationships between *core category* and other categories. Glaser emphasizes on induction or emerging theory without forcing [24] and suggests several theoretical coding families to assist the GT researchers to develop substantive theory.

3.2 Participants Recruitment

This study was conducted with 45 voluntary Agile experts from 13 different countries across the world. Figure 2 shows the detail information of the participants. Several interviews conducted with the participants and were voice recorded and transcribed subsequently for further use in data analysis. Due to space limitation, only a few participants' quotes are provided to clarify the researchers' interpretations.

P#	SD exp. (years)	Agile exp. (years)	Agile position	Agile methods	Country	Projects domain	Branches	Transformation period (months)
P1	14	8	HDD	XP, Scrum, Kanban	Finland	Mobile software	1	12+, ongoing
P2	25	15	AC	Scrum, Kanban	USA	All domains	4	12+
P3	7	7	PM	XP, Scrum, Kanban	USA	Web applications	3	6+, ongoing
P4	10	2	PM	XP, Scrum, Kanban	Bulgaria	Banking	1	6+, ongoing
P5	10	2	PM	Scrum, Kanban	Iran	Financing	7	12+, ongoing
P6	11	8	CON	Scrum, Kanban, FDD	Australia	All domains	7	12-15
P7	6	2	DEV	Scrum	Greek	Office Auto.	1	12+, ongoing
P8	10	5	PM	Scrum, Kanban	Germany	Web applications	5	8+, ongoing
P9	20	10	HDD	Scrum	Spain	Custom Development	3	24
P10	20	3	SM	Scrum, Kanban	Spain	Transportation	4	24+, ongoing
P11	10	4	AC, SM	XP, Scrum, Kanban	India	Publication	1	+6, ongoing
P12	16	2	HDD	Scrum, Kanban	USA	Life and insurance	16	6+, ongoing
P13	14	6	AC, CON	Scrum, Kanban	Finland	Consultancy services	1	3-30
P14	15	3	MGR	Scrum, Kanban	Iran	HRMS, CMMS	1	12
P15	10	2	CON	Scrum	Indonesia	MIS	1	3+, ongoing
P16	21	10	PM	Kanban	USA	SCM	1	12
P17	19	5	PM	Scrum, Kanban	Sweden	Financing	2	24+, ongoing
P18	8	2	DEV	Scrum	Sweden	HRMS, MIS	1	24
P19	13	6	PM	Scrum	India	Banking	3	18 in USA, 24 in India
P20	11	3	HDD, PM	Scrum, Kanban	USA	Insurance, Banking	16	6+, ongoing
P21	16	7	SM	XP, Scrum	USA	Financing, HR	3	18
P22	11	5	AC	Scrum, Kanban	France	Banking	100+	12+, ongoing
P23	16	8	AC	XP, Scrum, Kanban	USA	All domains	7	6-24
P24	15	7	SM	Scrum, XP	USA	All domain	3	12
P25	8	4	DEV	Scrum, XP	USA	HRMS	1	6+, ongoing
P26	13	6	AC	Scrum, XP	India	Medical systems	3	15+
P27	14	5	SM	Scrum, Kanban	USA	Mobile applications	2	12+
P28	15	6	AC	Scrum, Kanban	Germany	Advisory/consulting	2	15+
P29	10	1	PM	Scrum	Norway	Web Agency	2	12+
P30	35	1	DEV	Scrum	USA	Finance, Banking	3	6+, ongoing
P31	17	4	QA, PM	Scrum	USA	All domains	3	12
P32	25	2	AC	Scrum, Scrumban	USA	All domains	2	12, ongoing
P33	41	3	MGR	Scrum, Kanban	USA	All domains	17	15+, ongoing
P34	7	2	QA	Scrum, Kanban	USA	Retail systems	3	9+, ongoing
P35	20	8	HDD	Scrum, XP	USA	Life and Annuity	50+	15
P36	17	7	MGR	Scrum, Scrumban	USA	Investment Services	21	18
P37	13	1	HDD	Scrum	USA	Finance, Banking	3	6, ongoing
P38	7	2	PM	Scrum	USA	Web Applications	1	12+, ongoing
P39	35	10	AC	Scrum, XP, Kanban	USA	Consultancy services	1	15
P40	30	15	AC	Scrum, Kanban	USA	Consultancy services	1	12
P41	11	3	PM	Scrum, Kanban	India	Mobile applications	2	12+, ongoing
P42	14	4	DEV	Scrum	Spain	Banking, HR	14	15+, ongoing
P43	18	6	AC	Scrum, XP	USA	Banking	2	6+, ongoing
P44	17	5	SM	Scrum, Kanban	Norway	HRMS	2	9+
P45	11	6	PM	Scrum	USA	Mobile applications	4	12+

Fig. 2. Demography of the participants (HDD: Head of development dept., AC: Agile coach, PM: Project manager, CON: Consultant, SM: Scrum master, MGR: Manager, DEV: Agile Developer, QA: Quality assurance)

4 Results: Managerial Concerns in Agile Transition

Data analysis discovered the main concern of the participants, '*Agile transition and adoption process*', and showed that '*Managerial concerns in Agile transition*' is one of the its related categories. This section describes this category in form of the Glaser Strategy Family, suggested by Glaser [25]. Figure 3 shows the emergent theory following the Weber's recommendations for illustration of relationships [30].

4.1 Focusing on People

The participants explained the importance role of people involved in the transition. The reality of transition is changing people behaviours and mindsets. Hence, focusing on them is so crucial within the transition.

"At Toyota this is also known as 'Building people before building cars', this is the fact behind any successful Agile transition... " P1, HDD.
"Focusing on people is success key of transition to Agile, it is all about people and their behaviours." P21, SM.

Some of the participant believed that companies and managers only need to help people to change themselves while empowering them.

"The technical implementation of Agile is usually the easiest; it's the people transformation which is difficult. Managers need to help them to change themselves and give them enough authority in their work." P6, CON.

Encouraging people, motivating and supporting them during the transition help them to adapt to their new responsibilities easier.

"Supporting and helping them, managers can do this, and persuade them [involved people] in transition. People need good supporters while they are changing their behaviours." P16, PM.

Almost all of the participants emphasized on the critical role of the people and believed that most of the transition efforts are related to people.

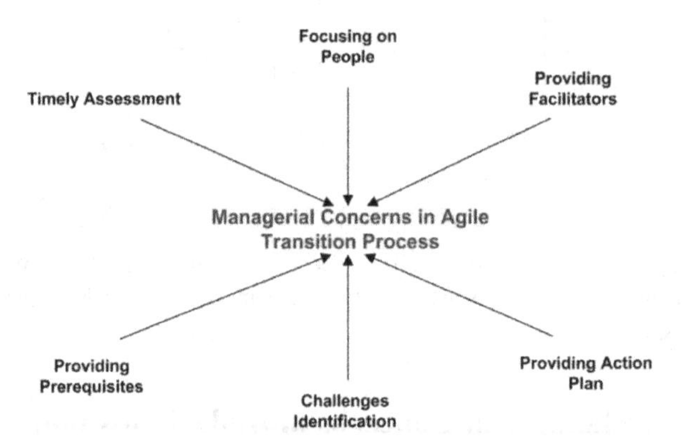

Fig. 3. Emergent of theory of Managing Agile Transition from underlying concepts

4.2 Providing an Action Plan

Agile transition should be considered as an organizational project. Hence, it needs to be planned carefully. Such a plan needs to cover all the transition issues and steps.

"Companies need to take it [transformation] seriously, not just like a momentary fashion. It should be done based on an action plan". P10, SM.

Companies can prepare a framework that considers all the required activities and highlights their business goals.

"We did a primary analysis on our needs and goals, and then prepared a plan or framework for it. Although it took a considerable time, but it helped us in all steps." P33, MGR.

Clearly, all the involved people need to be familiar with prepared action plan, its goals, and its milestones carefully.

> *"Companies need to make sure that the framework is in place and has been communicated before beginning an Agile transition."* P26, AC.

Finally, almost all of the involved Agile coaches emphasized on having an action plan before inception of the transition and addressed lack of it as a potential risk.

4.3 Transition Challenges Identification

During the interviews, almost all of the participants warned the researchers about the potential challenges during transition. They stressed that all members need to be aware about potential challenges.

> *"They [managers] should ensure that all members are aware about the transformation process and its challenges."* P19, PM.

Having enough knowledge assists all members to know the challenges and be ready to cope with them or reduce the likelihood of their occurrence.

> *"Make people strong enough before going Agile. If they know the challenges they'll be facing, it is much easier to provide a solution..."* P28, AC.

Furthermore, some of the participants explained the potential challenges that among them, 'lack of training', 'people resistance', 'people wrong mindset', 'customer issues', 'cultural issues', etc. were the most important ones. Some of them addressed 'lack or weak transition management' as a serious challenge as well (P1, P39).

4.4 Providing Prerequisites

Most of participants emphasized on providing transition prerequisites before starting the transition in order to facilitate the transition process.

> *"There are several prerequisites that I always ask my clients to provide them. I strongly believe that without them there will be no effective transition. For instance, if management is not committed to the change, transition would be meaningless."* P2, AC.

Some of the participants strongly believed on defining a preparation phase in order to provide the transition prerequisites prior to the transition.

> *"A distinct phase is needed to be defined before going Agile. I see most of companies who are not serious about it, while it is so critical and helps them in next phases"*, P43, AC.

The participants addressed several prerequisites too including 'people buy-in', 'training', 'pilot project selection', 'having convincing reasons for Agile', and 'defining business goals'.

> *"After many years experience, I strongly advise managers to provide the required prerequisites to have a smooth transition."* P39, AC.

4.5 Providing Facilitators

Due to wide range of challenges that software companies and teams may be faced with during the transition, the participants emphasized on providing transition facilitators that help teams to deal with the challenges.

> *"During the early meetings with my clients, I always ask senior managers to provide change facilitators; to me, this responsibility is the most important responsibility of managers during Agile transformation."* P13, AC.

Some of the participants believed that good facilitators can guarantee the success of the transition and lack of them causes trouble for involved people.

> *"... There are many factors and facilitators that existence or lack of them has a great impact on the success of project [Agile transition]... motivation and incentives, experienced coaches ... are some of them."* P35, HDD.

The respondents addressed several facilitators including, 'on-site coaching', 'management buy-in', 'team members buy-in', 'Agile champions', 'incentive factors', etc.

In general, the main focus of the facilitators should be on involved people and creating a positive atmosphere for the change.

4.6 Timely Assessment

The participants emphasized on conducting assessments within transformation in several points.

> *"We had a mistake in transformation; I recommend others to define measurement criteria for assessing the transformation process over time and identify their progress, challenges, weaknesses, and strengths."* P8, PM.

Although the participants addressed different points for conducting assessment, they emphasized on the role of assessment in increasing chance of transition success.

> *"The assessment of current situation, and existing assets, before starting the change will lead companies to prerequisites"* P2, AC.

Due to the importance impact of assessment on transition, many of the participants addressed pre-start up assessment as one of the critical prerequisites.

> *"Only by conducting a self-assessment before starting the project companies can decide about their required action plan and its step."* P40, AC.

Such an assessment was emphasized for various parts of the transition such as, 'pilot project selection', 'team selection', 'practice adaptation', 'defining goal', etc.

5 Discussion

Several research studies have been conducted to study Agile transition process, but yet there is no standard or popular roadmap to deploy Agile methods in software companies. Most of the studies in this context have focused on the journey of transition, lesson learned, and specific aspects of transition. Conducting a GT study showed that Agile transition and adoption process needs to be managed carefully to result in a successful transition. This study discovered the six most important managerial attentions to be considered during the transition, as explained in the above section. Conducting a major literature review supported the findings of this study strongly.

Conboy et al. by investigating on the challenges that are faced by software companies during the transition emphasized on the role of people and their skills in the transition process [7]. Lalsing et al. addressed people factors in Agile transition and put emphasis on focusing on people in this process [31]. Cockburn and Highsmith believed that since "Agile processes are designed to capitalize on each individual and each team's unique strengths", Agile transition must consider human factors carefully [32]. Focusing on people also has been emphasized by other researchers [33, 34].

Providing an action plan or framework was also stressed in the literature. Boehm and Turner suggested a risk based plan for using Agile approach in software projects [35]. Furthermore, some frameworks were suggested for Agile adoption [17, 36], however, they were not enough successful in the industry [18]. Also, a few experience-based roadmaps to going Agile [37].

Identification of the potential transition challenges was also emphasized by the literature. Conboy et al. believed that finding the challenges should be the first step of transition [7]. There are several research studies which by explaining the challenges, put emphasis on finding and handling the transition challenges before their occurrence [13, 23]. Furthermore, several studies described the challenges and problems they faced with during Agile migration and warned others about them.

Providing Agile transition prerequisites was another finding of this study. Esfahani focused on Agile pre-adoption and explained the necessity of providing prerequisites of the transition [38]. Other studies also focused on impacts of providing various prerequisites of the transition including 'defining business goals of transition' [39], 'initial training' [40], 'selecting an appropriate pilot project' [21, 41], 'appropriate team set up' [42] and so on. Almost all of them believed that these prerequisites directly impact the success of the transition.

Besides the prerequisites, software companies need to consider several change facilitators during the transition. Previous studies addressed various change facilitators including 'on-site training' [7], 'management and team members buy-in' [41, 43], 'right people selection and empowering team' [7], 'champions' [42], 'frequent communications'[44] , and 'incentive factors' [45].

Finally, the literature supported conducting timely assessment as one of the critical managerial concerns within Agile transition. Conducting assessment prior to starting the transition, method selection, during adaptation, and after adoption were emphasized by previous research studies [17, 46].

It should be noted that, beside the above brief discussion, some parts of findings of this study have explained previously in several distinct article or are under publishing process [6, 9, 21, 22, 27, 47].

6 Limitation

All the emergent codes, concepts, and categories in this study came from data directly, therefore the findings are grounded enough in substantive contexts [29]. Nonetheless, this article cannot claim that its findings are universal, since access to resources was limited to the participants of this study, but, it claims that its findings have characterized and described the context studied [48].

7 Conclusion and Future Work

The study showed various managerial concerns to be considered for managing Agile transition process. Most of them are related to the people and supporting them during this process. This reflects the critical role of the involved people throughout the transition. Preparing a perfect action plan, identifying potential challenge areas, providing prerequisites and facilitators to support the change, and conducting required assessment helps companies to manage the transition process well.

Conducting a quantitative research study and indentifying impact of the findings of this study on success of transformation is suggested for future work.

Acknowledgements. The authors would like to express their thanks to all the participants who voluntarily intended in this study. This research study was financially supported by the UPM International Graduate Research Fellowship (IGRF), Malaysia.

References

1. http://www.agilemanifesto.org
2. Cohen, D., Lindvall, M., Costa, P.: 62. Advances in Computers 62, 1–66 (2004)
3. Laanti, M., Salo, O., Abrahamsson, P.: Agile methods rapidly replacing traditional methods at Nokia: A survey of opinions on agile transformation. Information and Software Technology 53, 276–290 (2011)
4. Fulgham, C., Johnson, J., Crandall, M., Jackson, L., Burrows, N.: The FBI gets agile. IT Professional 13, 57–59 (2011)
5. http://www.versionone.com/state-of-agile-survey-results/
6. Gandomani, T.J., Zulzalil, H., Ghani, A.A.A., Sultan, A.M., Nafchi, M.Z.: Obstacles to moving to agile software development; at a glance. Journal of Computer Science 9, 620–625 (2013)
7. Conboy, K., Coyle, S., Wang, X., Pikkarainen, M.: People over process: Key challenges in agile development. IEEE Software 28, 48–57 (2011)
8. Nerur, S., Mahapatra, R., Mangalaraj, G.: Challenges of migrating to agile methodologies. Communications of the ACM 48, 72–78 (2005)

9. Javdani Gandomani, T., Zulzalil, H., Abd Ghani, A.A., Sultan, A.B.M., Khaironi, Y.S.: How Human Aspects Impress Agile Software Development Transition and Adoption. International Journal of Software Engineering and Its Applications (in-press 2014)
10. Highsmith, J.A.: Agile Software Development Ecosystems. Addison-Wesley Professional, Boston (2002)
11. Srinivasan, J., Lundqvist, K.: Using agile methods in software product development: A case study, pp. 1415–1420 (2009)
12. Schatz, B., Abdelshafi, I.: Primavera gets Agile: A successful transition to Agile development. IEEE Software 22, 36–42 (2005)
13. Srinivasan, J., Lundqvist, K.: Agile in India: Challenges and lessons learned. In: 3rd India Software Engineering Conference, ISEC 2010, pp. 125–130. ACM, New York (2010)
14. Williams, L.: What agile teams think of agile principles. Communications of the ACM 55, 71–76 (2012)
15. Patil, S.B., Rao, S., Patil, P.S.: Agile principles as a leadership value system in the software development: Are we ready to be unleashed?, pp. 765–766 (2011)
16. Qumer, A., Henderson-Sellers, B., McBride, T.: Agile adoption and improvement model. In: 4th European and Mediterranean Conference on Information Systems, EMCIS 2007, pp. 21–29 (2007)
17. Sidky, A., Arthur, J., Bohner, S.: A disciplined approach to adopting agile practices: the agile adoption framework. Innovations in Systems and Software Engineering 3, 203–216 (2007)
18. Rohunen, A., Rodriguez, P., Kuvaja, P., Krzanik, L., Markkula, J.: Approaches to agile adoption in large settings: a comparison of the results from a literature analysis and an industrial inventory. In: Ali Babar, M., Vierimaa, M., Oivo, M. (eds.) PROFES 2010. LNCS, vol. 6156, pp. 77–91. Springer, Heidelberg (2010)
19. Tolfo, C., Wazlawick, R.S., Ferreira, M.G.G., Forcellini, F.A.: Agile methods and organizational culture: Reflections about cultural levels. Journal of Software Maintenance and Evolution 23, 423–441 (2011)
20. Vijayasarathy, L., Turk, D.: Drivers of agile software development use: Dialectic interplay between benefits and hindrances. Information and Software Technology 54, 137–148 (2012)
21. Javdani Gandomani, T., Zulzalil, H., Abd Ghani, A.A., Md. Sultan, A.B., Sharif, K.Y.: Exploring Key Factors of Pilot Projects in Agile Transformation Process Using a Grounded Theory Study. In: Skersys, T., Butleris, R., Butkiene, R. (eds.) ICIST 2013. CCIS, vol. 403, pp. 146–158. Springer, Heidelberg (2013)
22. Javdani Gandomani, T., Zulzalil, H., Abd Ghani, A.A., Sultan, A.B.M., Sharif, K.Y.: Exploring Facilitators of Transition and Adoption to Agile Methods: a Grounded Theory Study. Journal of Software (in-press 2014)
23. Gandomani, T.J., Zulzalil, H., Ghani, A.A.A., Sultan, A.B.M.: Towards comprehensive and disciplined change management strategy in agile transformation process. Research Journal of Applied Sciences, Engineering and Technology 6, 2345–2351 (2013)
24. Glaser, B.: Basics of Grounded Theory Analysis: Emergence Vs. Forcing. Sociology Press, CA (1992)
25. Glaser, B.G.: Theoretical Sensitivity: Advances in the Methodology of Grounded Theory. The Sociology Press, Mill Valley (1978)
26. Corbin, J.M., Strauss, A.C.: Basics of Qualitative Research: Techniques and Procedures for Developing Grounded Theory (3e). SAGE Publications Inc., CA (2008)
27. Gandomani, T.J., Zulzalil, H., Ghani, A.A.A., Sultan, A.B.M., Sharif, K.Y.: How Grounded Theory can facilitate research studies in context of Agile software development. Science International-Lahore 25, 1131–1136 (2013)

28. Hoda, R., Noble, J., Marshall, S.: Using Grounded Theory to study the human aspects of Software Engineering. In: 2nd Workshop on Human Aspects of Software Engineering, HAoSE 2010 (2010)
29. Glaser, B.: Doing Grounded Theory: Issues and Discussions. Sociology Press, CA (1998)
30. Weber, R.: Evaluating and developing theories in the information systems discipline. Journal of the Association for Information Systems 13 (2012)
31. Lalsing, V., Kishnah, S., Pudaruth, S.: People factors in agile software development and project management. International Journal of Software Engineering & Applications (IJSEA) 3, 117–137 (2012)
32. Cockburn, A., Highsmith, J.: Agile software development: The people factor. Computer 34, 131–133 (2001)
33. Korsaa, M., Johansen, J., Schweigert, T., Vohwinkel, D., Messnarz, R., Nevalainen, R., Biro, M.: The people aspects in modern process improvement management approaches. Journal of Software: Evolution and Process 25, 381–391 (2013)
34. Moore, D.M., Crowe, P., Cloutier, R.: The balance between methods and people. CrossTalk 24, 11–14 (2011)
35. Turner, R., Boehm, B.: Balancing Agility and Discipline: A Guide for the Perplexed. Addison-Wesley/Pearson Education, Boston (2003)
36. Qumer, A., Henderson-Sellers, B.: A framework to support the evaluation, adoption and improvement of agile methods in practice. Journal of Systems and Software 81, 1899–1919 (2008)
37. Chan, F.K.Y., Thong, J.Y.L.: Acceptance of agile methodologies: A critical review and conceptual framework. Decision Support Systems 46, 803–814 (2009)
38. Esfahani, H.C.: Transitioning to Agile: A Framework for Pre-Adoption Analysis using Empirical Knowledge and Strategic Modeling. Graduate Department of Computer Science. PhD. University of Toronto, Canada (2012)
39. Holtsnider, B., Wheeler, T., Stragand, G., Gee, J.: Agile Development & Business Goals: The Six Week Solution. Morgan Kaufmann, MA (2010)
40. Wang, X., Conboy, K., Pikkarainen, M.: Assimilation of agile practices in use. Information Systems Journal 22, 435–455 (2012)
41. Pikkarainen, M., Salo, O., Kuusela, R., Abrahamsson, P.: Strengths and barriers behind the successful agile deployment-insights from the three software intensive companies in Finland. Empirical Software Engineering 17, 675–702 (2012)
42. Senapathi, M., Srinivasan, A.: Understanding post-adoptive agile usage: An exploratory cross-case analysis. Journal of Systems and Software 85, 1255–1268 (2012)
43. Chow, T., Cao, D.B.: A survey study of critical success factors in agile software projects. Journal of Systems and Software 81, 961–971 (2008)
44. Mishra, D., Mishra, A., Ostrovska, S.: Impact of physical ambiance on communication, collaboration and coordination in agile software development: An empirical evaluation. Information and Software Technology 54, 1067–1078 (2012)
45. O'Connor, C.P.: Anatomy and physiology of an Agile Transition. In: Agile Conference, Agile 2011, pp. 302–306. IEEE Computer Society (2011)
46. Cao, L., Mohan, K., Xu, P., Ramesh, B.: A framework for adapting agile development methodologies. European Journal of Information Systems 18, 332–343 (2009)
47. Gadnomani, T.J., Zulzalil, H., Abdul Ghani, A.A., Sultan, A.B.M.: Important considerations for agile software development methods governance. Journal of Theoretical and Applied Information Technology 55, 345–351 (2013)
48. Adolph, S., Hall, W., Kruchten, P.: A methodological leg to stand on: lessons learned using grounded theory to study software development. In: 2008 Conference of the Center for Advanced Studies on Collaborative Research: Meeting of Minds, pp. 166–178. ACM, Ontario (2008)

A Learning Automata-Based Version of SG-1 Protocol for Super-Peer Selection in Peer-to-Peer Networks

Shahrbanoo Gholami, Mohammad Reza Meybodi, and Ali Mohammad Saghiri

Department of Computer Engineering and Information Technology,
Amirkabir University of Technology, Tehran, Iran
{sh.gholami,mmeybodi,a_m_saghiri}@aut.ac.ir

Abstract. Super-peer topologies have been found efficient and effective in heterogeneous peer-to-peer networks. Due to dominant position of super-peers, super-peer selection requires a protocol that is aware of peer capacities. Lack of global information about other peers' capacity and dynamic nature of peer-to-peer networks are two major challenges that impose uncertainty in decision-making. SG-1, is a well-known super-peer selection protocol considering peer capacities, but lack of an appropriate decision-making mechanism makes this protocol slow at convergence and imposes overhead of client transfer between selected super-peers. In this paper, we propose an improved version of SG-1 that uses learning automata as an adaptive decision-making mechanism. For this purpose, each peer is equipped with a learning automaton which is used locally in the decisions taken by that peer. Simulations show effectiveness of proposed protocol in terms of convergence time, scalability, capacity utilization, behavior towards super-peer failure and communication cost, compared to SG-1.

Keywords: peer-to-peer network, super-peer, learning automata.

1 Introduction

Peer-to-Peer (P2P) overlay networks are distributed systems formed on top of the underlying physical network. Main purpose of these networks is sharing information and resources (i.e., CPU cycles, memory, storage space and bandwidth). Recent years, research attention increasingly focused on various aspects of P2P networks. Wide variety in distributions of peer characteristics such as the uptime, bandwidth, or available storage space demonstrates heterogeneity in P2P networks. As a result, pure form of P2P architectures can lead in poor performance because all peers without respect to their capacities act equal in all operations. Specially, when size of network increases such situation may lead in overloading of low capacity peers. Super-peer topologies developed in order to defeat this scalability limitation [1]–[3].

In super-peer topologies, peers are taken one of super-peer or client role. Super-peers are connected to each other and form an overlay like pure P2P systems. On the other hand, each client without taking part in search process just sends queries to its super-peer and receives results from it [4]. Efficient performance which is resulted by reducing the time needed for search, lead in increasing attention to super-peer systems

S. Boonkrong et al. (eds.), *Recent Advances in Information and Communication Technology*, Advances in Intelligent Systems and Computing 265,
DOI: 10.1007/978-3-319-06538-0_19, © Springer International Publishing Switzerland 2014

such as Gnutella vs. 6 [5] and Skype [6]. However protocols used in these large-scale distributed applications have no dynamic mechanism to adapt them when it is needed to rebuild the relationships between clients and super-peers. Shortcomings of super-peer topologies are pointed out by [4]. Moreover they propose super-peer redundancy to improve reliability in super-peer topologies. However they do not address super-peer selection problem.

Super-peers have a dominant position in super-peer topologies, so selecting subset of peers as super-peer requires an approach that considers capability of peers for providing service. Moreover, lack of global information about other peers' capacity and dynamic nature of peer-to-peer networks make this problem more challenging. SG-1 is a super-peer selection protocol that considers some of peer characteristics according to applications need, called capacity. SG-1 aims at forming a super-peer set with minimum cardinality to cover all other peers. For this purpose, SG-1 tries to select high capacity peers and utilizes capacity of selected super-peers in a greedy manner. As a result, in a network in which a larger number of peers have a relatively high capacity, SG-1 can lead in demoting some high capacity peers. Furthermore, SG-1 doesn't employ any appropriate decision-making mechanism. Therefore, inappropriate decisions taken by SG-1, makes this protocol slow at convergence and imposes overhead of client transfer between super-peers. In this paper, we propose an improved version of SG-1 that uses learning automata as an adaptive decision-making mechanism. We aim at minimizing number of super-peers to cover all other peers according to current network condition. Learning automata [7], [8] have been found [9], [10] to be useful in dynamic environments with uncertainty in decision-making. Therefore, it can be used to solve problems in P2P networks, where peers frequently join and leave and operate within a dynamic environment.

The rest of this paper is organized as follows. In Section 2, we review the related work. Learning automata will be discussed in Section 3. Section 4, represents our proposed protocol. Section 5, gives the performance evaluation of the proposed method, and finally Section 6 concludes the paper.

2 Related Work

Adaptive super-peer selection requires to consider a criterion that shows eligibility of peers to act as super-peer. SG-1 [11] is a well-known super-peer selection protocol considering peer capacities. Capacity is defined as a general concept that can be calculated by some peer properties, such as bandwidth and computational capabilities. In SG-1 every node maintains different views that each of them contains neighboring peers with some special characteristics. In their proposed protocol, information exchange with randomly selected neighbors lead in capacity comparison and rearranging the topology. The capacity parameter which is used in [12], [13] is similar to SG-1. In [12] the authors propose Myconet, a bio-inspired approach to model relationships between peers in a super-peer topology with focusing on self-healing ability. But, there is no report on the overhead evaluation of Myconet. [13] presents SPS protocol that mainly focuses on quickly construction of a super-peer overlay. In

SPS, each peer periodically rebuilds its set of super-peer candidates that imposes communication overhead. In protocols such as SG-1, Myconet and SPS each client is assigned to only one super-peer. [14] proposes another variation of this basic design that ordinary peers are allowed to be connected with each other and to be clients of more than one super-peer. They obtain efficiency in search operations but they do not provide any mechanism for super-peer election problem.

[15]–[19] considers lifetime of peer besides the concept of capacity. SPSI [15] introduce self-information theory to select super-peers. The mechanism which is used by ERASP [16] is similar to SG-1, but in ERASP client peers are allowed to have multiple super-peers. DLM [17] uses weighted metric for super-peer selection and presents a workload model to find the desirable number of super-peers adaptively. However, DLM is limited to file sharing applications. The same problem is defined in [18] as an optimization problem. In [18], the authors try to overcome shortcomings of DLM by particle swarm optimization. Some studies [20]–[22] focus on connections between super-peers and propose some special topologies. They do not address super-peer selection problem but [23] suggests gradient topology besides using a utility function that covers the concept of peer capacity.

In some applications, besides capacity the latency between super-peer and client is of great importance. In [24], authors define this problem as multiple colored domination and propose H2O, a super-peer selection protocol for unstructured P2P networks. But, H2O uses flooding mechanism that imposes considerable overhead. Natural evolution of SG-1 is SG-2 protocol [25] with bio-inspired approach for super-peer selection. SG-2 takes advantage of network proximity for building a super-peer overlay. The same problem is considered by [26], that proposes an algorithm inspired by a standard neural network learning algorithm. As another example, authors in [27], [28] aim at minimizing distance between super-peers and their clients, as well as between connected super-peers. In [27] they consider this problem as a hub-location problem and prove that it is NP-hard. After that in [28] they present a heuristic algorithm which is close to optimal, but requires a global view. Recently, [29] model super-peer selection problem in P2P streaming systems using game theory. Their proposed model improves system performance, by enabling upload capacity of entire network.

3 Overview of Learning Automata

learning automaton (LA) [7], [8] is an adaptive decision-making unit that has been shown to perform well in computer networks [9], [10]. LA can improve its performance by learning how to choose the optimal action from a finite set of allowed actions through repeated interactions with a random environment. At each iteration, the action is chosen at random based on a probability distribution kept over the action-set and at each instant the given action is served as the input to the random environment. The environments in which the reinforcement signal can only take two binary values 0 and 1 are referred to as p-model. Each LA belongs to one of fixed or variable structure category [8]. Later case are represented by a triple <

$\beta, \alpha, L >$, where β is the set of inputs, α is the set of actions, and L is learning algorithm that is a recurrence relation which is used to modify the action probability vector.

Let $\alpha_i(k) \in \alpha$ and $p(k)$ denote the action selected by LA and the probability vector defined over the action set at instant k, respectively. Let a and b denote the reward and penalty parameters and determine the amount of increases and decreases of the action probabilities, respectively. Let r be the number of actions that can be taken by LA. At each instant k, the action probability vector $p(k)$ is updated by the linear learning algorithm given in (1), if the selected action $\alpha_i(k)$ is rewarded by the random environment, and it is updated as given in (2), if the taken action is penalized. If $a = b$, the recurrence equations (1) and (2) are called linear reward-penalty (L_{RP}) algorithm. More information can be found in [7], [8].

$$p_j(n + 1) = \begin{cases} p_j(n) + a\left(1 - p_j(n)\right), & j = i \\ (1 - a)\, p_j(n), & \forall j, \ j \neq i \end{cases} \quad (1)$$

$$p_j(n + 1) = \begin{cases} (1 - b)\, p_j(n), & j = i \\ \frac{b}{r-1} + (1 - b)\, p_j(n), & \forall j, \ j \neq i \end{cases} \quad (2)$$

4 Proposed Method

This section contains overview of SG-1 protocol [11] and detailed description of our proposed protocol, SG-LA. At first, we introduce common concepts and notations used in these protocols: Given a network with n peers, each peer p is associated with a parameter $capacity(p)$, shows the maximum number of clients that p can handle if elected as super-peer. At any given time, $load(p)$ denotes the current number of client peers currently managed by super-peer p. If $load(p)$ is less than $capacity(p)$, p is set as under-loaded. In SG-LA, client c is associated to super-peer s, if and only if s is under-loaded and has more capacity. Being under-loaded is common constraint, but in SG-1it is allowed for a super-peer and its client to be equal in capacity. In both SG-1 and SG-LA, each client is assigned to only one super-peer.

4.1 SG-1 Protocol

In SG-1 every node has four neighbor sets: $\text{view}_{\text{connected}}$, $\text{view}_{\text{super-peer}}$, $\text{view}_{\text{under-loaded}}$ and $\text{view}_{\text{client}}$. Neighbor items are also called node descriptors, including the identifier, capacity, current role (super-peer or client) and other useful property of a peer. $\text{view}_{\text{connected}}$ contains the neighbors forming the underlying topology and it is maintained by an underlay peer sampling service (Newscast [30]). $\text{view}_{\text{super-peer}}$ and $\text{view}_{\text{under-loaded}}$ in each peer p respectively contains a random sample of super-peers and under-loaded super-peers in the system, that are known to p. Finally, $\text{view}_{\text{client}}$ for each super-peer, consists of clients currently associated with it. SG-1 assigns each node the initial role of super-peer. Each super-peer p periodically examines members in its under-loaded neighbor set and tries to find a suitable candidate for client transfer. If such a candidate s exists, transferring process is done from p to s, according to Fig. 1 (here C denotes the number of clients that are allowed to be accepted by s). If p misses all of its clients, and s is not exhausted, peer p becomes client of s. Otherwise p remains in super-peer role and the comparison will be made between capacity of p and that of highest capacity client of s, denoted as r. If r's capacity is higher than capacity of p, role swapping is made between p and r. During this process, whenever each client loses its super-peer, it becomes a super-peer by itself.

4.2 SG-LA Protocol

In our proposed protocol, SG-LA, we apply learning automata in order to train nodes during gossip-based algorithms. Each peer p in the network is equipped with a learning automaton $Role\_LAp$ that uses L_{RP} algorithm to help the node for making decision about its suitable role according to network condition. Therefore the action set of each $Role\_LA$ consists of two actions; Super and Client. It should be noted that, two terms "action" and "role" are different from each other: each peer in the network introduces itself according to its current role and takes the corresponding responsibilities. But, the action set of each peer is related to its own $Role\_LA$. Each peer makes use of its selected action, locally, to be able to make decisions about its role.

When each peer p joins the system, SG-LA assigns client role to p. So, initially $\text{view}_{\text{under-loaded}}$ in each peer contains random sample of participating peers. This view is updated in each round by a gossiping protocol. As running SG-LA and arising super-peers in the network, $\text{view}_{\text{under-loaded}}$ rapidly converged to desirable neighbor set which is used in our algorithms.

After joining the system, each peer for the first time, performs its learning phase according to Fig. 2. In learning phase, each peer p updates its action probability vector according to "its selected action" and "the comparison which is made between its capacity and that of neighbor in its under-loaded view". Then peer p randomly chooses an action again. This process is repeated for each neighbor in under-loaded view of peer p. It is clear that our policies for getting reward or penalty are based on the goal of maximizing capacity of selected super-peers. The last selected action can be considered as the output of learning phase, denoted as p.LPH.

DoTransfer(super-peer ***p***, super-peer *s*)

set *C* with *min*(capacity(*s*)-load(*s*), load(***p***))
Transfer *C* from ***p*** to ***s***
If load(***p***) = 0 and load(*s*) < capacity(*s*)
    p becomes client of *s*
Else
    set ***r*** with the highest capacity client of *s*
    If capacity(***r***) > capacity(***p***)
        swap role of ***r*** and ***p***

Fig. 1. DoTransfer method in SG-1

Learning phase for each peer *p*

For each $s \in p.\text{view}_{\text{underloaded}}$ **do**
    Peer ***p*** randomly chooses an action by ***p***.Role_LA
    If ***p***.Role_LA.*action* = client
      if $capacity(s) > capacity(p)$
        Update action probability vector by Equation (1)
      else if $capacity(s) < capacity(p)$
        Update action probability vector by Equation (2)
    Else if ***p***.Role_LA.*action* = super-peer
      if $capacity(s) > capacity(p)$
        Update action probability vector by Equation (2)
      else if $capacity(s) < capacity(p)$
        Update action probability vector by Equation (1)
Peer ***p*** randomly chooses an action by ***p***.Role_LA → ***p*.LPH**

Fig. 2. Learning phase for each peer *p*

Since under-loaded view of each peer is updated periodically, each peer performs learning phase in each round except for attached clients that perform this phase on demand. In other words learning phase is the starting step in our designed algorithms for unattached clients and super-peers but attached clients perform this phase if a request received. Since joining the system each peer periodically performs an algorithm, according to its current role. As shown in Fig. 3: unattached clients perform an algorithm shown in part (a) and super-peers act as part (b). Attached clients perform some trivial task that is discussed at the end of this section. At the rest, we introduce notations and give a detailed description of our algorithms. Notations used as follow: **SP**: super-peer, **USP**: under-loaded super-peer, **AC**: attached client, **UC**: unattached client, **super(s)**: super-peer of s, **p.LPH**: output of current learning phase running by peer p, and **p. HigherSet/p. LowerSet** consists of neighboring nodes whose capacity is higher/lower than that of p. There are some key rules applied in our algorithms:

1. Each UC or AC can promote to super-peer role if and only if its LPH is super.
2. Each USP is allowed to accept more load if and only if its LPH is super.
3. Each UC can become AC if and only if its LPH is client.
4. SG-1.Dotransfer(p,s) method can be performed if and only if p and s are SPs, s has higher capacity and s.LPH is super.
5. Each SP, p is allowed to swap role of its highest capacity client with one of lower capacity SPs in $p.\text{view}_{\text{under-loaded}}$ if and only if p.LPH is super.

According to part (a) of Fig. 3, in first step, each UC such as p performs its learning phase. According to rule (1), if p.LPH is super, p itself promotes to super-peer role. Otherwise p considers its HigherSet: In each iteration, p picks its highest capacity neighbor, s, and sends it a request message. If s is an USP or it is an UC, this message made s to use its own LPH. If s is an USP and rule (2) is obeyed by s, p becomes its client. If s is an UC and follows rule (1), firstly s promotes as SP then according to rule (2), p becomes its client. If s is an AC, at first p considers $super(s)$, instead of s. If $super(s)$ is an USP and obeys rule (2), p becomes its client. Otherwise, s utilizes its own LPH, if rule (1) is followed by s, firstly s promotes as super-peer then according to rule (2), p becomes its client. Above process continues until p becomes AC or no more peer exists in HigherSet of p.

According to part (b) of Fig. 3, at first step, each SP such as p performs its learning phase. Then p utilizes its LPH to check rule (2). If p.LPH is super, peer p is eligible to absorb more clients. So, if p is USP, it picks the lowest capacity neighbor x from p.LowerSet. If x is an UC, it uses its own LPH, if x follows rule (3), x is attached to p. Otherwise, if x is a SP, rule (4) is investigated to perform SG-1.DoTransfer(x,p) (as shown in Fig. 1). This loop continues until p becomes fully-loaded or no more peer exists in LowerSet of p. Then SP p, examines its view to check whether exists any super-peer m in its under-loaded view whose capacity is lower than that of p's highest capacity client, r, or not. If there is such super-peer, according to rule (5) swapping role is made between m and r.

Then as shown in part (b) of Fig. 3 (line 11), each SP such as p, regardless of its LPH, forms its HigherSet. In this way, SPs can compete with other higher capacity peers for absorbing clients. In each iteration, p picks its highest capacity neighbor, s.

If s is a SP, rule (4) is investigated to perform SG-1.DoTransfer(p,s). Otherwise If s is an UC, and follows rule (1) firstly s promotes as super-peer then rule (4) is investigated to perform SG-1.DoTransfer(p,s). Finally, If s is an AC, p considers $super(s)$ instead of s. Then rule (4) is investigated to perform SG-1.DoTransfer(p,$super(s)$). Selecting candidates in order to perform SG-1.DoTransfer method continues until SP p absorbed by other super-peer or when there is no more peer in its HigherSet.

	(a) Local: client peer ***p***, **Remote:** peer ***s***	**(b) Local:** super-peer ***p***, **Remote:** peers ***x***, ***s***
01	***p*** performs its learning phase according to Fig. 2	***p*** performs its learning phase according to Fig. 2
02	**If *p*. LPH** = super	**If *p*. LPH** = super
03	***p*** promotes to super-peer role	**While *p*.**LowerSet is not empty **and *p* is USP**
04	**Else**	pick ***x***, lowest capacity peer in ***p*.**LowerSet
05	**While *p*.**HigherSet is not empty **and** Found=false	**if *x* is UC and *x*. LPH** = client
06	pick ***s***, highest capacity peer in ***p***.HigherSet	***x*** becomes client of ***p***
07	**If *s* is USP** and ***s*. LPH** = super	**Else if *x*** is a **SP**
08	***p*** becomes client of ***s*** //Found = true	**SG-1.DoTransfer(*x*, *p*)**
09	**Else if *s*** is **AC**	check eligibility of its highest capacity client to swap
10	**If super(*s*)** is **USP** and **super(*s*).LPH** = super	**While *p*.**HigherSet is not empty **and *p* is SP**
11	***p*** becomes client of **super(*s*)**//Found = true	pick ***s***, highest capacity peer in ***p*.**HigherSet
12	**Else**	**If *s*** is a **SP and *s*. LPH** = super
13	checkAttached = true	**SG-1.DoTransfer(*p*, *s*)**
14	**If** checkAttached = true **or *s*** is an **UC**	**Else if *s*** is an **AC and super(*s*). LPH** = super
15	**If *s*. LPH** = super	**SG-1.DoTransfer(*p*, super(*s*))**
16	***s*** promotes to super-peer role	**Else if *s*** is **UC and *s*. LPH** = super
17	***p*** becomes client of ***s*** //Found = true	***s*** promotes to super-peer role
18	**End While**	**SG-1.DoTransfer(*p*, *s*)**

Fig. 3. Algorithm corresponding to (a): unattached clients, (b): super-peers

Every round, each AC such as p forms a candidate list including its neighboring nodes whose capacity is higher than that of $super(p)$. Then, p sends this list by candidate message to its SP. As a result, more opportunities are given to SPs, without imposing significant responsibility to ACs.

5 Performance Evaluation

In this section we have developed our protocol using PeerSim simulator [31] to evaluate efficiency of SG-LA in comparison with the well-known approach, SG-1. Two different distributions, power-law (with parameter á = 2) and uniform are applied to initialize capacity of nodes. In general, power-law distribution, maximum capacity of 500, partial view of size 30 and a network size of 10^5 is used as a default configuration, unless noted explicitly. Results are averaged over 25 experiments with a=b=0.1 as reward and penalty parameters. At the rest of this section, we have focused on the following major aspects: 1) Convergence time, 2) Capacity utilization, 3) Robustness in the face of super-peers' failure, 4) Scalability, and 5) Communication cost.

Let convergence time be the number of rounds needed to converge in an overlay with a stable number of super-peers that no more clients join or leave them. Fig. 4 shows number of selected super-peers as simulation proceeds, (for two different capacity distributions). According to Network column in part (a) of Table 1, when using power-law distribution only a small number of peers have a comparably high capacity but according to Network column in part (b), in uniform distribution capacity values assigned uniformly. Number of selected super-peers grouped by capacity values and number of rounds needed to find each subset are extracted from experiments of Fig. 4 and shown in Table 1. According to part (a) and part (b) of Table 1, in both capacity distributions SG-LA and SG-1 only select their super-peers from highest capacity subsets. But the number of rounds needed to find super-peers included in each subset, is comparably lower in SG-LA. In other words, high capacity peers in SG-LA learn about their eligibility faster than SG-1. Because in SG-LA each peer is able to learn its eligibility compared to other participating peers in adaptive manner. However, in SG-1 each peer joins as super-peer and role changes are dependent on capacity comparison with a randomly selected neighbor. According to part (a) of Table 1, when capacity parameter follows power-law distribution both of SG-1 and SG-LA obtain the same number of super-peers which is close to minimum possible. When capacity parameter follows uniform distribution, number of participating peers included in two first subsets is close to each other. In such situation, majority of second highest subset are demoted as clients by SG-1. But, in SG-LA higher number of peers from second highest subset, find that they are eligible with respect to current network condition. The behavior of SG-1 can be explained as a result of greedy manner of this protocol in utilizing capacity of super-peers.

Fig. 5 shows capacity utilization of SG-1 and SG-LA as simulation rounds increase, (for both capacity distributions).

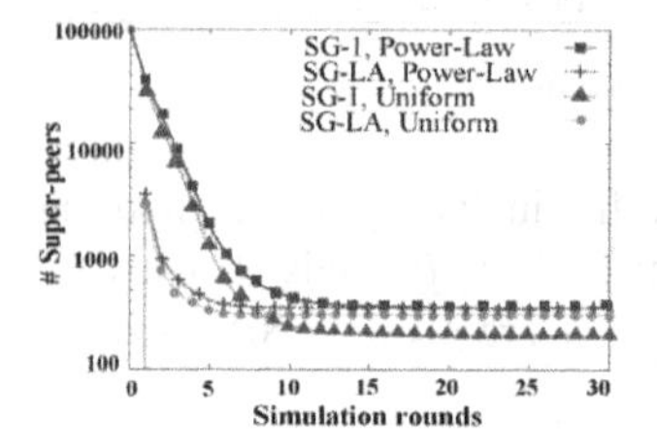

Fig. 4. Number of selected super-peers as simulation proceeds

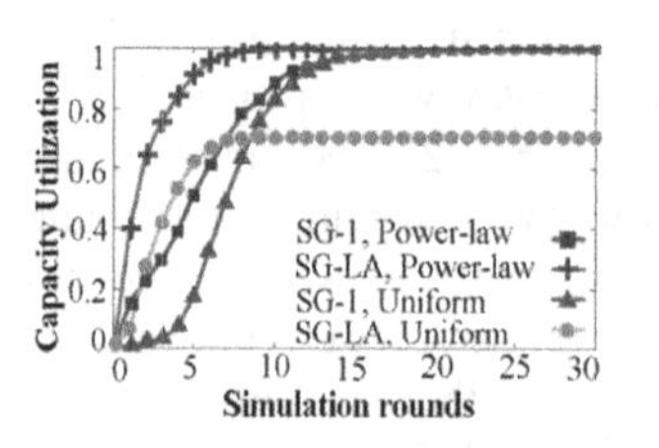

Fig. 5. Capacity utilization

Table 1. Number of selected super-peers grouped by capacity values and number of rounds needed to find each subset, SG-LA compared with SG-1. (a): Power-law, (b): Uniform.

(a)	Network	SG-LA		SG-1	
capacity	# peers	# supers	# rounds	# supers	# rounds
[401,500]	49	49	1	49	1
[301,400]	85	85	1	84	9
[201,300]	156	156	2	155	13
[101,200]	496	55	11	57	25
[1,100]	99214	0	-	0	-
(b)	**Network**	**SG-LA**		**SG-1**	
capacity	# peers	# supers	# rounds	# supers	# rounds
499	199	199	1	192	20
498	200	102	10	8	22
≤497	99601	0	-	0	-

At any given time capacity utilization (*CU*) can be defined by ratio of current number of attached clients to total capacity provided by super-peers, as given in (3) (let *S* denotes number of selected super-peers). According to Fig. 5, when power-law distribution is applied, SG-LA can reach 99% *CU* in only 9 rounds, but it takes SG-1 for 20 rounds.

$$CU = \#\ attached\ clients\ / \sum_{i=1}^{S} Capacity(i) \tag{3}$$

In SG-LA, each peer learns about its eligibility in its learning phase therefore high capacity peers are able to take super-peer role in first few rounds. But, SG-1 makes each decision by only one capacity comparison. Due to large number of role changes needed in SG-1, transferring clients between selected super-peers is more probable. As a result, it takes SG-1 a long time to utilize capacity of selected super-peers. In the case of uniform distribution, until round 9, SG-LA has higher *CU*, and finally it converges to 70% *CU*. Because according to part (b) of Table 1, in SG-LA higher number of peers find themselves eligible with respect to current network condition. So, as far as possible, SG-LA prevents high capacity peers from demotion. However, SG-1 utilizes capacity of super-peers in a greedy manner. So, until round 19, SG-1 tries to get 99% *CU*. In this way, some of high capacity peers are demoted by SG-1.

To evaluate robustness of our protocol in the face of super-peers' failure, Fig. 6 shows number of selected super-peers in a scenario in which 50% of super-peers removed at round 30, in the case of power-law distribution. The same scenario is evaluated with uniform distribution and *CU* of these experiments is shown in Fig. 7. Important results of these evaluations gathered in Table 2 and Table 3. According to Table 3 when using power-law distribution, in SG-LA after crash 50% of clients change into unattached statue. Then, all of them change into super-peer role, so in worst case SG-1 has experience of 16% *CU* (Fig. 7). Finally as shown in Table 2, SG-1 converges after 27 rounds. In the same situation, after crash, SG-LA starts with 17% unattached clients. Each of these clients performs its learning phase and makes a decision for its role according to the network situation after crash. As shown in Fig. 7, SG-LA reaches 99% utilization faster than SG-1 and in worst case never goes below 50% utilization. Finally, SG-LA converges in only 15 rounds (Table 2) and the minimum possible number of super-peers (546) is obtained.

According to Table 3, when uniform distribution is applied, on average SG-LA selects higher number of super-peers (281) compared with SG-1 (200). It should be noted that, according to Table 2, SG-LA selects these super-peers only from three highest capacity subsets (499,498,497) and selection from each subset is limited to scenarios in which other higher subsets have been selected completely. After crash in SG-LA, unattached clients start performing their learning phase. A high number of them in first 4 rounds find that they are eligible to be a super-peer, so SG-LA starts with low *CU* (Fig. 7). Then by repetition of learning phase and going into competition with other higher capacity peers for client transfer, in next 4 rounds, some of these super-peers demote and number of super-peers became stable. At the same time, *CU* increase and converges to 73% (Fig. 7). Due to larger number of high capacity peers in uniform distribution, this amount of *CU* is reasonable to prevent eligible super-peers from demotion. According to Table 3, this behavior of SG-LA resulted in lower client transfer compared with SG-1. According to Fig. 7, after crash, there is a sharp fall in *CU* of SG-1. Because all of peers promote as super-peer, except for small number of remaining attached clients. In next 18 rounds, SG-1 utilizes capacity of super-peers in a greedy manner which is resulted in higher number of client transfers compared with SG-LA (Table 3).

To evaluate scalability of our protocol, Table 4 shows number of rounds needed to converge, with both capacity distributions in different network sizes. SG-1 assigns each node the initial role of super-peer. So, by growth in the size of overlay, the number of role changes needed to obtain the target topology increases. On the other hand SG-1 dose not employ any appropriate mechanism for making decision about demotions and promotions. But, in SG-LA each peer joins as unattached client and takes one of super-peer or client role with respect to the action chosen in its learning phase. Due to adaptive manner of SG-LA, it is less affected by growth in network size. According to Table 4, smaller deviation in number of rounds needed to converge as the network size increases, shows scalability of SG-LA compared with SG-1.

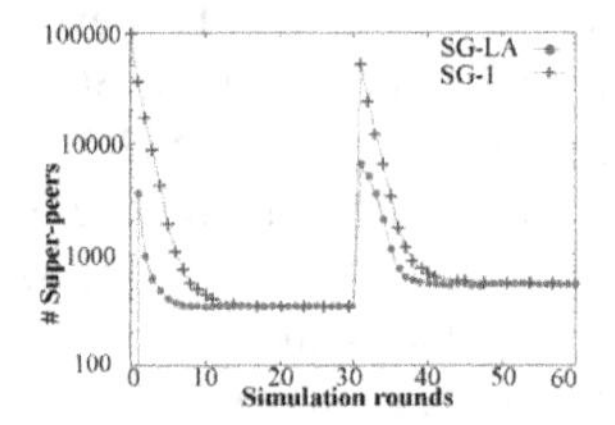

Fig. 6. Failure scenario, Power-law distribution

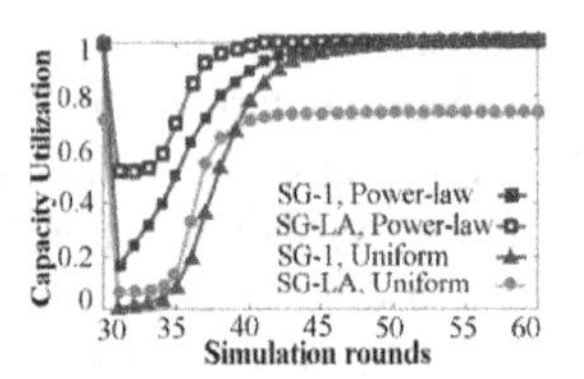

Fig. 7. Capacity utilization in failure scenario

Table 2. Percentage of selected super-peers grouped by capacity values and number of rounds needed to find each subset (after crah), SG-LA compared with SG-1. (a): Power-law, (b): Uniform

(a)	SG-LA		SG-1	
capacity	% supers	# rounds	% supers	# rounds
[401,500]	100	1	100	1
[301,400]	100	1	100	1
[201,300]	100	1	100	2
[101,200]	85	15	86	27
[1,100]	0	-	0	-
(b)	**SG-LA**		**SG-1**	
capacity	% supers	# rounds	% supers	# rounds
499	100	1	99.95	9
498	98.8	7	47.9	20
497	16	12	0	-
≤496	0	-	0	-

Table 3. Failure scenarios, SG-1 compared with SG-LA

	Unattached clients		# supers		# rounds		# Client transfers	
	SG-1	SG-LA	SG-1	SG-LA	SG-1	SG-LA	SG-1	SG-LA
Power-law	%52	%17	552	546	27	15	16.33	9.07
Uniform	%37	%21	200	281	22	13	16.59	4.5

Overhead includes number of messages sent to ask for information, candidate messages, and number of client transfers. According to Fig. 8, number of Request messages per peer in SG-LA and SG-1 is close to each other. Candidate message is dedicated to SG-LA and for each peer it is less than one. It is the only type of message sent by attached clients so it does not impose significant responsibility to them. Moreover, candidate messages resulted in less convergence time, because they provide more opportunities for super-peers as higher capacity candidates to compete in absorbing clients. Fig. 9 shows number of client transfers per peer in different network sizes. As shown in Fig. 9, even in a network with 10^5 peers, number of client transfers in SG-LA is less than SG-1. Because in SG-LA each peer makes its decisions based on the action which is selected in its learning phase but in SG-1 these decisions are made by comparing capacity with only one of randomly selected neighbors. On the other hand, in SG-LA only super-peers whose chosen action is "super", are allowed to accept more clients. As a result, client transfer in SG-LA is less probable. As shown in Fig. 8 and Fig. 9, total communication cost of SG-LA is less than that of SG-1.

Table 4. NS: Network Size, SD: Standard Deviation, (a): power-law, (b) uniform

	SG-1		SG-LA	
	(a)	(b)	(a)	(b)
NS=10^3	10	8	5	4
NS=10^4	18	13	9	7
NS=10^5	25	24	11	10
SD	7.5	8.1	3.05	3

Fig. 8. Overhead of transmitted messages

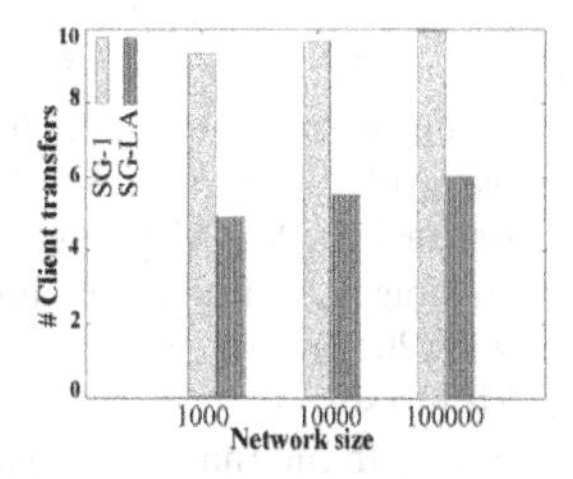

Fig. 9. Overhead of client transfer

6 Conclusion

This paper presents SG-LA, a learning automata based version of SG-1 protocol. We aim at minimizing number of super-peers according to network condition. For this purpose, each peer in the network makes use of its own learning automaton to be able to make decision about its eligibility. According to simulations, in both power-law and uniform capacity distributions, SG-LA can converge considerably faster compared with SG-1. Specifically in the case of power-law distribution, SG-LA can reach maximum capacity utilization faster than SG-1. And when capacity parameter follows uniform distribution, despite of SG-1, SG-LA prevents relatively high capacity peers from demotion, as far as possible. As shown in a catastrophic failure scenario, above results is not limited to normal situation. Furthermore, SG-LA is less affected by super-peer failure and it is able to recover faster. On the other hand, SG-LA reduces overhead of client transfers and shows better scalability compared with SG-1.

Acknowledgements. This work is supported by Iran Telecommunication Research Center (ITRC) grant.

References

1. Androutsellis-Theotokis, S., Spinellis, D.: A survey of peer-to-peer content distribution technologies. ACM Computing Surveys (CSUR) 36(4), 335–371 (2004)
2. Lua, E.K., Crowcroft, J., Pias, M., Sharma, R., Lim, S.: A survey and comparison of peer-to-peer overlay network schemes. IEEE Communications Surveys & Tutorials 7(2), 72–93 (2005)

3. Meshkova, E., Riihijärvi, J., Petrova, M., Mähönen, P.: A survey on resource discovery mechanisms, peer-to-peer and service discovery frameworks. Computer Networks 52(11), 2097–2128 (2008)
4. Yang, B., Garcia-Molina, H.: Designing a super-peer network. In: Proceedings of 19th International Conference on Data Engineering, pp. 49–60 (2003)
5. Gnutella, `http://rfc-gnutella.sourceforge.net`
6. Skype, `http://www.skype.com/en/`
7. Najim, K., Poznyak, A.S.: Learning automata: theory and applications. Pergamon Press, Inc. (1994)
8. Narendra, K.S., Thathachar, M.A.: Learning automata: an introduction. Printice-Hall Inc., Englewood Cliffs (1989)
9. Akbari Torkestani, J., Meybodi, M.R.: An intelligent backbone formation algorithm for wireless ad hoc networks based on distributed learning automata. Computer Networks 54(5), 826–843 (2010)
10. Esnaashari, M., Meybodi, M.R.: Deployment of a mobile wireless sensor network with k-coverage constraint: a cellular learning automata approach. Wireless Networks 19(5), 945–968 (2013)
11. Montresor, A.: A robust protocol for building superpeer overlay topologies. In: Proceeding of Fourth International Conference on Peer-to-Peer Computing, pp. 202–209 (2004)
12. Snyder, P.L., Greenstadt, R., Valetto, G.: Myconet: A fungi-inspired model for superpeer-based peer-to-peer overlay topologies. In: 3rd International Conference on Self-Adaptive and Self-Organizing Systems, pp. 40–50 (2009)
13. Liu, M., Harjula, E., Ylianttila, M.: An efficient selection algorithm for building a super-peer overlay. Journal of Internet Services and Applications 4(4), 1–12 (2013)
14. Singh, A., Haahr, M.: Creating an adaptive network of hubs using Schelling's model. Communications of the ACM 49(3), 69–73 (2006)
15. Gao, Z., Gu, Z., Wang, W.: SPSI: A hybrid super-node election method based on information theory. In: 14th International Conference on Advanced Communication Technology, pp. 1076–1081 (2012)
16. Liu, W., Yu, J., Song, J., Lan, X., Cao, B.: ERASP: an efficient and robust adaptive superpeer overlay network. In: Progress in WWW Research and Development, pp. 468–474 (2008)
17. Xiao, L., Zhuang, Z., Liu, Y.: Dynamic layer management in superpeer architectures. IEEE Transactions on Parallel and Distributed Systems 16(11), 1078–1091 (2005)
18. Sachez-Artigas, M., Garcia-Lopez, P., Skarmeta, A.F.G.: On the feasibility of dynamic superpeer ratio maintenance. In: International Conference on Peer-to-Peer Computing, pp. 333–342 (2008)
19. Chen, N., Hu, R., Zhu, Y., Wang, Z.: Constructing fixed-ratio superpeer-based overlay. In: 3rd International Conference on Computer Science and Information Technology, vol. 7, pp. 242–245 (2010)
20. Li, J.S., Chao, C.H.: An Efficient Superpeer Overlay Construction and Broadcasting Scheme Based on Perfect Difference Graph. IEEE Trans. Parallel Distrib. Syst. 21(5), 594–606 (2010)
21. Tan, Y., Lin, Y., Yu, J., Chen, Z.: k-PDG-Based Topology Construction and Maintenance Approaches for Querying in P2P Networks. J. Comput. Inf. Syst. 7(9), 3209–3218 (2011)
22. Teng, Y.H., Lin, C.N., Hwang, R.H.: SSNG: A Self-Similar Super-Peer Overlay Construction Scheme for Super Large-Scale P2P Systems. In: 17th International Conference on Parallel and Distributed Systems, pp. 782–787 (2011)
23. Sacha, J., Dowling, J., Cunningham, R., Meier, R.: Using aggregation for adaptive super-peer discovery on the gradient topology. In: Keller, A., Martin-Flatin, J.-P. (eds.) SelfMan 2006. LNCS, vol. 3996, pp. 73–86. Springer, Heidelberg (2006)

24. Lo, V., Zhou, D., Liu, Y., GauthierDickey, C., Li, J.: Scalable supernode selection in peer-to-peer overlay networks. In: Second International Workshop on Hot Topics in Peer-to-Peer Systems, HOT-P2P 2005, pp. 18–25 (2005)
25. Jesi, G., Montresor, A., Babaoglu, O.: Proximity-aware superpeer overlay topologies. IEEE Trans Network Serv. Manage 4(2), 78–83 (2007)
26. Dumitrescu, M., Andonie, R.: Clustering Superpeers in P2P Networks by Growing Neural Gas. In: 20th Euromicro International Conference on Parallel, Distributed and Network-Based Processing, pp. 311–318 (2012)
27. Wolf, S.: On the complexity of the uncapacitated single allocation p-hub median problem with equal weights. In: Internal Report 363/07, University of Kaiserslautern, Kaiserslautern, Germany (2007)
28. Wolf, S., Merz, P.: Evolutionary local search for the super-peer selection problem and the p-hub median problem. In: Bartz-Beielstein, T., Blesa Aguilera, M.J., Blum, C., Naujoks, B., Roli, A., Rudolph, G., Sampels, M. (eds.) HM 2007. LNCS, vol. 4771, pp. 1–15. Springer, Heidelberg (2007)
29. Chen, J., Wang, R.M., Li, L., Zhang, Z.H., Dong, X.S.: A Distributed Dynamic Super Peer Selection Method Based on Evolutionary Game for Heterogeneous P2P Streaming Systems. In: Mathematical Problems in Engineering (2013)
30. Jelasity, M., Kowalczyk, W., Van Steen, M.: Newscast computing. Technical Report IR-CS-006, Vrije Universiteit Amsterdam, Department of Computer Science, Amsterdam, The Netherlands (2003)
31. Jelasity, M., Montresor, A., Jesi, G.P., Voulgaris, S.: PeerSim P2P Simulator (2009), http://peersim.sourceforge.net/

A New Derivative of Midimew-Connected Mesh Network

Md. Rabiul Awal[1,*], M.M. Hafizur Rahman[1], Rizal Bin Mohd. Nor[1], Tengku Mohd. Bin Tengku Sembok[2], and Yasuyuki Miura[3]

[1] Department of Computer Science, KICT, IIUM, Malaysia
rabiulawal1@gmail.com, {hafizur,rizalmohdnor}@iium.edu.my
[2] Cyber Security Center, National Defense University Malaysia, Malaysia
tmtsembok@gmail.com
[3] Graduate School of Technology, Shonan Institute of Technology, Japan
miu@info.shonan-it.ac.jp

Abstract. In this paper, we present a derivative of Midimew connected Mesh Network (MMN) by reassigning the free links for higher level interconnection for the optimum performance of the MMN; called Derived MMN (DMMN). We present the architecture of DMMN, addressing of nodes, routing of message and evaluate the static network performance. It is shown that the proposed DMMN possesses several attractive features, including constant degree, small diameter, low cost, small average distance, moderate bisection width, and same fault tolerant performance than that of other conventional and hierarchical interconnection networks. With the same node degree, arc connectivity, bisection width, and wiring complexity, the average distance of the DMMN is lower than that of other networks.

Keywords: Massively Parallel Computers, Interconnection Network, DMMN, MMN, and Static Network Performance.

1 Introduction

Interconnection network is one of the crucial parameters for modern high performance computing. After the introduction of packet switching [1], it has become the "Performance determining factor" for massively parallel computers (MPC) [2] and dominates the performance of a computing system [3,4]. Current research suggests that MPCs of next decade will contain 10 to 100 millions of nodes [5] with computing capability at the tens of petaflops or exaflops level. With this huge amount of nodes conventional topologies for MPC possess a large diameter, hence completely infeasible for next generation MPCs. The hierarchical interconnection network (HIN) provides a cost-effective way in which several network topology can be integrated together [6]. Therefore, HIN is a plausible alternative way to interconnect the future MPC [6] systems. A variety of hypercube based HINs found in the literature, however, its huge number of physical

* Corresponding author.

S. Boonkrong et al. (eds.), *Recent Advances in Information and Communication Technology*, Advances in Intelligent Systems and Computing 265,
DOI: 10.1007/978-3-319-06538-0_20,

links make it difficult to implement. To alleviate this problem, several k-ary n-cube based HIN have been proposed [7,8]. Nevertheless, the performance of these networks does not yield any obvious choice of a network for MPC. No one is clear winner in all aspect of MPC design. As the performance improvement of an interconnection network is likely related to smaller diameter, the problem of designing interconnection network with low diameter with scalability of network size is still desirable [8,9]. A TESH network [10,11] is a k-ary k-cube HIN aiming for large-scale 3D MPC systems, consisting of multiple basic modules (BMs) which are 2D-mesh networks. The BMs are hierarchically interconnected by a 2D-torus network to build higher level networks. Additionally, MInimal DIstance MEsh with Wrap-around links (midimew) network is an optimal topology [12,13]. With this key motivation, to find a network which is suitable for interconnecting a large number of nodes while keeping small diameter, we have replaced the higher level 2D-torus of a TESH network by a 2D midimew network. To use the free ports in the periphery of the 2D-mesh network for higher level interconnection, we kept the basic module as 2D-mesh network same as TESH network. Hence the TESH network becomes Midimew-connected Mesh Network (MMN). The horizontal free ports of the BM are distributed for symmetric tori connection and vertical free links are used for diagonal wrap-around connection. This new HIN, thus allowing exploitation of computation locality, as well as providing scalability up to a million of nodes. In our previous research [15] we arranged the free links of Basic Module (BM) in a specific manner. To the thirst of more efficient way to interconnect the higher level networks for better performance, we derived the free links of BMs in a particular way. Hence we call it Derived MMN (DMMN).

The remainder of the paper is organized as follows. In Section 2, we present the basic architecture of the DMMN. Addressing of nodes and the routing of messages are discussed in Section 3 and Section 4, respectively. The static network performance of the DMMN is discussed in Section 5. Finally, in Section 6, we conclude this paper.

2 Architecture of the DMMN

Derived Midimew connected Mesh Network (DMMN) is a hierarchical interconnection network consisting of multiple basic modules (BM) that are hierarchically interconnected to form a higher level network. Basically the DMMN has two major parts of its architecture, the basic module (BM) and higher level networks. The BMs act as the basic building blocks of DMMN whereas higher level networks determines the construction of DMMN from BMs.

2.1 Basic Module of DMMN

Basic Module of DMMN is a 2D-mesh network of size $(2^m \times 2^m)$. BM consists of 2^{2m} processing elements (PE) with 2^m rows and 2^m columns, where m is a positive integer. Considering $m = 2$, a BM of size (4×4) is portrayed in Figure 1.

Each BM has $2^{(m+2)}$ free ports at the contours for higher level interconnection. The usability of free ports of DMMN is defined by the number of higher levels and denoted by q. All ports of the interior nodes are used for intra-BM connections. All free ports of the exterior nodes, either one or two, are used for inter-BM connections to form higher level networks. In this paper, BM refers to a Level-1 network.

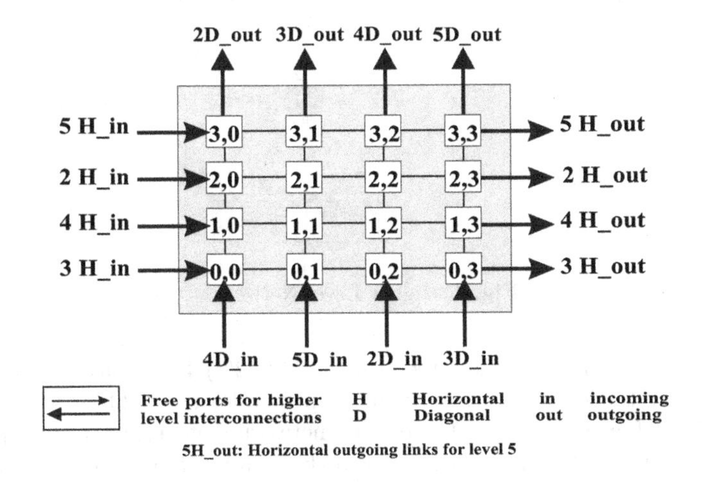

Fig. 1. Basic Module

2.2 Higher Level DMMN

Successive higher level networks are built by recursively interconnecting 2^{2m} immediate lower level subnetworks in a $(2^m \times 2^m)$ midimew network. In a midimew network, one direction (either horizontal or vertical) is symmetric tori connected and other direction is diagonally wrap-around connected. We have assigned the vertical free links of the BM for symmetric tori connection and horizontal free links are used for diagonal wrap-around connection. As portrayed in Figure 2, considering (m = 2) a Level-2 DMMN can be formed by interconnecting $2^{(2\times 2)} = 16$ BMs. Similarly, a Level-3 network can be formed by interconnecting 16 Level-2 sub-networks, and so on. Each BM is connected to its logically adjacent BMs. It is useful to note that for each higher level interconnection, a BM uses $4 \times (2^q) = 2^{q+2}$ of its free links, $2^{(2q)}$ free links for diagonal interconnections and $2^{(2q)}$ free links for horizontal interconnections. Here, $q \in \{0, 1, ..., m\}$,, is the inter-level connectivity. $q = 0$ leads to minimal interlevel connectivity, while $q = m$ leads to maximum interlevel connectivity. For example the (4×4) BM has $2^{(2\times 2)} = 16$ free ports as shown in Figure 1. If we chose $q = 0$, then $4 \times (2^0) = 4$ of the free ports and their associated links are used for each higher level interconnection, 2 for horizontal and 2 for diagonal interconnection. Among these 2 links, one is used for incoming link and another one for used for outgoing link, i.e., a single links is used for diagonal in, diagonal out, horizontal in, and horizontal out.

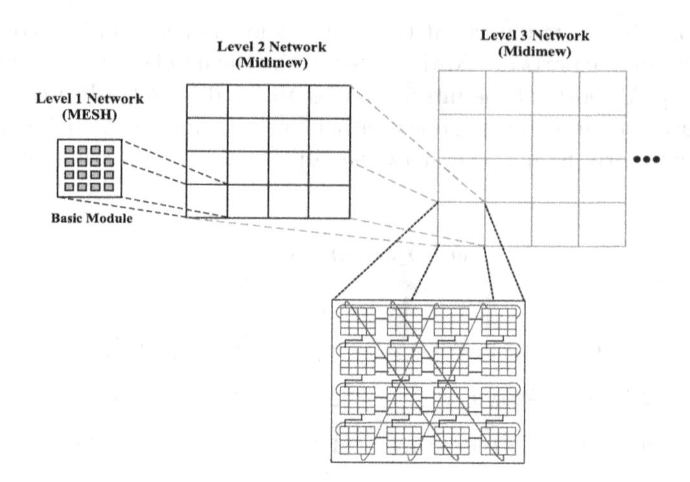

Fig. 2. Higher Level Network

A DMMN(m, L, q) is constructed using $(2^m \times 2^m)$ BMs, has L levels of hierarchy with inter-level connectivity q. In principle, m could be any positive integer value. However, if $m = 1$, then the network degenerates to a hypercube network. Hypercube is not a suitable network, because its node degree increases along with the increase of network size. If $m = 2$, then it is considered the most interesting case, because it has better granularity than the large BMs. If $m \geq 3$, the granularity of the family of networks is coarse. If $m = 3$, then the size of the BM becomes (8×8) with 64 nodes. Correspondingly, the Level-2 network would have 64 BMs. In this case, the total number of nodes in a Level-2 network is $N = 2^{2\times3\times2} = 4096$ nodes, and Level-3 network would have 262144 nodes. Clearly, the granularity of the family of networks is rather coarse.In the rest of this paper we consider m = 2, therefore, we focus on a class of DMMN(2,L,q) networks.

The highest level network which can be built from a $(2^m \times 2^m)$ BM is $L_{max} = 2^{m-q} + 1$ with $q = 0$ and $m = 2$, $L_{max} = 5$, Level-5 is the highest possible level. The total number of nodes in a network having $(2^m \times 2^m))$ BMs is $N = 2^{2mL}$. If the maximum hierarchy is applied then number of total nodes which could be connected by DMMN(m,L,q) is $N = 2^{2m(2^{m-q}+1)}$. For the case of (4×4) BM with $q = 0$, a DMMN network consists of over 1 million nodes.

3 Addressing of Nodes

Nodes in the BM are addressed by an address block, consisting of two digits, the first is representing the horizontal coordinate and the next is representing the vertical coordinate. The address of the nodes are expressed by the base-4 numbers. In case of higher levels, 1 address block is used for each level.

Again the blocks are consists of two digits with base-4 numbers. More generally, in a Level-L MMMN, the node address is represented by:

$$\begin{aligned} A &= A^L A^{L-1} A^{L-2} \ldots \ldots A^2 A^1 \\ &= a_{n-1}\, a_{n-2}\, a_{n-3}\, a_{n-4} \cdots \cdots a_3\, a_2\, a_1\, a_0 \\ &= a_{2L-1}\, a_{2L-2}\, a_{2L-3}\, a_{2L-4} \cdots \cdots a_3\, a_2\, a_1\, a_0 \\ &= (a_{2L-1}\, a_{2L-2})\, (a_{2L-3}\, a_{2L-4}) \ldots\ldots (a_3\, a_2)\, (a_1\, a_0) \end{aligned} \quad (1)$$

Here, the total number of digits is $n = 2L$, where L is the level number. A^L is the address of level L and $(a_{2L-1}a_{2L-2})$ is the co-ordinate position of Level-$(L-1)$ for Level-L network. Pairs of digits run from group number 1 for Level-1, i.e., the BM, to group number L for the L-th level. Specifically, l-th group $(a_{2l-1}a_{2l-2})$ indicates the location of a Level-$(l-1)$ subnetwork within the l-th group to which the node belongs; $1 \le l \le L$. In a two-level network the address becomes $A = (a_4a_3)(a_1a_0)$. The first pair of digits (a_4a_3) identifies the BM to which the node belongs, and the last pair of digits (a_1a_0) identifies the node within that BM.

The address of a node n^1 encompasses in BM_1 is represented as $n^1 = (a^1_{2L-1}\, a^1_{2L-2}\, a^1_{2L-3}\, a^1_{2L-4} \cdots \cdots a^1_3\, a^1_2\, a^1_1\, a^1_0)$. The address of a node n^2 encompasses in BM_2 is represented as $n^2 = (a^2_{2L-1}\, a^2_{2L-2}\, a^2_{2L-3}\, a^2_{2L-4} \cdots \cdots a^2_3\, a^2_2\, a^2_1\, a^2_0)$. In DMMN, the node n^1 in BM_1 and n^2 in BM_2 are connected by a link if the following condition is satisfied.

$$\begin{aligned} &\exists i\{a^1_i(a^2_i \pm 1) mod\, 2^m \wedge \forall j(j \neq i \rightarrow a^1_j = a^2_j)\} \\ &where\; i\%2 = 0, i, j \geq 2\,; \\ &\exists i\{a^1_i = (a^2_i \pm 1) \wedge \forall j(j \neq i \rightarrow a^1_j = a^2_j)\} \\ &where\; a^1_i = 2^m - 1, i\%2 = 1, i, j \geq 2\,; \\ &\exists i\{a^1_i = (a^2_i \pm 1) mod\, 2^m \wedge \forall j(j \neq i \rightarrow a^1_j = a^2_j + 2)\} \\ &where\; i\%2 = 1, i, j \geq 2 \end{aligned}$$

The assignment of inter-level ports for the higher level networks has been done quite carefully so as to minimize the higher level traffic through the BM. The address of a node n1 encompasses in BM1 is represented as . The address of a node n2 encompasses in BM2 is represented as . The node n1 in BM1 and n2 in BM2 are connected by a link if the following condition is satisfied.

4 Routing of DMMN

Routing of messages in the DMMN is simple, top to bottom fashion in order[10,11]. Routing of highest level is done first, the lower level routing at last. BM has outlet/inlet port for higher levels. When a particular transaction of packet is set up from a source to destination, first the shortest path is calculated. Based on the shortest path, outlet port for source and inlet port for destination are fixed. The packet uses the outlet port to reach at highest level

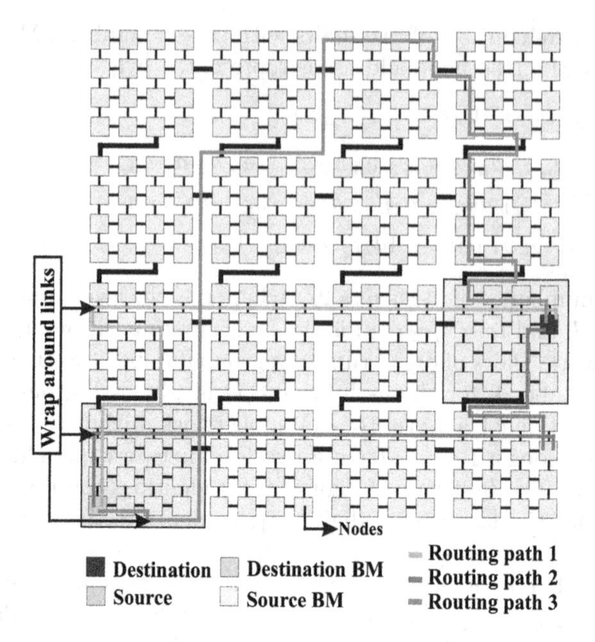

Fig. 3. Routing of DMMN

sub-destination and continue to move through the sub-network to lower level sub-destination until it reaches its final destination. Horizontal routing is performed first, once the packet matches the destination column then diagonal routing starts.

Figure 3 illustrates a routing between level-2 DMMN. Let us consider, a packet is to be routed from source node 0000 to destination node 1323. For this transaction, three routing path is shown. For routing path 1, first the packet moves to outlet node 0030 for level 2. Then it enters to node 1002 and the horizontal BM address is matched. Now the packet will move to the vertical direction using a wraparound link and match vertical BM address by entering destination node 1323. Hence the order of routing is followed by deterministic strategy. The other routing paths are followed by same manner. 9 hops, 11 hops and 22 hops are needed for the packet to reach destination through routing path 1, 2 and 3 respectively. Thus the routing path 1 is shortest and will be followed by packet for routing. Here we assumed that, all the links of DMMN are bidirectional full-duplex links.

5 Static Network Performance

Several topological properties and performance metrics of interconnection network are closely related to many technological and implementation issues. The static network performances do not reflect the actual performance but they have a direct impact on network performance. In this section we discuss about several performance metrics. For the performance evaluation, we have considered mesh, torus, TESH network, MMN and the proposed DMMN. Some performance metrics like diameter and average distance of MMN, DMMN and TESH were evaluated by simulation, the other metrics like Wiring Complexity, cost were evaluated by their corresponding equations.

5.1 Node Degree

Node degree is the maximum number of neighbor nodes are directly connected with a node. It refers to the number of links at a node. Constant node degree is preferable for networks. Network with constant degree is easy to expand. Also the cost is related to the node degree proportionally. For fair comparison, we have consider degree 4 network. It is shown in Table 1 that the degree of the mesh, torus, TESH, MMN, and DMMN are equal, it is 4 are independent of network size.

5.2 Diameter

Diameter refers to the maximum distance between any pair of source and destination. In other words the number of maximum links to cross for any transaction with a pair of nodes in a given network. Diameter indicates the locality of the network. Latency and message passing time depend on the diameter. Small diameter gives better locality to the network. Hence smaller diameter is convenient. We have evaluated the diameter of the TESH, MMN and DMMN network by simulation and mesh and torus network by their static formula and the results are presented in Table 1. Clearly, the DMMN has a much smaller diameter than that of TESH and mesh networks; equal to MMN and a slightly large diameter than that of torus networks.

5.3 Cost

Cost is one of the important parameter for evaluating an interconnection network. Node degree and diameter effect the performance metrics of the interconnection network including message traffic density, faulttolerance and inter-node distance. Therefore, the product ($diameter \times node\, degree$) is a good criterion to indicate the relationship between cost and performance of a parallel computer systems [16]. The cost of different networks is plotted in Table 1. The DMMN is less costly than mesh and TESH, same to MMN and a slightly higher than torus network.

5.4 Average Distance

The average distance is the average of all distinct paths in a network. Average distance reflects the ease of communication within the network i.e. average network latency. A small average distance results small communication latency.

Table 1. Comparison of Static Network Performance of Various Networks

Network	Node Degree	Diameter	Cost	Average Distance	Ark Connectivity	Bisection Width	Wiring Complexity
2D-Mesh	4	30	120	10.67	2	16	480
2D-Torus	4	16	64	8	4	32	512
TESH(2,2,0)	4	21	84	10.47	2	8	416
MMN(2,2,0)	4	17	68	9.07	2	8	416
DMMN(2,2,0)	4	17	68	8.56	2	8	416

In store and forward communication which is sensitive to the distance, small average distance tend to favor the network [17]. But it is also crucial for distance-insensitive routing, such as wormhole routing, since short distances imply the use of fewer links and buffers, and therefore less communication contention. We have evaluated the average distances for DMMN, MMN, and TESH network by simulation and mesh and torus networks by their corresponding formulas and the results are tabulated in Table 1. It is shown that the average distance of DMMN is lower than that of MMN, mesh and TESH networks, and slightly higher than that of torus networks.

5.5 Bisection Width

The Bisection Width (BW) refers to the minimum number of communication links that must be removed to partition or segment the network into two equal halves. Small BW impose low bandwidth between two parts. Nevertheless large BW requires lots of wires and is difficult for VLSI design. Hence moderate BW is highly desirable. BW is calculated by counting the number of links that must to be eliminated from Level-L DMMN. Table 1 is showing that, BW of the DMMN is exactly equal to that of the MMN and TESH network and lower than that of mesh and torus network.

5.6 Arc Connectivity

The arc connectivity of a network suggest the minimum number of arcs that must be removed from the network to break it into two disconnected networks. It measures the robustness of a network and the multiplicity of paths between nodes over the network. High arc connectivity improves performance during normal operation, and also improves fault tolerance. A network is maximally fault-tolerant if its connectivity is equal to the degree of the network. From Table 1 it is clear that for DMMN, MMN and TESH, the arc connectivity is exactly equal. Nonetheless arc connectivity of torus is equal to its degree, thus more fault tolerant than others.

5.7 Wiring Complexity

The wiring complexity of an interconnection network refers to the total number of links required to form the network. It has a direct correlation to hardware cost and complexity. A (16 × 16) 2D-mesh and 2D-torus networks have $\{N_x \times (N_y 1) + (N_x 1)Ny = 16 \times (161) + (161) \times 16 = 480\}$ and $(2 \times N_x \times N_y = 2 \times 16 \times 16) = 512$ links, respectively. Ni represents the number of nodes in the ith dimension. The wiring complexity of a Level-L DMMN, MMN, and TESH networks is $[\# \ of \ links \ in \ a \ BM \times k_{2(L1)} + \sum_{x=2}^{L} 2(2^q) \times k^{2(L1)i}]$. Considering, $m = 2$, a BM of DMMN, MMN, and TESH network have 24 links. Hence the total number of links of a Level-2 DMMN, MMN, and TESH are 416. Table 1 is showing that the total number of links of DMMN is lower than that of mesh and torus network and exactly equal to that of MMN and TESH network.

The static network performance is claiming that torus network is better than DMMN except in the term of wiring complexity. Now, torus network has $N_x + N_y$ long wrap-around links, where $N_x \times N_y$ is the network size. In case of DMMN, from Level-2 to Level-L, each level contains $(2^m/2) + (2^m/2)$ wrap-around links. Also the wrap-around links of DMMN do not increase with network size, instead they increase with higher levels. But in torus they increase with network size. Hence the implementation of DMMN is easier than torus.

6 Conclusion

A new derivative of MMN, called DMMN, is proposed for the future generation MPC systems. The architecture of the DMMN, addressing of nodes, and routing of message have been discussed in detail. We have evaluated the static network performance of the DMMN, as well as that of several other networks. From the static network performance, it has been shown that the DMMN possesses several attractive features, including constant node degree, small diameter, low cost, small average distance, and better bisection width. We have seen that with the same node degree, arc connectivity, bisection width, and wiring complexity, the average distance of the DMMN is lower than that of MMN, TESH, and mesh networks. Also, DMMN has slightly higher diameter and average distance to that of torus network. The DMMN yields better static network performance with low cost, which are indispensable for next generation MPC systems. This paper focused on the architectural structure and static network performance. Issues for future work include the following: (1) evaluation of static network performance considering different value of m; L; and q and (2) evaluation of dynamic communication performance using dimension order routing algorithm.

Acknowledgments. The authors are grateful to the anonymous reviewers for their constructive comments which helped to greatly improve the clarity of this paper. This work is partly supported by FRGS13-065-0306, Ministry of Education, Government of Malaysia.

References

1. Baran, P.: On distributed communications networks. IEEE Transactions on Communications Systems 12, 1–9 (1964)
2. Wu, C.L., Feng, T.Y.: Tutorial: Interconnection networks for parallel and distributed processing. IEEE Computer Society, Los Alamitos (1984)
3. Yang, Y., Funahashi, A., Jouraku, A., Nishi, H., Amano, H., Sueyoshi, T.: Recursive diagonal torus: an interconnection network for massively parallel computers. IEEE Transactions on Parallel and Distributed Systems 12, 701–715 (2001)
4. Rahman, M.H., Jiang, X., Masud, M.A., Horiguchi, S.: Network performance of pruned hierarchical torus network. In: 6th IFIP Int'l Conf. on Network and Parallel Computing, pp. 9–15 (2009)
5. Beckman, P.: Looking toward exascale computing. In: 9th Int'l Conf. on Parallel and Distributed Computing, Applications and Technologies, p. 3 (2008)
6. Abd-El-Barr, M., Al-Somani, T.F.: Topological properties of hierarchical interconnection networks: a review and comparison. J. Elec. and Comp. Engg. 1 (2011)
7. Lai, P.L., Hsu, H.C., Tsai, C.H., Stewart, I.A.: A class of hierarchical graphs as topologies for interconnection networks. J. Theoretical Computer Science, Elsevier 411, 2912–2924 (2010)
8. Liu, Y., Li, C., Han, J.: RTTM: a new hierarchical interconnection network for massively parallel computing. In: Zhang, W., Chen, Z., Douglas, C.C., Tong, W. (eds.) HPCA 2009. LNCS, vol. 5938, pp. 264–271. Springer, Heidelberg (2010)
9. Camarero, C., Martinez, C., Beivide, R.: L-networks: A topological model for regular two-dimensional interconnection networks. IEEE Transactions on Computers 62, 1362–1375 (2012)
10. Jain, V.K., Ghirmai, T., Horiguchi, S.: TESH: A new hierarchical interconnection network for massively parallel computing. IEICE Transactions on Information and Systems 80, 837–846 (1997)
11. Jain, V.K., Horiguchi, S.: VLSI considerations for TESH: A new hierarchical interconnection network for 3-D integration. IEEE Transactions on Very Large Scale Integration (VLSI) Systems 6, 346–353 (1998)
12. Beivide, R., Herrada, E., Balcazar, J.L., Arruabarrena, A.: Optimal distance networks of low degree for parallel computers. IEEE Transactions on Computers 40, 1109–1124 (1991)
13. Puente, V., Izu, C., Gregorio, J.A., Beivide, R., Prellezo, J., Vallejo, F.: Improving parallel system performance by changing the arrangement of the network links. In: Proceedings of the 14th Int'l Conf. on Supercomputing, pp. 44–53 (2000)
14. Lau, F.C., Chen, G.: Optimal layouts of midimew networks. IEEE Transactions on Parallel and Distributed Systems 7, 954–961 (1996)
15. Awal, M.R., Rahman, M.H., Akhand, M.A.H.: A New Hierarchical Interconnection Network for Future Generation Parallel Computer. In: 16th Int'l Conf. on Computers and Information Technology (2013)
16. Kumar, J.M., Patnaik, L.M.: Extended hypercube: A hierarchical interconnection network of hypercubes. IEEE Transactions on Parallel and Distributed Systems 3, 45–57 (1992)
17. Lonka, O., Naralchuk, A.: Comparison of interconnection networks. LUT (2008)

Modeling of Broadband over In-Door Power Line Network in Malaysia

Kashif Nisar*, Wan Rozaini Sheik Osman, and Abdallah M.M. Altrad

International Telecommunication Union ITU-UUM
Asia Pacific Centre of Excellence for Rural ICT Development,
School of Computing, College of Arts and Sciences,
Universiti Utara Malaysia, 06010 UUM Sintok, Malaysia
{kashif,rozai174}@uum.edu.my, abdaltrad85@gmail.com

Abstract. Malaysia is considered the eighth Asian country out of the top 15 countries in household broadband penetration at 34.5%. Users in rural areas who cannot receive Digital Subscriber Line (DSL) or cable modem services. In addition, owing to the high cost of Information and Communication Technology (ICT) infrastructure deployment in the rural areas, delay in broadband services is being experienced. Therefore, the Power Lines Communication (PLC) technology could have the potential to provide a broadband access through the entire electricity grid. Broadband PLC uses power lines as a high-speed digital transmission channel. This paper investigates low voltage Channel Transfer Function (CTF) of PLC technology for the purpose of data transmission regarding Malaysia electrical cable specification by using Matlab-Simulink simulation tool. Since that could help Malaysia licensable activities to provide Broadband PLC service out of harm's way of other data communication networks.

Keywords: Information and Communication Technology, Power Lines Communication, Low Voltage, Channel Transfer Function.

1 Introduction

There are four broadband services in Malaysia namely, traditional or fixed broadband, mobile broadband, wireless broadband, and satellite broadband. The services are operated mostly by Telekom Malaysia (TM [10]. According to the Malaysia Internet Usage Statistics and Marketing Report on 2010, the number of subscribers in Malaysia was 2.9 million in 2004, in 2005 it increased to 3.5 million, and in 2006 the number of subscribers was close to five million. This is an encouraging growing trend, and most of the Internet subscribers were eyeing for high-speed broadband infrastructure. Malaysia has achieved a household broadband penetration rate of almost 61.5% and the target is 75% by end 2015, spurred on by the National Broadband Initiative (NBI). In order to provide high transmission rates, a lot of broadband technology trails have been applied over time in several countries to

* Corresponding author.

S. Boonkrong et al. (eds.), *Recent Advances in Information and Communication Technology*, Advances in Intelligent Systems and Computing 265,
DOI: 10.1007/978-3-319-06538-0_21, © Springer International Publishing Switzerland 2014

produce new broadband technology that increase household Internet access [9]. Power line network is a possible infrastructure for Information and Communication Technology (ICT) services provision through Power Lines Communication (PLC) technology which is a valid medium for managing data communication system's Quality of Service (QoS). In addition, PLC network is able to achieve the quality of service for QoS requirements in data transmission [13].

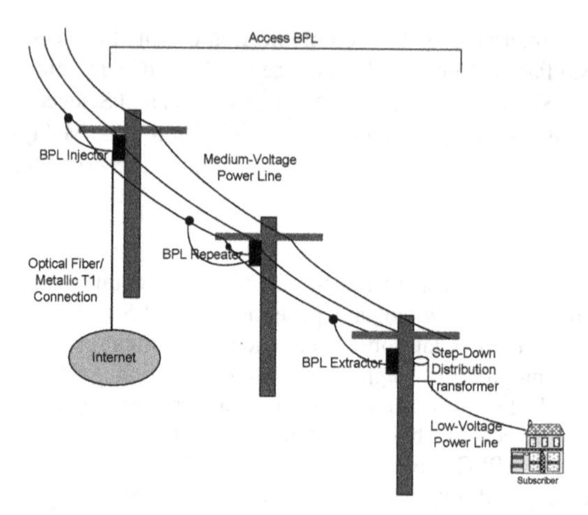

Fig. 1. General Broadband over Power Line Access Infrastructure

There are several successful installed projects that make the data transmission over power lines very possible, namely Broadband over Power Lines (BPL), PLC, Digital Power Line (DPL) or Power Lines Transmission (PLT). Broadband PLC works by transmitting high frequency signals through the same power cable used for carrying electricity to households [6] see Fig. 1.

PLC has been used for many decades, but a variety of new services and applications require more reliability and higher data rates [1]. However, the power line channel is very hostile. Channel characteristics and parameters vary with frequency, location, time and the type of equipment connected to it. The lower frequency regions from 10 kHz to 200 kHz are especially susceptible to interference. Recently, PLC has grown rapidly due to its potential of substantial cost savings, it is found clearly in several applications of data communication such as vehicle data communications, advanced control metering and home area network [4]. However, with the purpose of develop efficient Broadband PLC network, it is required to fully characterize the power line specifications. These studies are primarily focused on frequencies up to 30 MHz.

Unfortunately, the Broadband PLC technology still faces transmission channel modeling problems. Many efforts have been put in determining accurate channel models for the power line, but there is no widely accepted data transmission model yet [11]. Furthermore, several Broadband PLC projects have been developed but without an accurately transmission modeling due to the power network structure differences between countries, which affect the performance and characteristics of Broadband PLC technology, for reasons of different numbers of branches, power load impedances, and power line length [7], [12], and [14].

Channel transfer modeling which has been studied in several studies basically follows two approaches; a top-down and/or a bottom-up approach. The top-down approach considers the power line channel as a black box, and defines the multipath propagation through an echo model and obtains the results from real measurements, while the bottom-up approach starts with the theoretical analyzing of power line [2], [3] and [4]. In order to form channel transfer function which is unfortunately not standardized yet.

Channel coding or decoding techniques refer to error detection and correction which can be classified into two broad categories: Automatic Retransmission Query (ARQ) or Automatic Repeat Request and Forward Error Correction (FEC) [8]. In ARQ systems, errors due to the channel are controlled via error detection and retransmission. ARQ appends redundant bits to allow error detection at the receiver. FEC systems on the other hand, detect and correct symbol errors at the receiver without the need to request for symbol retransmission. In a Broadband PLC system where real-time high data rate transmission is required, ARQ techniques are not practical because in the presence of impulsive noise, large number of data packets need to be retransmitted which results in long delays and reduces both power and bandwidth efficiencies [5].

2 Malaysia Broadband over Power Line Regulations

With the improvements in telecommunications technology such as BPL, Malaysian Communications and Multimedia Commission (MCMC) identify the possible advantages that could affect not only corporate structures and businesses, but also to the individual user of this technology too. MCMC has a certainty where BPL has the potential of carrying affordable broadband access to businesses and households, and also brings broadband connectivity to the underserved rural communities (last-mile). The shortage of alternative telecommunications infrastructure has been a barrier to broadband penetration in this country and BPL is a potential alternative for rapid low cost access solution to bridge the digital divide.

MCMC had executed a consultation in order to forming rules toward BPL technology. Results of consultation have introduced a roll-out of BPL and policy must be taken into consideration when allowing BPL to be deployed in Malaysia under limited conditions, where it can be installed in households (residence PLC appliances). In other word, BPL is only in the home environment where is no licensing required. Second area can utilize BPL is private network facilities and or service. BPL is installed on the infrastructure of a corporation/business entity for its own consumption. The operation of broadband over power line system in Malaysia should be subjected to the following conditions:

2.1 Frequency Band Blocking and Power Adjustment

BPL system must have frequency notching, frequency band blocking and power adjustment. Which will enhance the service performance and reliability, and also to inhibit possible interferences to other frequencies users. MCMC focuses on electromagnetic compatibility; BPL system must work with surrounding equipment without effect their performance or causes any harm interference. The radiated emission limits for BPL installation should follow the Federal Communications Commission (FCC) Part 15 & 15.209 specified in Table 1.

Table 1. Radiated emission limits

Frequency Bandwidth	Radiated Emission Limits
1 to 30 MHz 9 KHz	30 μV/m(29.59 dBμV/m) at 30 meters
30 to 88 MHz 120 KHz	100 μV/m(40 dBμV/m)at 3 meters

3 LV Channel Transfer Function Results

There are two mathematical approaches in order to model any communication medium; first method follows a Top-down approach where the channel transmission model is derived from an analytical expression whose parameters are extracted from determine statistics as mention in previous section. Second approach is Bottom-up channel modeling, which exploits Transmission Line (TL) theory to derive the channel transmission model from the knowledge of the network topology, cable parameters and appliance impedance that are connected to the network.

This research focuses on the modeling PLC technology based on low voltage power cable specification in Malaysia specifically, on its channel transfer function using Matlab/Simulink software. The idea is to create the power line channel transfer function, with considering the LV scenario at a transmission band from 1 to 30 MHz. The transmission process depends on the power line networks components and its topology structure. Basically there are a few forms of topologies in network such as ring, radial and hybrid.

Power lines show a tough environment for high frequency communication signals. Noise, impedance and attenuation are there during transmission data with differences based on high variable and time, frequency and location. In order to overcome these channel impairment issues, a typical indoor power network with radial topology connected between a transmitter and receiver units shown in Fig. 2 which represents a simplified indoor power line channel. Port 1 is the transmitter where the signal is sent, and port 2 is the receiver where the signal strength is measured.

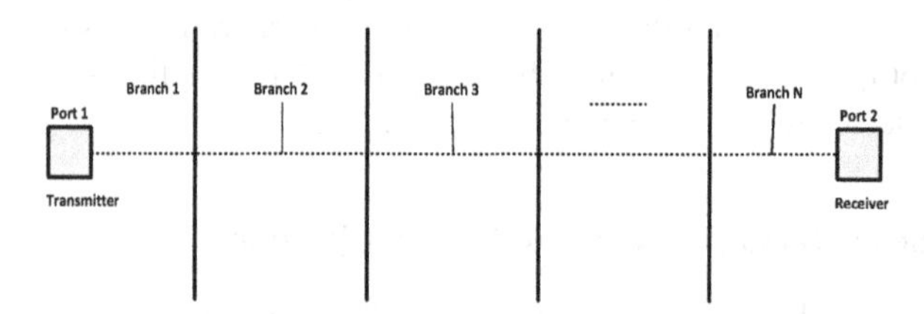

Fig. 2. Indoor Radial Power Line Channel

Fig.3 displays a two-wire transmission line is produced in Southern cable Manufacture, Kedah, Malaysia with its physical characteristics which could help determine the channel transfer function and the relative permittivity and permeability of the wires. Matlab/Simulink software assumes the relative permittivity and permeability are uniform. The resistance (R), inductance (L), conductance (G), and capacitance (C) per unit length (meters) are as follows:

$$R = \frac{1}{\pi a\, \delta_{cond}\, \sigma_{cond}} \tag{1}$$

$$L = \frac{\mu}{\pi} a\cosh\frac{d}{2a} \tag{2}$$

$$G = \frac{\pi \varpi \varepsilon''}{a\cosh\left[\frac{D}{2a}\right]} \tag{3}$$

$$C = \frac{\pi \varepsilon}{a\cosh\left[\frac{D}{2a}\right]} \tag{4}$$

Where:

σ_{cond} is conductivity in the conductor.
$\delta cond = \sqrt{1/\pi f\, \mu \sigma_{cond}}$ is the skin depth.
μ is the permeability of the dielectric.
ε is the permittivity of the dielectric.
ε'' is the imaginary part of ε, $\varepsilon'' = \varepsilon_0, \ \varepsilon_r \tan\delta$

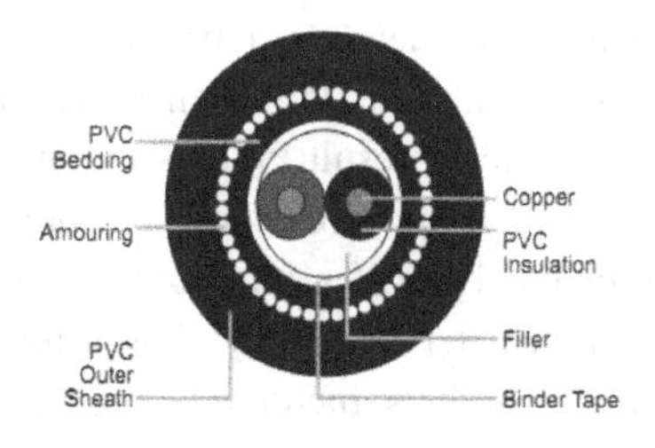

Fig. 3. Cross Sectional View of Two Wire Transmission Line

Table 2 depicts the parameters of the low voltage cable which is produced in Jalan Merbau Pulas, Kawasan Perindustrian Kuala Ketil Kuala Ketil, Baling, Kedah, Malaysia manufacture for simulation purpose:

Table 2. Power line specification parameters

Diameter (mm^2)	Radius (mm^2)	Conductivity (C^0)	Permeability(μ)	Permittivity (ε)	Line Length (m)	Separation Distance (m)
1.5	0.75	12.1 (12.1e-1)		0.8	5	
	(0.00075m)			0.17	10	1.2e-3
	(0.75e-4)			0.36	20	
2.5	1.25	7.41 (7.41e-1)		0.8	5	
	(0.00125m)			0.17	10	1.4e-3
	(0.12e-3)			0.36	30	
4	2 (0.002m)	4.61 (4.61e-1)		0.8	5	
	(0.2e-3)		1	0.17	10	1.6e-3
				0.36	30	
6	3 (0.003m)	3.8 (3.8e-1)		0.8	5	
	(0.3e-3)			0.17	10	1.6e-3
				0.36	30	

The frequency response of powerline communication system is presented by using magnitude versus frequency and phase versus frequency plots for representing a CTF. The transfer function phasor can then be simply defined at any frequency. A magnitude of an output is the magnitude of a phasor representation of the transfer function at a given frequency, multiplied by the magnitude of the input. The phase of the output is the phase of the transfer function added to the phase of the input. To get the transfer function whose frequency versus magnitude (dB), plot shows the attenuation in the signal strength and angle radian versus frequency plot gives the phase distortion or delay see depicted result Figures below which based on effect of varying line length. Fig. 4(a), 4(b) and 4(c) show the presence of deep notches at specific frequencies in the channel transfer function. To establish a communication between two access points, the carrier frequency chosen must not drop at these deep notches. Also, when the line length is increased, the number of those notches is increased proportionally.

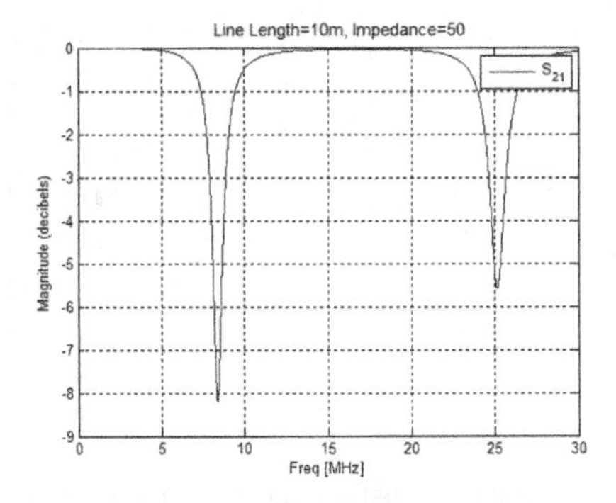

Fig. 4(a). Magnitude of Transfer Function (Line Length=10 m)

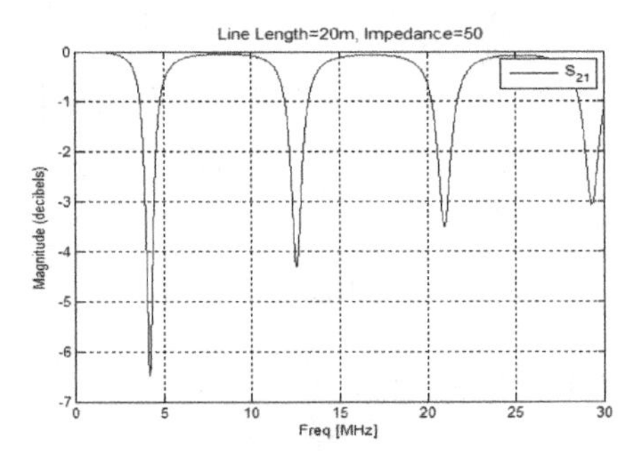

Fig. 4(b). Magnitude of Transfer Function (Line length= 20 m)

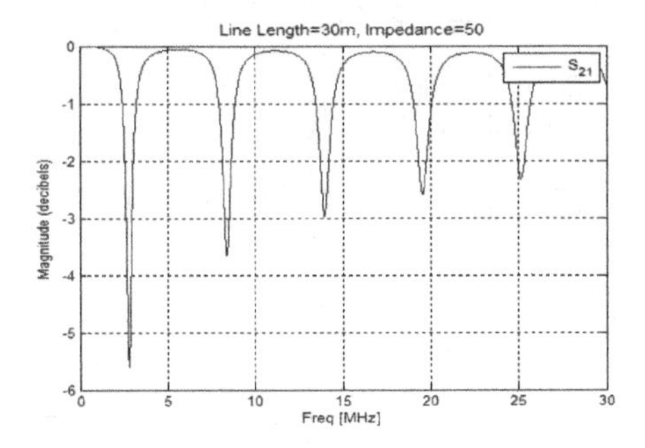

Fig. 4(c). Magnitude of Transfer Function (length= 30 m)

In addition, Fig. 5(a), 5(b) and 5(c) display that there are deep notches at certain frequencies in the channel transfer function for example there is a discontinuity in the phase characteristics leading to phase distortion or delay. Again, the number of such notches is proportional to the line lengths.

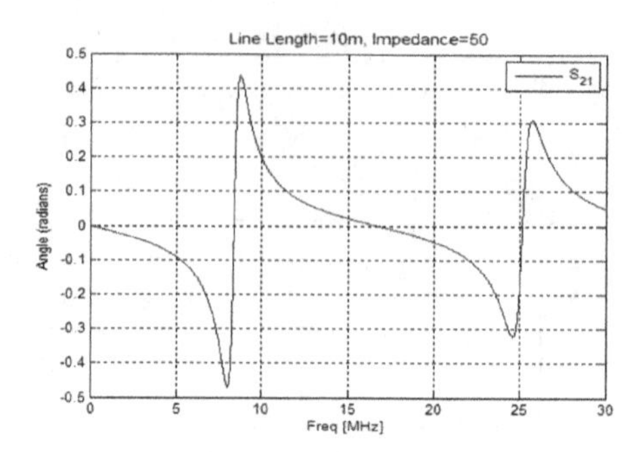

Fig. 5(a). Phase of Transfer Function (Line Length=10 m)

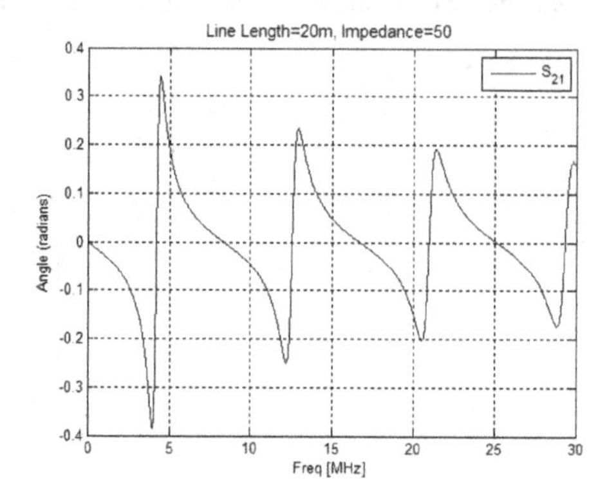

Fig. 5(b). Phase of Transfer Function (Line length= 20 m)

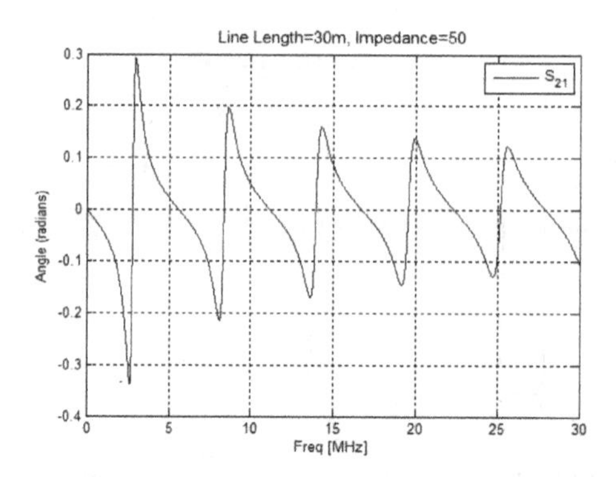

Fig. 5(c). Phase of Transfer Function (Line length= 30 m)

Generally, on the figures above we can see for example at 10 rad/sec with 10m cable length the phasor representation of the transfer function has a magnitude of 0.5 and a phase of 0.2°. This means that at 10 rad/sec the magnitude of the output will be 0.5 times the magnitude of the input and the output will lag the input by 0.2° and so on.

The basic application of PLC used to accessing telecommunication networks in order to finally analyse the advantages and disadvantages. From the economic viewpoint, it is very sensible to use a pre-installed power line grid instead of using new cables. It surely reduces a lot of money and time and so this is the biggest advantage of the technology. In several countries, PLC is becoming a reliable high speed source of Ethernet. In addition, PLC gratefully made it likely to benefit internet connections especially in last-mile and remote areas, PLC characteristics are quite different than the conventional wirings. Comparatively, it is a harsh medium whereas data transmission over it can generate a lot of harms. Continuous plugging and unplugging of electronic devices such as televisions, washing machines etc, makes power line prone to an unpredictable interference and noise on the transmission path.

4 Conclusion

The PLC technology still faces transmission channel modeling problems. Many efforts have been put in determining accurate channel models for the power line, but there is no widely accepted data transmission model yet. Furthermore, several projects have been developed but without an accurately transmission modeling due to the power network structure differences between countries, which affect the performance and characteristics of PLC technology. There are two main issues behind transmission channel modeling namely channel transfer function modeling and channel coding or decoding Techniques. Our result on the modeling of powerline for communication specifically, on its channel transfer function using Matlab/Simulink software, shows that there are deep notches appeared on injected frequency. In our future work, model will be implemented through more simulation based on testbed environment.

Acknowledgment. This work was carried out in the framework of the PLC project Ref: UUM/ITU/P-30 and financed by the International Telecommunication Union, Unversiti Utara Malaysia ITU-UUM Asia Pacific Centre of Excellence for Rural ICT Development, Universiti Utara Malaysia 06010 Sintok, Kedah, Malaysia. The authors would like to thanks Prof. Dr.-Ing. habil. Unger, Herwig, Fern University in Hagen, Germany for his insightful comments on final draft of this research paper.

References

1. Anatory, J., Theethayi, N., Thottappillil, R., Mvungi, N.H.: A Broadband Power-Line Communication System Design Scheme for Typical Tanzanian Low-Voltage Network. IEEE Transactions Power Delivery, 1218–1224 (2009)
2. Arora, S., Chandna, V.K., Thomas, M.S.: Modeling of Broadband Indoor Power Line Channel for Various Network Topologies. In: Innovative Smart Grid Technologies (ISGT) Conference, pp. 229–235. IEEE PES, India (2011)
3. Chandna, V.K., Zahida, M.: Effect of Varying Topologies on the Performance of Broadband over Power Line. IEEE Transactions Power Delivery, 2371–2375 (2010)

4. Chong, A.Y.-L., Darmawan, N., Ooi, K.-B., Lin, B.: Adoption of 3G Services Among Malaysian Consumers: an Empirical Analysis. Journal of Mobile Communication 8, 129–149 (2010)
5. Ghanim, T., Valenti, M.C.: The Throughput of Hybrid-ARQ in Block Fading Under Modulation Constraints. In: 40th Annual Conference of the Information Sciences and Systems, pp. 253–258. Princeton, New Jersey (2006)
6. Held, G.: Understanding Broadband over Power Line. Taylor and Francis Group, New York (2006)
7. Real world field test results homepluge technology, `http://www.homeplug.org/tech/whitepapers/devolo_presentation.pdf`
8. Hrasnica, H., Haidine, A., Lehnert, R.: Broadband Powerline Communications. In: Network Design. Wiley and Sons, West Sussex (2005)
9. Ibrahim, Z., Ainin, S.: The Influence of Malaysian Telecenters on Community Building. Electronic Journal of e-Governmen. 71, 77–86 (2009)
10. Jusoff, K., Hassan, Z., Razak, N.: Bridging the Digital Divide: an Analysis of the Training Program at Malaysian Telecenters. In: International Conference of Applied and Theoretical Mechanics WSEAS, pp. 15–23. Universiti Kebangsaan Malaysia Selangor, Malaysia (2010)
11. Kellerbauer, H., Hirsch, H.: Simulation of Powerline Communication with OMNeT++ and INET-Framework. In: IEEE International Symposium on the Power Line Communications and its Applications (ISPLC), pp. 213–217. IEEE Symp, Udine (2011)
12. Li, H., Sun, Y., Yao, Y.: The Indoor Power Line Channel Model Based on Two-Port Network Theory. In: 9th International Conference on the Signal Processing, Beijing, China, pp. 132–135 (2008)
13. Mlynek, P., Koutny, M., Misurec, J.: Power Line Modelling for Creating PLC Communication System. International Journal of Communications 1, 13–21 (2010)
14. Philipps, H.: Development of a Statistical Model for Powerline Communication Channels. In: 4th International Symposium on the Power-Line Communications and its Applications (ISPLC), Glasgow, Scotland, pp. 153–160 (2000)

Improving Performance of Decision Trees for Recommendation Systems by Features Grouping Method

Supachanun Wanapu[1], Chun Che Fung[2], Jesada Kajornrit[2], Suphakit Niwattanakula[1], and Nisachol Chamnongsria[1]

[1] School of Information Technology, Suranaree University of Technology, Nakhonratchasima 30000, Thailand
[2] School of Engineering and Information Technology, Murdoch University, South Street, Murdoch, WA 6150, Australia
supachanun@g.sut.ac.th, L.Fung@murdoch.edu.au, j_kajornrit@hotmail.com, {suphakit,nisachol}@sut.ac.th

Abstract. Recently, recommendation systems have become an important tool to support and improve decision making for educational purposes. However, developing recommendation systems is far from trivial and there are specific issues associated with individual problems. Low-correlated input features is a problem that influences the overall accuracy of decision tree models. Weak relationship between input features can cause decision trees work inefficiently. This paper reports the use of features grouping method to improve the classification accuracy of decision trees. Such method groups related input features together based on their ontologies. The new inherited features are then used instead as new features to the decision trees. The proposed method was tested with five decision tree models. The dataset used in this study were collected from schools in *Nakhonratchasima* province, Thailand. The experimental results indicated that the proposed method can improve the classification accuracy of all decision tree models. Furthermore, such method can significantly decrease the computational time in the training period.

Keywords: Recommendation Systems, Decision Trees, Features Grouping, Learning Object, Ontologies.

1 Introduction

Recently, recommendation systems have gained much attraction in their ability to support and improve decision making for educational purposes. In developing countries such as Thailand, recommendation systems can assist academic staff in their decisions to determine appropriate approaches for their delivery and enhancement of their teaching. One of the applications of academic recommendation systems is to discover the relationship between appropriate Learning Objects (LOs) and learner characteristics of each student [1-4]. Decision Trees (DTs) are one of the popular techniques to handle such task [4, 5]. Among several classification models, DTs are transparent [5-7] and explainable, also easy for deployment.

S. Boonkrong et al. (eds.), *Recent Advances in Information and Communication Technology*, Advances in Intelligent Systems and Computing 265,
DOI: 10.1007/978-3-319-06538-0_22,

However, efficiency of DT is subject to the characteristics of the input features applied to the model. In some cases, if the relationships between input features are weak, DTs may provide poor classification accuracy [8, 9]. This problem possibly exists in the datasets that have a few input features, of course as in the case of the dataset used in this study. To address this problem, this study introduces a features grouping method to group related input features together before feeding to a DT model. The hypothesis of this study is that if input features are grouped in an appropriate way, the classification accuracy of DTs could be increased while training time should be decreased.

This paper is part of a study on the development of a framework for the management of Learning objects in alignment with individual learner styles. This is a report on some of the preliminary progress. The rest of the paper is organized as follows. Section 2 provides the background on the dataset used in this study. Section 3 presents the proposed grouping method. Section 4 shows the experimental results and Section 5 provides discussion on the progress. The conclusion is finally presented in Section 6.

2 Datasets and Input Features

The dataset used in this study were collected through surveys from high school students in *Nakhonratchasima* province, Thailand. This study selected this targeted location because it is the second most populated province in Thailand. Questionnaires were used in the collection process. The items in the questionnaires consist of three parts as follows.

- *Part A* (General Information) collects general background information of the students. The features in this part are *Gender*, *Grade Level* and *Science Favorite.* The type of these features is binary.
- *Part B* (Leaning Styles) relates to the learning style and characteristics of the students. This part has one feature with three possible choices, that is, *Participant* (P), *Collaborative* (C) and *Independent* (I).
- *Part C* (Learning Objects) contains the characteristics of the learning objects that are preferred by the learners. The type of feature is binary set (Y = preferred, N = not preferred) and the number of the features is nineteen, $\{LOs\_C_1, LOs\_C_2,\ldots, LOs\_C_{19}\}$. They are, *Video*, *Animation*, *Graphic*, *Simulation*, *Overview*, *Chapter*, *Summary*, *Fact*, *Process*, *Procedure*, *Demo*, *Example*, *Exercise*, *Description*, *Time*, *Schedule*, *Sound*, *Place* and *Assignment* respectively.

The features in Part B are adopted from part of the Grasha-Riechmann Learning Style Scale (GRSLSS) [10]. These criteria were selected because they were especially designed to be used with high school students [11] and they were most often used by previous studies for the classification of learning styles in Thailand [12, 13]. In Part C, the features adopted were: (i) multimedia type features from metadata standards

[14, 15], (ii) content object features from ALOCOM ontology [16-18] and (iii) the environment features involved with the learning procedures from Dunn and Dunn's learning styles model [19]. Figure 1 shows an example of the dataset.

ID	*Learning Style*	*Gender*	*Grade Level*	*Science Favorite*	**Learning Objects Characteristics**				
					LOs_C_1 Video	LOs_C_2 Animation	LOs_C_3 Graphic	...	LOs_C_{19} Assignment
1	I	F	S	M	Y	N	Y	...	Y
2	C	M	H	F	N	N	N	...	N
3	P	F	H	F	Y	N	Y	...	N
...	...	...	...	...	...	...	...	...	...
1,290	P	M	S	F	Y	N	Y	...	Y
No. of classes (C)	$C_{D1}=3$	$C_{D2}=2$	$C_{D3}=2$	$C_{D4}=2$	$C_{C1}=2$	$C_{C2}=2$	$C_{C3}=2$		$C_{C19}=2$

Fig. 1. Matrix of dataset before grouping

3 Definition of Learning Objects and Grouping Procedure

Learning Objects (LOs), in general, are referred as tools that assist academic staff to support or enhance students' leaning, such as power point presentation. Wiley [20] attempted to define LOs as "*any digital resource that can be reused to support learning*". In 2002, the definition of LOs by the IEEE is given as "*any entity, digital or non-digital, that may be used for learning, education or training*". Since then, LOs characteristics have been defined in many ways and approaches for practical implementation. Several educational metadata standards can be found at IEEE LOM [15], Dublin Core [14], IMS consortium [21], and SCORM [22].

At the same time, development of LOs based on ontologies and web standards such as XML, RDF, OWL, etc. have also been carried out for the management of learning environments. This ontology-based definition is an appropriate approach to clearly characterize the relationships of LOs. This definition is adopted in this study to group the features in Part C of the dataset. According to Figure 1, nineteen LOs features can be grouped into finer groups based on ontology-based definition as follows and shown in Figure 2.

- Group 1 (LOs_G_1) is multimedia type {Video, Animation, Graphic, Simulation}.
- Group 2 (LOs_G_2) is narrative content type {Overview, Chapter, Summary}.
- Group 3 (LOs_G_3) is cognitive content type {Fact, Process, Procedure, Demo}.
- Group 4 (LOs_G_4) is supporting content type {Example, Exercise, Description}.
- Group 5 (LOs_G_5) is environment content type {Time, Schedule, Sound, Place, Assignment}.

ID	*Learning Style*	*Gender*	*Grade Level*	*Science Favorite*	**Group of Learning Object Characteristics**				
					LOs_G_1	LOs_G_2	LOs_G_3	LOs_G_4	LOs_G_5
1	I	F	S	M	YYYN	NNN	YYYY	NYY	YYYNY
2	C	M	H	F	NNNN	NYN	YYYN	NNN	NYNNY
3	P	F	H	F	NYYN	NNY	NNNN	YYY	NNNNN
...	...	...	...	...	...	...	...	...	...
1,290	P	M	S	F	NNYY	YYY	NYNN	YNY	YYYYY
No. of classes (C)	$C_{D1}=3$	$C_{D2}=2$	$C_{D3}=2$	$C_{D4}=2$	$C_{G1}=16$	$C_{G2}=8$	$C_{G3}=16$	$C_{G4}=8$	$C_{G5}=32$

Fig. 2. Matrix of dataset after grouping

Group 1 is a group of features defined in metadata standard [14, 15]. Group 2 to 4 are the features grouped by ALOCOM ontology [16-18]. Group 5 is a group of features defined in Dunn and Dunn's learning styles model [19].

Once the features have been grouped, encoding for the new features has to be carried out. In Group 1, the number of binary features is four; the number of possible combination is sixteen or 2^4. Then new possible values assigned to this feature is a set of sixteen {YYYY, YYYN, YYNN,..., NNNN}. In the same manner, this binary encoding is applied to the other four groups. Therefore, the numbers of members in the sets of the other four groups are eight, sixteen, eight and thirty-two respectively. Therefore, the size of input vector to DTs is reduced from nineteen to five.

4 Experiment and Results

In the experiment, the dataset of 1,290 records were randomly divided into two sets: 60 percent of dataset (774 records) were used as training set and the remaining 40 percent (516 records) were used as the testing set for validation purpose. Five DT algorithms have been used in this experiment, namely, Random Tree, J48 Tree, NB Tree, BF Tree and Simple Cart Tree.

To evaluate the proposed grouping methods, DTs were training with two input types, before grouping (19 features) and after grouping (5 features). Output of the DTs was Learning Style (data in part B). In other words, the number of classes is three (*Participant*, *Collaborative* and *Independent*). Table 1 and Table 2 show the results of this experiment.

Table 1. Accurcy of DTs before grouping (Learning Style)

Model	Accuracy (%)			
	Independent	Participant	Collaborative	Total
Random Tree	38.93 %	54.88 %	50.59 %	**49.42 %**
J48 Tree	41.22 %	56.28 %	41.76 %	**47.67 %**
NB Tree	29.01 %	63.72 %	34.12 %	**45.16 %**
BF Tree	32.06 %	48.84 %	41.76 %	**42.25 %**
Simple Tree	34.35 %	51.63 %	35.88 %	**42.05 %**

Table 2. Accuracy of DTs after grouping (Learning Style)

Model	Accuracy (%)			
	Independent	Participant	Collaborative	Total
Random Tree	53.44 %	66.05 %	44.71 %	**55.81 %**
J48 Tree	42.75 %	66.98 %	41.18 %	**52.33 %**
NB Tree	41.22 %	72.56 %	51.76 %	**57.75 %**
BF Tree	34.35 %	65.58 %	32.35 %	**46.71 %**
Simple Tree	34.35 %	62.79 %	40.59 %	**48.26 %**

According to the results as shown in the tables above, it can be seen that the total accuracy of all five DTs have been improved. The accuracy of Random tree improved 6.40 percent, J48 tree improved 4.65 percent, NB tree improved 12.60 percent, BF tree improved 4.46 percent and Simple tree improved 6.20 percent. Overall, the average improvement based on all of the five DTs was 6.86 percent.

This implies that the grouping method is able to represent the inner-relationship between the input features before they were applied to the DTs, and this approach has caused the accuracies of all five DTs to improve. Furthermore, it seems that such grouping method is generally applicable and they have improved results from all the DT algorithms.

With the positive results, this study then expanded the proposed technique to other types of outputs. The expanded experiments were conducted on the three features in Part A, namely, *Gender*, *Grade Level* and *Science Favorite*. The experimental results are showed in Table 3 to Table 8.

Table 3. Accurcy of DTs before grouping (Gender)

Models	**Accuracy (%)**		
	Male	**Female**	**Total**
Random Tree	61.54 %	49.57 %	**56.20 %**
J48 Tree	65.38 %	45.65 %	**56.59 %**
NB Tree	70.98 %	54.78 %	**63.76 %**
BF Tree	76.22 %	35.65 %	**58.14 %**
Simple Tree	69.58 %	44.78 %	**58.53 %**

Table 4. Accurcy of DTs after grouping (Gender)

Models	**Accuracy (%)**		
	Male	**Female**	**Total**
Random Tree	76.92 %	57.83 %	**68.41 %**
J48 Tree	77.62 %	56.09 %	**68.02 %**
NB Tree	76.57 %	58.26 %	**68.41 %**
BF Tree	73.08 %	45.65 %	**60.85 %**
Simple Tree	65.73 %	54.35 %	**60.66 %**

Table 5. Accurcy of DTs before grouping (Grade Level)

Models	**Accuracy (%)**		
	High school	**Second school**	**Total**
Random Tree	70.07 %	48.35 %	**59.88 %**
J48 Tree	68.25 %	60.74 %	**64.73 %**
NB Tree	64.96 %	58.26 %	**61.82 %**
BF Tree	72.26 %	42.98 %	**58.53 %**
Simple Tree	64.23 %	59.92 %	**62.21 %**

Table 6. Accurcy of DTs after grouping (Grade Level)

Models	Accuracy (%)		
	High school	**Second school**	**Total**
Random Tree	66.79 %	61.16 %	**64.15 %**
J48 Tree	74.09 %	59.92 %	**67.44 %**
NB Tree	74.82 %	61.57 %	**68.60 %**
BF Tree	68.61 %	57.85 %	**63.57 %**
Simple Tree	64.96 %	60.74 %	**62.98 %**

Table 7. Accurcy of DTs before grouping (Favored Science)

Models	Accuracy (%)		
	Favorite	**Merely**	**Total**
Random Tree	62.73 %	53.06 %	**58.14 %**
J48 Tree	59.41 %	52.24 %	**56.01 %**
NB Tree	59.41 %	62.04 %	**60.66 %**
BF Tree	54.24 %	57.55 %	**55.81 %**
Simple Tree	53.51 %	60.41 %	**56.78 %**

Table 8. Accurcy of DTs after grouping (Favored Science)

Models	Accuracy (%)		
	Favorite	**Merely**	**Total**
Random Tree	69.00 %	64.08 %	**66.67 %**
J48 Tree	63.84 %	57.96 %	**61.05 %**
NB Tree	61.25 %	72.65 %	**66.67 %**
BF Tree	59.04 %	57.96 %	**58.53 %**
Simple Tree	56.83 %	58.37 %	**57.56 %**

In case of *Gender* output, the accuracy of Random tree improved 12.21 percent, J48 tree improved 11.43 percent, NB tree improved 4.65 percent, BF tree improved 2.71 percent and Sample tree improved 2.13 percent. Overall, the average improvement based on five DTs was 6.63 percent.

In case of *Grade level*, the improved accuracy of DTs in the same order are 4.26, 2.71, 6.78, 5.04, and 0.78 percent respectively. Overall, the average improvement based on five DTs was 3.91 percent. And finally in case of *Favored science*, the improved accuracy of DTs in the same order are 8.53, 5.04, 6.01, 2.71, and 0.78 percent respectively. Overall, the average improvement based on five DTs was 4.61 percent.

In term of training time, the grouping method satisfactory reduced the training time of all five DTs. Since the dimension of input vector was reduced, the training time was decreased, at least, in the case of this dataset. Figure 3 shows these outcomes. In addition, the experiment was conducted on a computing platform with an Intel Core i7 - 2.0 GH CPU, running under an OSX 10.9 operating system.

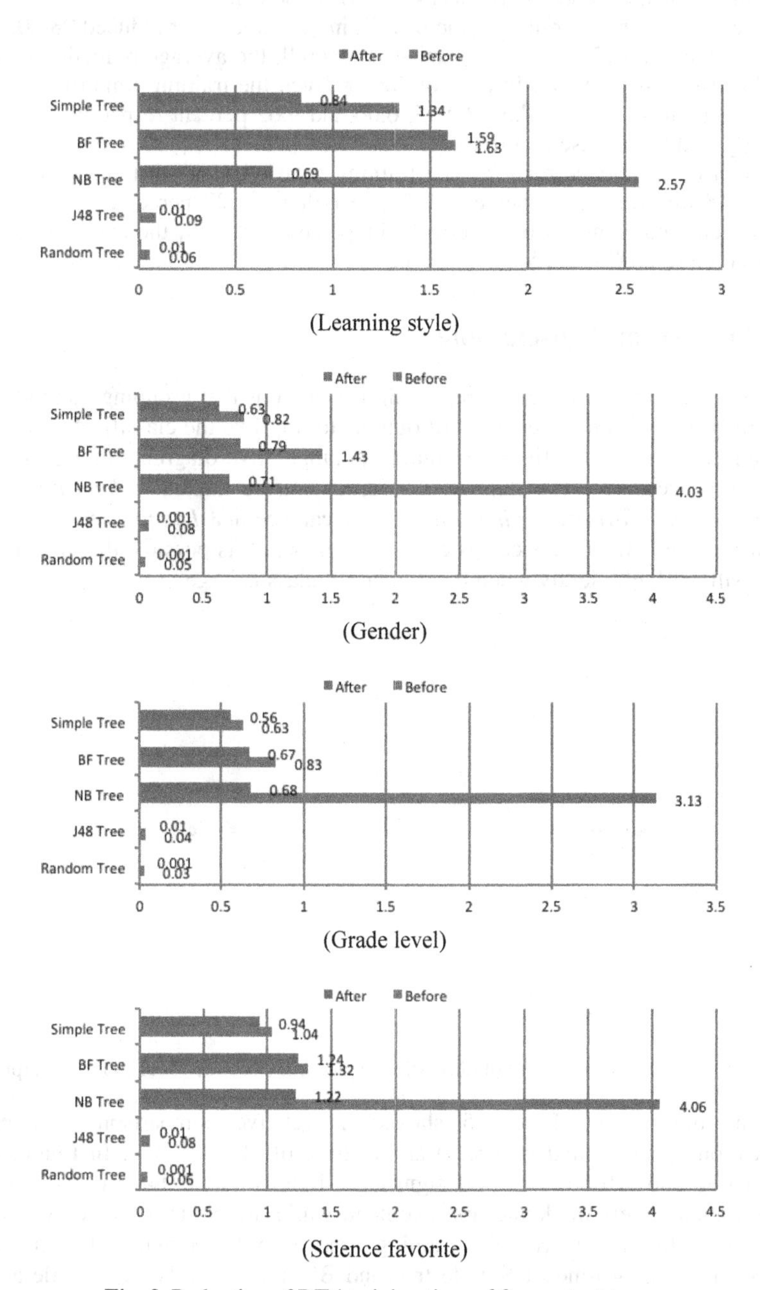

Fig. 3. Reduction of DTs' training time of four output types

For detailed results and in the case of *Learning style*, the training time of Random tree reduced 83.88 percent, J48 tree reduced 88.89 percent, NB tree reduced 73.15 percent, BF tree reduced 2.45 percent and Simple tree reduced 37.31 percent. Overall, the average reduction based on five DTs was 57.03 percent.

In case of *Gender*, the training time of DTs in the same order reduced 98.00, 98.75, 82.38, 44.76 and 23.17 percent respectively. Overall, the average reduction based on five DTs was 69.41 percent. In case of *Grade level*, the training time of DTs in the same order reduced 98.33, 87.50, 69.95, 6.06 and 9.62 percent respectively. Overall, the average reduction based on five DTs was 54.29 percent.

In the case of *Science favorite*, the training time of Random tree reduced 96.67 percent, J48 tree reduce 75 percent, NB tree reduce 78.27 percent, BF tree reduce 19.28 percent and Simple tree reduce 11.11 percent. Overall, the average reduction time based on five DTs was 56.07 percent.

5 Analysis and Discussions

Based on the experimental results reported, it is obvious that grouping method can be successfully extended to other types of outputs to improve the classification accuracy while reduced the training time. In Figure 4, comparative diagram among four types of outputs is presented. The improved accuracy can be ranked as *Learning style* > *Gender* > *Science favorite* > *Grade level*. One can see that *Leaning style* consists of three classes while the others consisted of two classes. It is possible that the grouping method will work efficiently when the number of class is large.

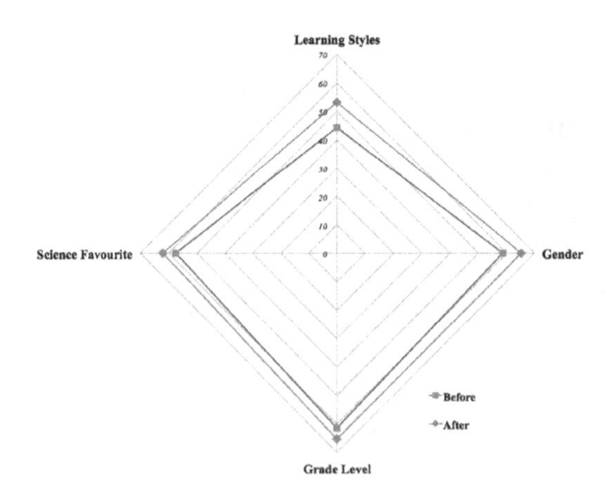

Fig. 4. Relative improvement of classification accuracy between four types of output

On the other hand, Figure 5 shows the relative comparison of improved classification accuracy and reduced training time of all five DTs. In Figure 5 (a), Random tree and NB tree showed significant improvement, J48 tree and BF tree provide comparatively moderate improvement, while Simple tree showed only slight improvement. In Figure 5(b), the training time of NB tree reduced significantly, whereas, the training time of Simple tree and BF tree was relatively moderate. For Random tree and J48 tree, the training time before grouping is rather small, the grouping method may not show any advantage. One possible reason is that when the number of input features decreases, the initial search space of Random tree and probability calculation space of NB tree also decrease. Consequently, classification accuracy and training time of such DTs are improved.

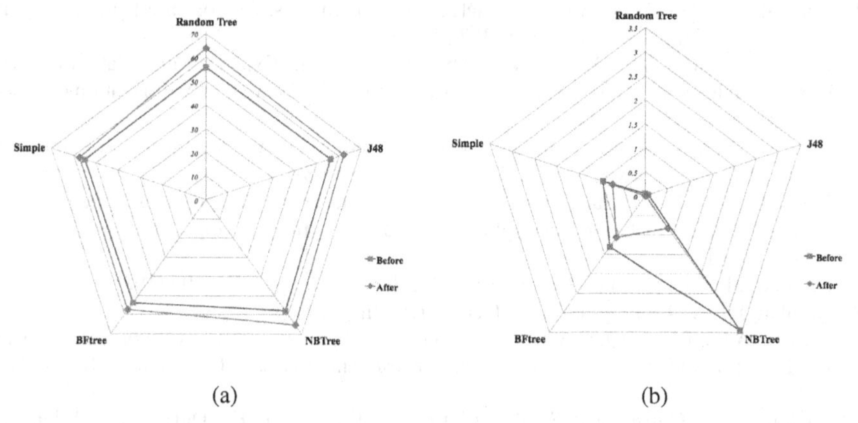

(a) (b)

Fig. 5. Relative comparisons based on all five decision trees of (a) increased accuracy and (b) reduced training time

6 Conclusion

Recently, recommendation systems have become a useful tool for academic staffs to support and enhance their teaching and delivery of lessons. However, establishing efficient recommendation systems is not a trivial task and it has to meet many issues with respect to specific problems. One specific problem is due to the weakness of relationship between the input features, and it causes the performance of the decision tree classifiers deteriorates. This study reports the use of a features grouping method to improve the performance of decision trees. The features grouping method is conducted based on the concept of ontology. The dataset used in this study was collected from schools in a rural area of Thailand. The experimental results indicated that if the input features are effectively grouped, classification accuracy will be increased. Furthermore, training time of decision trees will also be decreased. In the future, other advanced grouping techniques will be investigated.

Acknowledgment. I would like to thank Office of the Higher Education Commission, Thailand for providing me with grant under the higher education research promotion and national research university for Ph.D. programs, that enabled me to visit and carry out research at Murdoch university, Perth, Western Australia.

References

1. Klašnja-Milićević, A., Vesin, B., Ivanović, M., Budimac, Z.: E-Learning personalization based on hybrid recommendation strategy and learning style identification. Computers & Education 56(3), 885–899 (2011)

2. Chen, W., Persen, R.: Recommending collaboratively generated knowledge. Computer Science and Information Systems 9(2), 871–892 (2012)
3. Vesin, B., Ivanovic, M., Klasnja-Milicevic, A., Budimac, Z.: Ontology-based architecture with recommendation strategy in java tutoring system. Computer Science and Information Systems 10(1), 237–261 (2013)
4. Özpolat, E., Akar, G.B.: Automatic detection of learning styles for an e-learning system. Computers & Education 53(2), 355–367 (2009)
5. Ocepek, U., Bosnić, Z., Nančovska Šerbec, I., Rugelj, J.: Exploring the relation between learning style models and preferred multimedia types. Computers & Education 69, 343–355 (2013)
6. Chen, C.-M.: Intelligent web-based learning system with personalized learning path guidance. Computers & Education 51(2), 787–814 (2008)
7. Lin, C.F., Yeh, Y.-C., Hung, Y.H., Chang, R.I.: Data mining for providing a personalized learning path in creativity: An application of decision trees. Computers & Education 68, 199–210 (2013)
8. Quinlan, J.R.: Induction of decision trees. Machine Learning 1, 81–106 (1986)
9. Quinlan, J.R.: C.45: Programs for Machine Learning (1993)
10. Grasha, A.F., Riechmann, S.W.: A rational to developing and assessing the construct validity of a student learning styles scale instrument. Journal of Psychology 87(2), 213–223 (1974)
11. Baykul, Y.A., Gürsel, M., Sulak, H., Ertekin, E., Yazıcı, E., Dülger, O., Aslan, Y., Büyükkarcı, K.A.: A Validity and Reliability Study of Grasha-Riechmann Student Learning Style Scale. International Journal of Human and Social Sciences 5(3), 177–184 (2010)
12. Maneenil, S., Srisa-ard, B., Chookhampaeng, C.: Causal Factors Influencing Critical Thinking of Students in Different Learning Styles. Journal of Education, Mahasarakham University 4(4), 88–95 (2010)
13. Thonthai, T.: Learning Styles of Graduate Diploma Students in Teaching Profession at Princess of Naradhiwas University. Princess of Naradhiwas University Journal 1(3), 145–157 (2009)
14. Dublin Core Metadata Element Set, Version 1.1, `http://dublincore.org/documents/dces/`
15. Final 1484.12.1 LOM Draft Standard Document, `http://ltsc.ieee.org/wg12/20020612-Final-LOM-Draft.html`
16. Knight, C., Gašević, D., Richards, G.: Ontologies to integrate learning design and learning content. Journal of Interactive Media in Education 7, 1–24 (2005)
17. Verbert, K., Jovanović, J., Duval, E., Gašević, D., Meire, M.: Ontology-based learning content repurposing: The ALOCoM framework. International Journal on E-learning 5(1), 67–74 (2006)
18. Verbert, K., Klerkx, J., Meire, M., Najjar, J., Duval, E.: Towards a Global Component Architecture for Learning Objects: An Ontology Based Approach. In: Meersman, R., Tari, Z., Corsaro, A. (eds.) OTM-WS 2004. LNCS, vol. 3292, pp. 713–722. Springer, Heidelberg (2004)
19. Dunn, R., Dunn, K.: Teaching Secondary Students Through Their Individual Learning Styles Practical Approaches For Grades, pp. 7–12 (1993)
20. Wiley, D.A.: A Learning Object Design and Sequencing Theory. Brigham Young University, Provo (2000)
21. IMS learning resource meta-data information model, `http://www.imsproject.org/`
22. SCORM, 2nd ed., `http://www.adlnet.org`

Improving Service Performance in Cloud Computing with Network Memory Virtualization

Chandarasageran Natarajan

School of Science & Technology, Wawasan Open University, Malaysia
cnatarajan@wou.edu.my

Abstract. Resources abstractions have been very critical in securing performance improvement and higher acceptance by the end users in cloud computing. System virtualization, storage virtualization and network virtualization had been realized and has become as a part large systems abstraction in cloud computing. Memory high-availability in network environments is an added advantage in our presentation as an integral part in extending cloud computing multi various services to include network memory virtualization.

In this paper, we describe the virtualization of memory in cluster environment that could be applied universally using RDMA utility to map and access memory across the network. We suggested a combination of using the latency of remote memory and direct remote memory mapping facilities in our implementation. A low-level remote memory allocation and replacement technique is introduced to minimize page faulting and provide option to be more fault tolerance. We proposed a low level memory management technique in a network environment which would be able to support Service Oriented Architecture (SOA) and cloud computing.

Keywords: Network Memory, Virtualization, Cloud Computing and RDMA protocol.

1 Introduction

Emergence of cloud computing and related technologies had changed the way how computing are perceived today. Together with the grid computing and virtualization techniques the resources are being integrated in a hierarchical fashion in the cloud environment thus offering dramatic performance improvements over high-speed networks. In this proposal, we are exposing the possibilities of memory harvesting from idle machines on a network environment. Recent research studies shows most of the times; the machines in a network environment remain idle or at mostly available for a definite periods of time [1].

There is much research work done to support the idle memory utilization in cluster workstations, but this option of using the idle memory in the cloud resource virtualization for performance improvement is yet to be identified as a separate issue. Even though, some work was done as an overall initiative of resource management in grid computing and as remote memory resource. Studies show that in a small scale

S. Boonkrong et al. (eds.), *Recent Advances in Information and Communication Technology*, Advances in Intelligent Systems and Computing 265,
DOI: 10.1007/978-3-319-06538-0_23, © Springer International Publishing Switzerland 2014

networking environment this wastage of memory resources are quite “critical” and can be used for performance improvement without additional cost. The memory shortage crisis has been a continuous issue even when systems’ are being provided with more Ram after the cost of the RAM had drastically dropped over the years. This is due to the latest application and system software become more demanding and memory consuming to run efficiently. Consequently, it is a waste if we do not fully utilize the resources we have in a network environment without any additional cost to recruit the resources for our usage when they are idle.

We are proposing an in-depth analysis of the techniques and algorithms used in the effort of harvesting and utilizing the idle memory in a network using the cluster technology as the base. This proposal views the idle memory pool to be shared by the other participating workstations in the network environment for overall performance improvement. Thus, this will add an extra level of memory hierarchy in the cloud computing, which gives us the fast and slow remote memory access.

In our work here, we are proposing a low-level memory resource management algorithm option that had been implemented with the cluster technology as the basis to support the implementation of this network memory management. We applied our strategy by isolating the nodes in the environment as server node (only one), idle nodes and participating nodes.

2 Related Work

In providing a better understanding for the need to harvest these huge resources that are freely available in a network, we carried out our comparative study in a three-fold approach of identifying idle network memory availability, the reasons to exploit them and various methods deployed in performing the task.

A very early attempt to resolve memory shortage problem during page faulting was done by Markatos and Dramitinos [2], they capitalize the remote network memory available in their environment. Later a more comprehensive study was done to establish a Reliable Remote Memory Pager to support the idea [3], which provides an alternative memory paging, faster-than-disk paging option which bring it closer to utilization of the remote network memory. In both of these works, it was established the reliability of the remote memory that can be utilized which initially was seen as memory redundancy in a network environment. At the same time, Dhalin et. al. also used the remote memory to improve the file system performance [4], which boost the usage of idle network memory researches to support that the usage of remote network memory is a viable option to be explored further.

The utilization of ideal memory becomes an important concept to be exploited as a new research outcome. The idea of implementing a global memory management in a workstation clusters by Feeley et. al. [5], make a more promising end for the others to make use the idle memory in the workstation cluster. Feeley further establish the Global Memory System for a network environment in his work [6], where a complete analysis was done in handling the network memory as a global memory entity. He also provided prove that the global memory management algorithm that he implemented had better performance advantage over others in the past. An interesting research, similar to this with additional distributed workload sharing of CPU and memory resources for performance improvement was successfully done in [7] which capture the importance of memory usage tradeoff with computing power.

In view of all these development, a thorough study was done to justify the need to further the work on this idle remote memory in the network. Acharya and Setia [1] concluded in their work that huge amount of memory sits idle in any given network can be used without affecting the performance of the workstation in the network with certain memory recruitment policy. A user-level system, Dodo [8] was designed to harvest and exploit the idle memory space in workstation cluster that improve system throughput. On a different approach, Ioannidis et. al., improve the performance of transaction-based system in a similar way by using idle network memory [9]. In order to better understanding of a need for memory management in a network environment, we also identified the cost-benefit study to substantiate the claim [10]. This report shows that in long-term benefit based on economic principles, it is a compulsory for us to tap the idle memory of a network.

Further to the findings, a challenging approach was made with Network Memory File Systems. This is an example where network memory is been used as file system cache or used directly as the storage device. The speed of the applications can be easily improved by eliminating writing temporary files to the disk, instead this time we will write to the network memory. This new device is called the Network RamDisk (NRD). A NRD is a block device that unifies all the idle main memories in a workstation cluster under a (virtual) disk interface was well described in [11]. There are many applications, for instance Web servers, compilers and document processing systems, tend to use temporary files to process the required data and then delete them after finalizing the results would be benefitted using NRD concept.

In Flouris and Markatos an extensive discussion of the Network RAM are given with detailed description of the implementation and alternatives in the use of the idle memories in a network [12]. An idea of remote memory pager was described and implemented successfully to utilize network memory [13] with added flexibility. Xiao et. al., improve the Network RAM performance with incorporating the job migration scheme which provide effectiveness and scalability [14]. Other studies show the usage of the memories of cluster-based servers that improvised the web server performance [15, 16].

Later an interesting concept was introduced when Storage Area Network (SAN) were explored widely and used for storage purpose [17]. Hansen and Lachaize implement a prototype to use idle disks of a cluster for high performance storage [18] bring the concept of using idle resources in a network environment as a valuable asset. Thus, would enhance grid resource management system as a reality [19].

A more promising research emerge with the advancement of network theories, that integrate P2P and grid as internet distributed computing (IDC) [20] which also include the mobile resources to be as a part of the idea in resource aggregation and orchestration. The network memory would be easily used in the resource virtualization. Using virtualization techniques, Waldspurger [21] implemented a memory resource management in a virtual machine, VWware ESX server to establish efficient idle memory utilization in the server performance. In this study, three issues were addressed; memory virtualization, reclamation mechanisms (idle memory) and sharing memory.

In support of Software Oriented Architecture (SOA) and Cloud Computing, new foundation on virtualization has emerged vigorously [22, 23, 24]. The evolution processes that naturally seen moving toward extensive cloud computing, on-demand information technology services and products. Every focus is being directed at cloud computing advances and break-through that would cater wider usage of resources effectively. In our work, we had explicitly looked how memory in network environment can be exploited in the similar ways that we have seen above used for other resources especially on data storage models [4, 25] using DMA and RDMA [18, 26, 27]. The resource management is established with different ways, using servers is consider a more centralized techniques but without much overhead to improve on the memory latency when using network RAM as a number of research done had used similar approach [15, 21, 28, 29]. It is also recommended for the use of faster Ethernet facilities.

An implicit usage of Remote DMA (RDMA) in the research paradigm to establish communication channel between two nodes in the network is described for exploiting storage devices in support of computing processing [29, 30, 31]. This new technology standard is well established to make use of the idle resources in the network environment and become advantage for our purpose here in harvesting the idle memory. Remote Direct Memory Access (RDMA) is "also known as the PUT/GET or one-sided communication model – is a mechanism for transferring data over a network to (PUT) or from (GET) an address in another node's memory without the explicit involvement of the remote node's CPU" [40]. Pakin [32] also were able to reduce the extra sender-initiated message passing between the nodes by introducing the receiver-initiated message passing technique.

Finally, we also investigated the latest technology demand for memory related cloud computing and data intensive application usage to support our research objective to harvest the idle memory for virtualization. Service Oriented Architecture (SOA) had been around for sometimes with the idea of providing computing resources as services to the user [33]. Many researchers had given thought of this concept with the grid computing. Later on, convergence of multiple technologies abstraction and virtualization had initiated the clouds. Memory need would be always there for various reasons [34, 35] for which we had worked on network memory utilization.

Interestingly, the re-emergence of virtualization [36, 37] with its usage in the High Performance Computing (HPC) and grid environment show a great potential as a method for enhancing cloud computing performance. Vallee et. Al., explored in depth how virtualization is incorporated in HPC. Furthermore, in [38, 39, 40 and 41] RDMA protocol based investigation shows improvised data transfer in the Grid, Cloud and HPC environment.

3 Design Model Conceptualization

In our work, we provide a wider range of memory management solution for a cluster based network environment. The idea here is to utilize the idle network memory as much as possible to improve individual system's performance in a network. We had proposed a design concept model based on the following questions: How a user in a network could exploit his neighbor's memory when needed urgently? What is the percentage of unused memory in a network at a given time that is not utilized? What is the statistical implication of page faulting that happens in a network environment on individual PCs? What is PC's idle time in a network?

3.1 Resource Collection

In our model, we begin with the resource collection module, which are essential composition of the main function. We divide the task to three steps module. The first one is called the Idle Client Extension (ICE), which is activated on an idle node. The next one is known as Memory Server Extension (MSE) that resides in the server. The last one is the Network Memory Resource Table (NMRT), which is in the server and all the client nodes.

1. Idle Client Extension: A client extension program module is loaded at every participating node. This module is there to monitor each of the nodes' activities. When an idle situation is triggered the module would notify the server and added to the memory pool. The same module is responsible to withdraw the node from the pool if the statuses of the node toggle from idle to active.

2. Memory Server Extension: This module is only loaded on the server and act as the memory pool manager that keep track all the available idle memory nodes and participating nodes and their supporting nodes. It is responsible to provide an up-to-date real time situation of the network memory to all the nodes.

3. Network Memory Resource Table: This NMRT maintenance are done by the server and delivered to the entire participating nodes whenever any changes happen to the table. This is the table that has the information of the idle memory nodes, participating nodes and all the used idle memory nodes. In our work we also have the information of the mirror idle nodes which also dynamic. This provides our system an advantage so that there would always be an alternative path when any idle node's status changes from idle to active.

3.2 Memory Management Extension

This module is where the policy and decision making of the memory management algorithm are implemented. 1. What default criteria are set to choose the idle memory resource? 2. Is the participating node given the choice of preference of the remote memory that is idle? 3. Can a node have a reservation or prior booking to the memory pool? The module integrated with the MSE module and the NMRT provide data used in a process to have a reliable remote memory selection algorithm. Our main concern here would be the fault tolerance when choosing the idle memory client node. Scalability would be possible when the algorithm is able to adjust the availability of the memory pool.

The following Figure 1 shows the layered conceptual model that we describe above interact with the RDMA protocols.

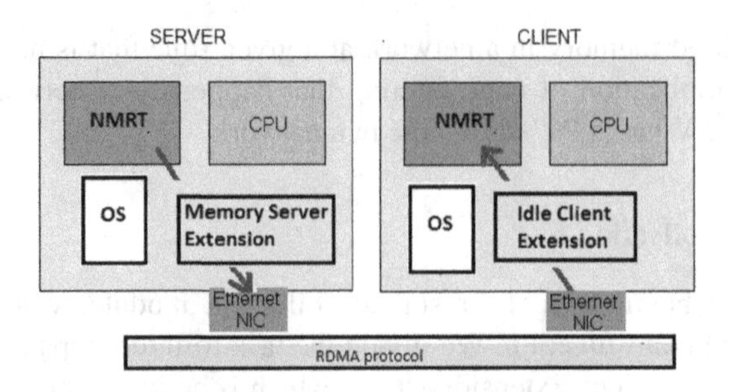

Fig. 1. Layered Network Memory Server-Client Model

4 Prototype Implementation

We divided our implementation into two different parts, as we need to address these two aspects separately in order to improve the overall performance. In the first stage, we address the state of nodes and state of paging activity of the client nodes. The server is responsible for both these functions, which enable the cluster-client architecture in our model. In the next stage, we address the state of memory request and protocol used by client nodes.

4.1 Algorithm Considerations

Here we describe briefly the basic algorithm condition that we use for the better performance. Our algorithm additions were made as two different parts so that we have a better control and distribution of the tasks. The first portion is used at server level to monitor idle memory and allocate them to participating nodes. It is also used to prevent any one node from monopolizing the idle memory pool. The other portion is used for handling the communication and the transfer protocols of memory data at the client node. The server is not involved with the details of the idle memory usage in any of the client node.

In order to enhance the performance of suggested algorithm, we had identified the consideration of the idle memory data collection in the following three fold sub-system as show in Figure 2 below.

a) *Segmentation of Workstations:* The segmentation is required to identify the grouping of the workstation according to either type, OS, configuration, distant to server or usage. This is done manually to group those machines with the same working environment background. We are using the physical location as the factor for segmentation of workstations.

The main reason physical location is chosen as the main factor is that the time required for memory swapping from one workstation to another workstation can ideally be minimized.

Besides, when the server intends to locate another idle workstation as a replacement for the client that turns from idle to busy, it can always start from the same segment that particular client resides in. This will certainly minimize the time to

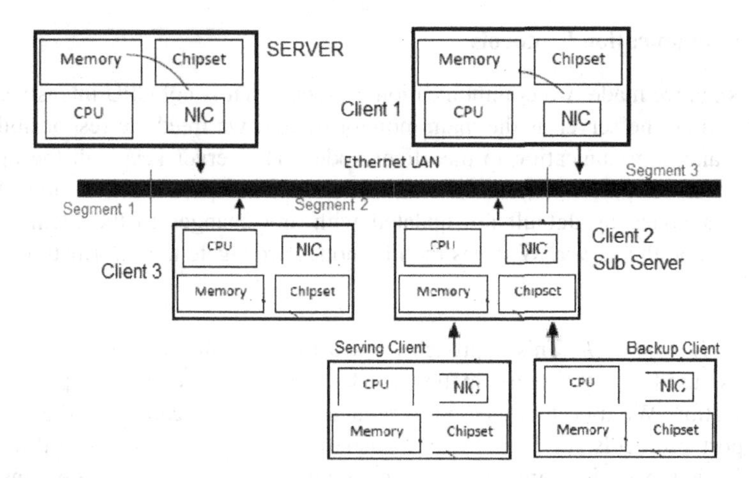

Fig. 2. Network segmentation showing the Server, Client and Sub server (client with Serving and backup client)

locate another idle workstation and eventually decrease the network traffic time to search for the content of memory resides in another workstation especially for those applications that refer to the content of memory frequently.

b) Memory Request: Memory request can also be divided into two different modes: -

i. Normal request – register with the server that this particular client is using local memory to carry out its tasks.
ii. Additional request – when local physical memory available for a particular client is not sufficient to carry out its tasks, it will request for additional network memory. By then, it will upgrade its status to become the "Sub Server" by monitoring possible two clients, i.e. the serving client and the backup client.

c) Idle Workstation Allocation: The server would perform an "intelligent search" to allocate an idle workstation to host guest data. It is done based on accumulated past history for that particular interval. As the server will continuously collects the data, therefore the result that it yields will reflect the "real time" status of the participating workstations.

4.2 Cluster-Client Architecture

In this architecture, the cluster server plays an important role in collecting information of idle nodes and remote memory usage in the network environment. A memory pool is established from the collection of idle memory. All these facts are disseminated on real time basis to the entire participating nodes which determine the course of action to be taken. The toggling mode of the node between idle and active state is crucial were addressed at this stage. Also some alternative arrangements were made to protect the client nodes when the participating idle nodes become active. We used this client-server architecture concept to emulate the benefits as been described in [21, 28]

4.3 Communication Protocols

In our design, we made two communication protocols that is not CPU intensive. Even though we used the server as the main monitoring tool we freed the responsibility of collection and communication to the client nodes. The server keeps all the updated information and supplies them to the client nodes upon request or by default. All the participating nodes by default are updated with any changes to the memory pool configuration at the server. Whereas the idle nodes propagate the information needed to the server.

1. *Proboscis Approach:* This approach is used to establish the connection between the nodes used as the participating nodes that act as point-to-point (P2P) communication. We used this method to reduce the communication overhead caused by transport protocols. In this way, we also minimized the network traffics. The communication extensions are built only on temporary basis when data are transferred to and from the server by the nodes. The Proboscis method used here would enable a P2P communication between the nodes to transfer the data smoothly [18] with establishing receiver side and sender side sockets.

2. *Direct Client Remote Memory Access:* This Direct Client Remote Memory Access is established when the memory of the idle node is used as the remote memory for a particular participating node. Only for this situation the communication extension are permanent as long as there are exchanges between these two nodes. We also used this method to provide a more local control of the idle memory to the participating node. We explore the usage of RDMA technology for greater reliability and faster data transfer between nodes in the network environments that participated in the memory sharing virtualization [31, 32].

5 Conclusion and Future Work

Computing resources are everywhere and if deployed properly would be able to enhance the computing performance tremendously. The network performance is undergoing a very dramatic improvement and thus reducing the processor/network performance ratio that changes the outlook of how this can be used in overall computing performance improvement in a network. The overall purpose of this research output can be applied directly for the grid and cloud computing where providing resources as a service is promoted for the computing community [19, 22, 24].

As a conclusion, idle network memories are there for us to harvest. But, we have to recruit them in a more meaningful way without much interference during the period the memory is hosting guest data. The most appropriate approach would be to identify the most suitable intervals to run this tool. The most preferred intervals would be different according to different environment.

Although we concluded here that there are definitely ways to recruit idle network memories, our suggestion would be one of the best approaches. However, we need to upgrade and enhance the tool in order to optimize its performance.

First of all, similar client utilities need to be developed on Linux platform to invite the Linux workstations to participate in this tool. Secondly, the allocation of idle workstations should be modified to justify the bias contribution of the busy frequency percentage.

A very promising trend is emerging to utilize this network memory by virtualization [32]. Our presentation here would contribute tremendously for cloud computing [34, 35] for application that are RAM and data intensive computing performance.

Acknowledgments. This project is funded by the Institute of Research and Innovation (IRI) Research Grant of Wawasan Open University (WOU), Penang, Malaysia.

References

1. Acharya, A., Setia, S.: Availability and utility of idle memory in workstation clusters. In: Proceedings of ACM SIGMETRICS Conference on Measuring and Modeling of Computer Systems, pp. 35–46 (1999)
2. Markatos, E.P., Dramitinos, G.: Implementation and Evaluation of a Remote Memory Pager. Technical Report FORTH/ICS 129 (1995)
3. Markatos, E.P., Dramitinos, G., Implementing, G.: of reliable remote memory pager. In: Proceedings of the 1996 Usenix Technical Conference, pp. 146–164 (1996)
4. Dahlin, M., Wang, R., Anderson, T., Patterson, D.: Cooperative Caching: Using Remote Memory to Improve File System Performance. In: Proceedings of the 1st Symposium on Operating System Design and Implementation, pp. 267–280 (1994)
5. Feely, M.J., Morgan, W.E., Pighin, F.H., Karlin, A.R., Levy, H.M.: Implementing global memory management systems in a Workstation Cluster. In: Proceedings of the 15th ACM Symposium on Operating System Principles, pp. 201–212 (1995)
6. Feeley, M.J.: Global Memory Management for Workstation Networks. PhD Thesis, University of Washington (1996)
7. Zhang, X., Qu, Y., Xiao, L.: Improving Distributed Workload Performance by Sharing Both CPU and Memory Resources. In: Proceedings of 20th International Conference on Distributed Computing Systems (ICDCS), pp. 10–13 (2000)
8. Kousshi, S., Acharya, A., Setia, S.: Dodo: A User-level System for exploiting idle memory in workstation clusters. Technical Report TRCS98-35, Department of Computer Science, University of California, Santa Barbara (1998)
9. Ioannidis, S., Markatos, E.P., Sevaslidou, J.: On using Network Memory to improve the performance of transaction-based System. Technical Report 190, ICS-FORTH, `http://www.ics.forth.gr/proj/avg/paging.html`
10. Amir, Y., Awerbuch, B., Borgstrom, R.S.: A Cost-Benefit Framework for Online Management of a Metacomputing System. In: Proceedings of 1st International Conference on Information and Computational Economy (ICE 1998), pp. 25–28 (1998)
11. Flouris, M.D., Markatos, E.P.: The Network RamDisk: Using Remote Memory on Heterogeneous NOWs. In: Progress Session of the USENIX 1999 Annual Technical Conference (1999)
12. Flouris, M.D., Markatos, E.P.: Network RAM. In: Buyya, R. (ed.) High Performance Cluster Computing, pp. 383–508. Prentice Hall, New Jersey (1999)

13. Markatos, E.P., Dramitinos, G.: Adding Flexibility to a Remote Memory Pager. In: Proceedings of the 4th International Workshop on Object Orientation in Operating Systems
14. Xiao, L., Zhang, X., Kubricht, S.A.: Incorporating Job Migration and Network RAM to Share Cluster Memory Resources. In: Proceedings of the 9th IEEE International Symposium on High-Performance Distributed Computing (HPDC-9), pp. 71–78 (2000)
15. Cuenca-Acuna, F.M., Nguyen, T.D.: Cooperative Caching Middleware for Cluster-Based Servers. In: Proceedings of the 10th IEEE International Symposium on High Performance Distributed Computing, HPDC-10 (2001)
16. Carrera, E.V., Bianchini, R.: Efficiency vs. Portability in Cluster-Based Network Servers. In: Proceedings of the 8th Symposium on Principle and Practice of Parallel Programming. ACM/SIGPLAN (2001)
17. Hansen, J.S.: Flexible network attached storage using remote DMA. In: Proceeding of Hot Interconnects 9, pp. 51–55. IEEE (2001)
18. Hansen, J.S., Lachaize, R.: Using Idle Disks in a Cluster as a High Performance Storage System. Technical Report, SARDES Project, SIRAC Laboratory, Grenoble, France (2002)
19. Buyya, R., Abramson, D., Giddy, J.: Grid Resource Management, Scheduling and Computational Economy. In: Proceedings of the 2nd International Workshop on Global and Cluster Computing (WGCC 2000), Tsukuba, Tokyo, Japan (2000)
20. Milenkovic, M., Robison, S.H., Knauerhase, R.C., Barkai, D., Garg, S., Tewari, V., Anderson, T.A., Bowman, M. (Intel): Toward Internet Distributed Computing, pp. 38–46. IEEE Computer Society (2003)
21. Waldspurger, C.A.: Memory Resource Management in VMware ESX Server. In: Proceedings of the 5th Symposium on Operating Systems Design and Implementation, OSDI 2002, Best paper award (2002)
22. Birman, K., Chockler, G., Renesse, R.: Towards a cloud computing research agenda. SIGACT News 40(2) (2009)
23. Vouk, M.A.: Cloud Computing – Issues, Research and Implementations. Journal of Computing and Information Technology (CIT) 16(4), 235–246 (2008)
24. Armbrust, M., Fox, A., Griffith, R., Joseph, A.D., Katz, R., Konwinski, A., Lee, G., Patterson, D., Rabkin, A., Stoica, I., Zaharia, M.: A view of cloud computing. Communications of the ACM 53(4), 50–58 (2009)
25. Anderson, E., Neefe, J.M.: An exploration of network ram. Technical Report (CSD98-1000) Dept. of Computer Science, University of California, Berkeley (1994)
26. Islam, N.S., Rahman, M.W., Jose, J., Rajachandrasekar, R., Wang, H., Subramoni, H., Murthy, C., Panda, D.K.: High performance RDMA-based design of HDFS over InfiniBand. In: Proceedings of the International Conference on High Performance Computing, Networking, Storage and Analysis, pp. 1–35. IEEE Computer Society Press (2012)
27. Chu, R., Xiao, N., Zhuang, Y., Liu, Y., Lu, X.: A Distributed Paging RAM Grid System for Wide-Area Memory Sharing. In: Proceedings of 20th International Parallel and Distributed Processing Symposium, IPDPS, pp. 1–10 (2006)
28. Gemikonakli, O., Mapp, G., Thakker, D., Ever, E.: Modelling and Performability Analysis of Network Memory Servers. In: Proceeding of the 39th Annual Simulation Symposium, ANSS 2006 (2006)
29. Balaji, P., Shah, H.V., Panda, D.K.: Sockets vs RDMA interface over 10-Gigabit networks: an in-depth analysis of the memory traffic bottleneck. In: RAIT Workshop 2004 (2004)

30. Mamidala, A.R., Vishnu, A., Panda, D.K.: Efficient shared memory and RDMA based design for mpi_allgather over InfiniBand. In: Mohr, B., Träff, J.L., Worringen, J., Dongarra, J. (eds.) PVM/MPI 2006. LNCS, vol. 4192, pp. 66–75. Springer, Heidelberg (2006)
31. Romanow, A., Bailey, S.: An Overview of RDMA over IP. In: Proceedings of the First International Workshop on Protocols for Fast Long-Distance Networks, PFLDnet 2003 (2003)
32. Pakin, S.: Receiver-initiated message passing over RDMA Networks. In: IEEE International Symposium on, Parallel and Distributed Processing, IPDPS 2008 (2008)
33. Perez-Conde, C., Diaz-Villanueva, W.: Emerging Information Tecnologies (II): Virtualization as Support for SOA and Cloud Computing. The European Journal of the Informatics Professional (CEPIS UPGRADE) 11(4), 30–35 (2010), http://cepis.org/upgrade
34. Hoff, T.: Are CloudBased Memory Architectures the Next Big Thing? Blog-Article (2009), http://highscalability.com/blog/2009/3/16/are-cloud-based-memory-architectures-the-next-big-thing.html
35. Ghalimi, I.C.: Cloud Computing is Memory Bound. An Intalio White Paper. Intalio, Inc., USA (2010)
36. Vallee, G., Naughton, T., Engelmann, C., Ong, H., Scott, S.L.: System-level virtualization for high performance computing. In: 16th Euromicro Conference on Parallel, Distributed and Network-Based Processing (PDP 2008), Toulouse, France (2008)
37. Subramoni, H., Lai, P., Kettimuthu, R., Panda, D.K.: High Performance Data Transfer in Grid Environment using GridFTP over Infiniband. In: 2010 10th IEEE/ACM International Conference on Proceeding of Cluster, Cloud and Grid Computing (CCGrid). IEEE (2010)
38. Woodall, T.S., Shipman, G.M., Bosilca, G., Graham, R.L., Maccabe, A.B.: High performance RDMA protocols in HPC. In: Mohr, B., Träff, J.L., Worringen, J., Dongarra, J. (eds.) PVM/MPI 2006. LNCS, vol. 4192, pp. 76–85. Springer, Heidelberg (2006)
39. Yu, W., Rao, S.V.N., Wyckoff, P., Vetter, J.S.: Performance of RDMA-capable storage protocols on wide-area network. In: Proceeding of Petascale Data Storage Workshop, PDSW 2008, 3rd edn. IEEE (2008)
40. Kissel, E., Swany, M.: Evaluating High Performance Data Transfer with RDMA Based Protocols in Wide Area Networks. In: Proceeding of 14th International Conference on High Performance Computing and Communications (HPCC 2012), Liverpool, UK (2012)
41. Yufei, R., Li, T., Yu, D., Jin, S., Robertazzi, T.G.: Middleware support for rdma-based data transfer in cloud computing. In: Parallel and Distributed Processing Symposium Workshops & PhD Forum (IPDPSW). IEEE (2012)

[illegible]

Tweet! – And I Can Tell How Many Followers You Have

Christine Klotz[1], Annie Ross[2], Elizabeth Clark[3], and Craig Martell[4]

[1] FernUniversität in Hagen, Fakultät für Mathematik und Informatik, Hagen, Germany
[2] Colorado State University, United States
[3] Middlebury College, United States
[4] Naval Postgraduate School, Department of Computer Science, Monterey, United States
c._klotz@web.de, {anniesross,eaclark07,craig.martell}@gmail.com

Abstract. Follower relations are the new currency in the social web. User-generated content plays an important role for the tie formation process. We report an approach to predict the follower counts of Twitter users by looking at a small amount of their tweets. We also found a pattern of textual features that demonstrates the correlation between Twitter specific communication and the number of followers. Our study is a step forward in understanding relations between social behavior and language in online social networks.

Keywords: Twitter, follower, user characteristics, text mining, n-grams, Naïve Bayes, tf-idf, online social networks.

1 Introduction

The rise of social media has created a new platform for viral and brand marketing, public relations and political activities. Therefore, "giving the right message to the right messengers in the right environment" [1] is essential for effective campaigns. The number of followers is often interpreted as sign of popularity and prestige, follower relations count as currency within the social web [2]. A marketplace for followers is already evolving [3].

From the perspective of information diffusion, Twitter users with high counts for followers can be categorized as 'information sources' or 'generators' and with high counts for followings as 'information seekers' or 'receptors' [4, 5]. Users with equal numbers of followers and followings are called 'acquaintances' to emphasize the reciprocity of the relationship [6]. Retweeting is associated with information sharing, commenting or agreeing on other peoples' messages and entertaining followers [7]. Tweet content may fall in the categories: 'Daily Chatter', 'Conversations' (tweets containing the @-symbol), 'Sharing Information' (tweets containing at least one URL) or 'Reporting News' (tweets via RSS feeds) [4].

User-generated content obviously plays an important role for tie formation on the social web. Hutto et al. [2] monitored Twitter users over a period of 15 month and concluded that "social behavior and message content are just as impactful as factors related to social network structure" for the prediction of follower growth rates. Kwak et al. [8] examined factors that impacted the stability of follower relationships. They

S. Boonkrong et al. (eds.), *Recent Advances in Information and Communication Technology*, Advances in Intelligent Systems and Computing 265,
DOI: 10.1007/978-3-319-06538-0_24, © Springer International Publishing Switzerland 2014

found that, beside a dense social network, actions of mutual behavior assured the stability of relations. These include retweeting and replying as well as the usage of similar hashtags in tweets and the inclusion of URLs. Both approaches needed a high amount of data points which were gathered over long periods of time.

In this study, we concentrate on a single point in time. We report an approach to predict the follower rates of randomly picked Twitter users by looking at a small amount of data. Therefore, we apply naïve Bayes classification on tweet content using character trigrams as feature sets. The contribution of this paper is three-fold. First, we show how to predict follower levels, and then compare the predictability of different data points that are often used for the detection of latent user attributes in a second step. Thirdly, we report an interesting pattern of textual features that shows the correlation between Twitter specific functionality (retweeting, replying, etc.) and the number of followers. Follower relations seem to be the new currency in the social web. Our findings show how tweets reflect the popularity of users. Understanding relations between behavior and language usage will lead to an improved understanding of the users.

2 Related Work

Kwak et al. [9] rank user influence with follower counts and retweet rates, and compare the results with PageRank values for the nodes. In a similar approach, Cha et al. [10] rank user influence in three categories by counting followers, retweet rate (forwarding of tweets) and mentions (referring to someone), respectively. Quercia et al. [11] extracted the numbers of followers, followings and lists (how many times a user appears on other reading lists) to predict personality with regression analysis. Klotz et al. [12] applied clustering algorithms on Twitter user data to identify personality types. Five features were chosen related to behavior (numbers of tweets and favorites) or social network (same as [11]). The results of those studies demonstrate personality prediction based on publicly available data. Golbeck et al. [13] used multiple feature sets including behavioral, structural and textual data for regression. The results showed best predictability for openness and worst for neuroticism, but the prediction tasks beat the baseline only slightly. Their low performance may be a result of the diffuse feature sets creating additional noise. Pennacchiotti et al. [14] compare different feature sets to predict socio-demographic attributes (ethnicity, political affiliation, etc.) with Gradient Boosted Decision Trees. In all three tasks, linguistic features performed well, with hashtags and topic words adding beneficial information. Liu et al. [15] developed a threshold classifier to predict gender on Twitter that takes first names as features, and adds other features only in case of low distinctiveness. This second feature set contains textual features, e.g. character trigrams, which show good performance.

Although many studies utilize usage data and user-generated content to identify user characteristics, little is known about the relations between these data points. For many applications, it is useful to target the right people in the social network. Our study is a step forward in understanding relations between social behavior and language in online social networks.

3 Experimental Setup

3.1 Data Collection

A sample of 1400 users was gathered by using random numbers with up to eight digits, which has been acknowledged as a range for valid user-IDs [6, 10]. All tweets posted by these users during a one-month-sampling period (19.06–21.07.2012) were stored as data collection 1. An extended data collection contained all tweets posted during a two-month-sampling period (19.06–29.08.2012). After filtering for users with at least 20 English tweets, data collection 1 contained a total of 527 users and 77,131 tweets, and data collection 2 contained 547 users and 131,575 tweets. Histograms plotted in log-log-space for the number of followers and friends/followings show an approximately normal distribution, while the histograms for favorites, lists, and tweets per day are Zipfian.

3.2 Classification Method and Feature Selection

For a pilot study a naïve Bayes classifier is a beneficial approach. The probabilistic model assumes that features are independent which simplifies the estimation [16]. Although this is often not true, i.e. in case of natural language, the classifier shows good performance and may even outperform more sophisticated models [16].

We did Laplace smoothing to account for trigrams that occur only within tweets of the testing set. The smoothing factor was $\alpha = 0.005$ and the size of vocabulary was $V = 256^3$, because a character trigram can be any permutation of three characters.

The equation for our classification process is as follows:

$$\underset{level_j}{argmax}\left(\ln\left(P\left(level_j\right)\right) + \textstyle\sum_i^N \ln\left(P(trigram_i|level_j)\right)\right) \quad (1)$$

$$with \quad P\left(level_j\right) = \frac{number\ of\ users\ in\ level_j}{total\ number\ of\ users}$$

$$P\left(trigram_i|level_j\right) = \frac{(number\ of\ times\ trigram_i\ occurs\ in\ level_j)+\alpha}{(total\ number\ of\ trigram\ tokens\ in\ level_j)+V}$$

Text mining approaches differ in the complexity of the classification method, and can also be conducted on varying sizes of feature sets. A feature set may be too simple to capture signal, but also too diffuse causing it to create additional noise. When performing natural language processing on social media text, it should be kept in mind that most techniques are designed for more grammatical and well-formed text sources than for short, noisy ones [17]. We chose character n-grams, in particular trigrams, as a simple, but well-performing feature set. Character n-grams account for uncommon textual features such as "Emoticons, abbreviations, and creative punctuation use (which) may carry morphological information useful in stylistic discrimination" [18]. They also provide good word coverage, and preserve word order. The latter compensates for elaborate text-processing. Character trigrams have been used in different prediction tasks on Twitter, e.g. as features for gender inference [15] and authorship attribution [18], and showed good performances. We used the frequencies of character trigrams as features in combination with techniques for noise reduction.

Messages on Twitter contain specific entities that refer to platform functionality. These are: the abbreviation 'RT' in retweets (forwarded messages), @-signs in replies and mentions, hashtags marking topics, and the inclusion of URLs. We filtered for these entities because they could create noise or provide false signaling. For example, including hashtag trigrams could result in subject matter detection, rather than user trait detection. However, the usage of platform functionality can be an important cue, so we replaced the entities by placeholders. This strategy led to the discovery of an interesting pattern for follower prediction (see Section 4.3).

3.3 Noise Reduction with Tf-idf

Tf-idf (term frequency-inverse document frequency) is a measurement for the representativeness of a token in a particular document in comparison with all documents of a corpus. The score increases with the number of times a token appears in the particular document (term frequency), but decreases with the number of times it appears in other documents of the corpus (inverse document frequency). The measurement is popular for keyword extraction, because high scores reflect representativeness of a term for the document, but low scores can be utilized for noise reduction.

We used augmented term frequency to prevent bias between users with many collected tweets (and therefore a long document) and users with few tweets. In this context, a document is the concatenation of all tweets we sampled for one user. As we tested our classifier with 10-fold cross-validation, we considered only the training set as corpus leaving out the documents of the test sample. This is the right way to perform cross-validation with multivariate classifier [16].

The scores are calculated as follows:

$$\textit{tf-idf-value} = TF * IDF \tag{2}$$

with

$$TF = 0.5 + \frac{0.5 * (\textit{raw count of trigram in document})}{(\textit{raw count of most common trigram in document})}$$

$$IDF = \ln\left(\frac{(\textit{number of documents in corpus})}{(\textit{number of documents containing trigram})}\right)$$

For varying cutoff percentages $p=\{0, 0.05,\ldots,0.3\}$, any trigram whose tf-idf-value fell under the threshold was excluded from the feature set. These can be either trigrams that occur seldomly in one particular document (low term frequency) accounting for uncommon words/misspellings or very frequently in the corpus (low inverse document frequency) accounting for stopwords. As the idf-score decreases faster than the tf-score, we mostly filtered for very frequent words that do not bear much information.

4 Results

4.1 Prediction of Followers

To predict follower rates, we performed a supervised binary classification with a sliding $\lambda = 0.1, 0.2, ..., 0.9$ resulting in two bins with the ranges $[0, maximum * \lambda]$ and $]maximum * \lambda, maximum]$. This division technique produces unbalanced bins, so we limited the number of users in the training set so that each bin would have approximately the same number of users, giving baseline around 50%. We worked on data collection 2, to provide sufficient data. The naïve Bayes classifier used character trigrams as features. Platform specific entities were filtered out. To reduce noise, we filtered for trigrams with low tf-idf-scores as described in Section 3.3. We tested our classifier with 10-fold cross-validation. Table 1 gives the results for three runs with different setups.

Table 1. Binary classification of follower counts including noise reduction

	Bin division $\lambda = 0.4$ Baseline Accuracy = 51.7		Bin division $\lambda = 0.5$ Baseline Accuracy = 51.3		Bin division $\lambda = 0.7$ Baseline Accuracy = 51.5	
Tf-idf cutoff	**Accuracy**	**Improvement above Baseline**	**Accuracy**	**Improvement above Baseline**	**Accuracy**	**Improvement above Baseline**
0.00	63.9	23.6%	60.6	18.1%	58.2	13.0%
0.05	63.7	23.2%	60.4	17.7%	58.2	13.0%
0.10	63.5	22.8%	60.6	18.1%	58.2	13.0%
0.15	65.1	25,9%	61.7	20.3%	60.8	18.1%
0.20	65.5	26.7%	61.7	20.3%	60.9	18.3%
0.25	65.3	26.3%	60.9	18.7%	59.3	15.1%
0.30	65.0	25,7%	60.4	17.7%	59.9	16.3%

The classifier performed over baseline for all λ and best for $\lambda = 0.4$. With very low or very high values for λ the performance drops. The good overall performance in predicting five levels of followers (Section 4.2) makes it unlikely that very high or very low numbers of followers are harder to predict. The drop in performance is more likely caused by the reduction of training data when balancing the bins.

Furthermore, the reduction of noise with tf-idf filtering improves the performance. Social media texts are often noisy because of non-linguistic fragments as well as spelling mistakes [17], so the removal of least occurring trigrams like misspellings as well as stopwords bearing no information is a valuable approach. We got best results for a threshold of $p = 0.2$. When increasing the cutoff further, the performance is slightly reduced. As Twitter messages are very short and full of slang [17], a further reduction may remove information rather than noise. Furthermore, using too high a percentage may fit the classifier too specifically to the training set and not be effective for classification of the users in the testing subset. This is called overfitting.

The results of our experiment demonstrate that there is signal inside tweets to classify users according to their level of followers. For future research, a more advanced classification method could improve the performance.

4.2 Comparing the Predictability of Different User Traits

Approaches that predict user attributes mine behaviour data from online social networks as these sources have been proven to bear valuable information [19]. In [12], the five-dimensional feature space contains the tweet rate for communicativeness, the numbers of follower and following representing the position within the social network, the number of lists as further measurement for participation on the platform as well as bookmarking in terms of the number of favorites. We attempt to predict these five traits with textual features as explorations suggested correlations between behaviour and language usage. Each trait was considered an independent experiment and no interaction between traits was investigated. We computed the trait ranges, and divided the scales in five equal sized levels for a 5-way classification task. Since each level has an equal range of values, the boundary of level_i is defined as:

$$\left[\frac{i}{5} * \text{maximum}, \frac{i+1}{5} * \text{maximum}\right) \tag{3}$$

where *i=0,1,…,5* and *maximum* is the maximum value for the specific trait. The task was performed as supervised 5-way classification with character trigrams as the features. Platform specific entities were filtered out, all other trigrams were preserved. We ran 10-fold cross-validation and used data collection 1, because the work reported in [12] is based on that sample. Table 2 shows the results in terms of accuracy (column 2) and improvement above baseline (column 4). Column 3 gives the accuracy baselines. Baselines are computed as the percentage of objects in the largest bin for each trait. The baselines differ, as the bins are equal sized in terms of range, but not in terms of number of objects, and the five traits possess different ranges. The prediction was successful for followers and for the number of lists. The other tasks failed.

Table 2. Results of 5-way-classification task for five different user traits

Trait	Accuracy	Baseline	Improvement above baseline
Followers	45.6	35.0	30%
Followings	56.8	58.4	-
Favorites	39.7	41.4	-
Lists	40.6	28.2	44%
Tweets Per Day	28.4	33.5	-

Textual features, in particular, character trigrams do not bear beneficial information for the prediction of followings, favorites and tweet rates. On the other hand, the content of tweets has signal to predict how many users will receive the tweet directly by following the blogger or indirectly by the number of lists the blogger is added to. The popularity of the blogger is obviously correlated with the content of the tweet.

4.3 Patterns of Specific Textual Features

The 5-way classification task for follower prediction uncovered an interesting pattern of textual features which may be a reason for the good classification results. The pattern is drawn in Figure 1. Whereas the average number of URLs in tweets increases with the level of followers, the correlation function between follower level and retweet rate is monotonically decreasing. URLs in tweets signal information sharing [4]. Retweeting behavior is associated with commenting or agreeing on messages, entertaining and showing presence, as well as information sharing [7], whereby the content is forwarded, not contributed. Our results indicate that users with lower follower counts are more likely promoters than originators of information. In other words, gaining followers is mainly achieved through producing original content.

At the same time, we found a nearly stable rate of @-references for all follower levels. Tweets containing the @-sign mark conversations [4], because the sign is used to explicitly reply to or mention someone else. The amount of @-signs did not differ significantly among the five follower levels, so the proportion of conversations is much likely the same for people with low follower vs. high follower rates.

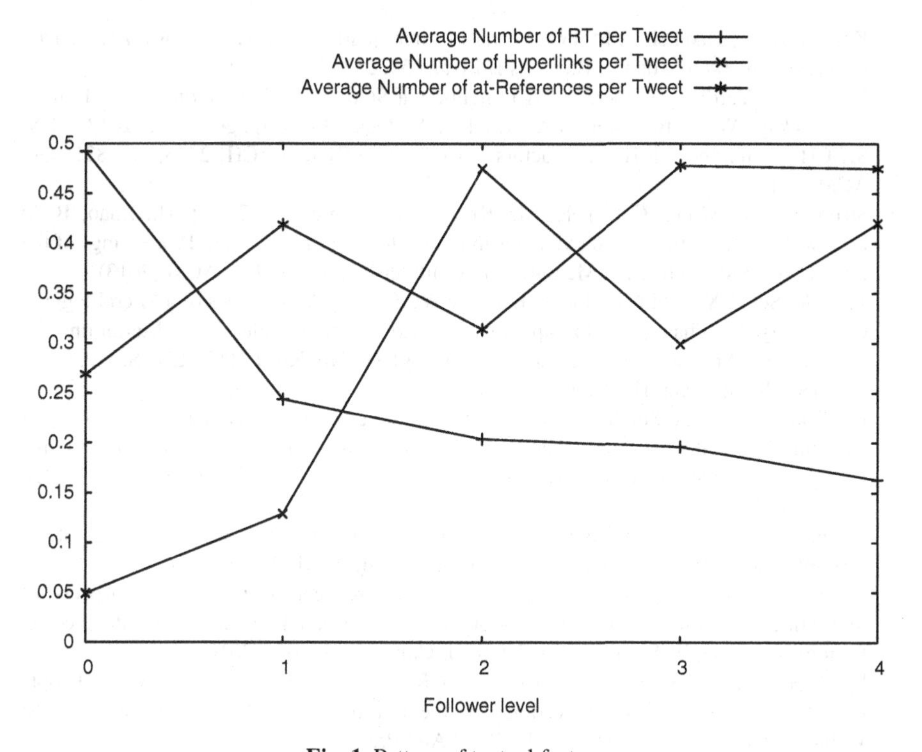

Fig. 1. Pattern of textual features

5 Conclusions

We reported an approach to predict the follower counts of Twitter users by looking at a small amount of their tweets. We used a naïve Bayes classifier with character trigrams as feature set, and reduced noise with tf-idf scores. Our results show that tweet content bears enough information for the prediction of followers and lists, but not for followings, favorites and tweet rates. The popularity of the blogger is obviously correlated with the content of the tweet. We also found a pattern of textual features that demonstrates the correlation between Twitter specific communication and the number of followers. Our results indicate that gaining followers is mainly achieved through producing original content and that the proportion of conversations is much likely the same for people with low follower vs. high follower rates. Our study is a step forward in understanding relations between social behavior and language in online social networks. In the future, more advanced classification methods could improve the performance, and uncover new relations between features and user classes.

References

1. Kaplan, A.M., Haenlein, M.: Two hearts in three-quarter time: How to waltz the social media/viral marketing dance. Business Horizons 54, 253–263 (2011)
2. Hutto, C.J., Yardi, S., Gilbert, E.: A Longitudinal Study of Follow Predictors on Twitter. In: Mackay, W.E., Brewster, S.A., Bødker, S. (eds.) Proceedings of the 2013 ACM SIGCHI Conference on Human Factors in Computing Systems (CHI 2013), pp. 821–830. ACM (2013)
3. Stringhini, G., Wang, G., Egele, M., Kruegel, C., Vigna, G., Zheng, H., Zhao, B.Y.: Follow the green: growth and dynamics in twitter follower markets. In: Proceedings of the 2013 Conference on Internet Measurement Conference, pp. 163–176. ACM (2013)
4. Java, A., Song, X., Finin, T., Tseng, B.: Why We Twitter: An Analysis of a Microblogging Community. In: Zhang, H., Spiliopoulou, M., Mobasher, B., Giles, C.L., McCallum, A., Nasraoui, O., Srivastava, J., Yen, J. (eds.) WebKDD/SNA-KDD 2007. LNCS, vol. 5439, pp. 118–138. Springer, Heidelberg (2009)
5. De Choudhury, M., Sundaram, H., John, A., Seligmann, D.: Dynamic prediction of communication flow using social context. In: Brusilovsky, P., Davis, H.C. (eds.) Proceedings of the Nineteenth ACM Conference on Hypertext and Hypermedia, pp. 49–54. ACM (2008)
6. Krishnamurthy, B., Gill, P., Arlitt, M.: A few chirps about twitter. In: Proceedings of the First Workshop on Online Social Networks, pp. 19–24. ACM (2008)
7. Boyd, D., Golder, S., Lotan, G.: Tweet, tweet, retweet: Conversational aspects of retweeting on twitter. In: Proceedings of the 43nd Hawaii International Conference on System Sciences (HICSS-43), pp. 1–10. IEEE Computer Society (2010)
8. Kwak, H., Moon, S., Lee, W.: More of a Receiver Than a Giver: Why Do People Unfollow in Twitter? In: Proceedings of the 6th International AAAI Conference on Weblogs and Social Media (ICWSM). The AAAI Press (2012)
9. Kwak, H., Lee, C., Park, H., Moon, S.: What is Twitter, a social network or a news media? In: Proceedings of the 19th International Conference on World Wide Web, pp. 591–600. ACM (2010)

10. Cha, M., Haddadi, H., Benevenuto, F., Gummadi, K.P.: Measuring user influence in Twitter: The million follower fallacy. In: Cohen, W.W., Gosling, S.D. (eds.) Proceedings of the Fourth International Conference on Weblogs and Social Media, ICWSM 2010, Washington, DC, USA, May 23-26, pp. 10–17. The AAAI Press (2010)
11. Quercia, D., Kosinski, M., Stillwell, D., Crowcroft, J.: Our Twitter Profiles, Our Selves: Predicting Personality with Twitter. In: 2011 IEEE International Conference on Privacy, Security, Risk, and Trust (PASSAT), and IEEE International Conference on Social Computing (SocialCom), pp. 180–185. IEEE (2011)
12. Klotz, C., Akinalp, C.: Identifying Limbic Characteristics on Twitter. In: Meesad, P., Unger, H., Boonkrong, S. (eds.) $IC^2IT2013$. AISC, vol. 209, pp. 19–27. Springer, Heidelberg (2013)
13. Golbeck, J., Robles, C., Edmondson, M., Turner, K.: Predicting Personality from Twitter. In: 2011 IEEE International Conference on Privacy, Security, Risk, and Trust (PASSAT), and IEEE International Conference on Social Computing (SocialCom), pp. 149–156. IEEE (2011)
14. Pennacchiotti, M., Popescu, A.-M.: A Machine Learning Approach to Twitter User Classification. In: Adamic, L.A., Baeza-Yates, R.A., Counts, S. (eds.) Proceedings of the Fifth International Conference on Weblogs and Social Media, Catalonia, Spain, July 17-21. The AAAI Press (2011)
15. Liu, W., Ruths, D.: What's in a Name? Using First Names as Features for Gender Inference in Twitter. In: Analyzing Microtext: Papers from the 2013 AAAI Spring Symposium. The AAAI Press (2013)
16. Hastie, T., Tibshirani, R., Friedman, J.H.: The elements of statistical learning. Data mining, inference, and prediction. Springer, Heidelberg (2009)
17. Baldwin, T., Cook, P., Lui, M., MacKinlay, A., Wang, L.: How Noisy Social Media Text, How Diffrnt Social Media Sources? In: Mitkov, R., Park, J.C. (eds.) Proceedings of the 6th International Joint Conference on Natural Language Processing (IJCNLP 2013), pp. 356–364. Asian Federation of Natural Language Processing (2013)
18. Boutwell, S.R.: Authorship Attribution of Short Messages Using Multimodal Features. Naval Postgraduate School, Monterey (2011)
19. Kosinski, M., Stillwell, D., Graepel, T.: Private traits and attributes are predictable from digital records of human behavior. Proceedings of the National Academy of Sciences 110, 5802–5805 (2013)

PDSearch: Using Pictures as Queries

Panchalee Sukjit[1], Mario Kubek[1], Thomas Böhme[2], and Herwig Unger[1]

[1] FernUniversität in Hagen, Faculty of Mathematics and Computer Science, Hagen, Germany
{panleek,dr.mario.kubek,herwig.unger}@gmail.com
[2] Technische Universität Ilmenau, Faculty of Mathematics and Natural Sciences, Ilmenau, Germany
thomas.boehme@tu-ilmenau.de

Abstract. Search engines usually deliver a large amount results for each topic addressed by a few (mostly 2 or 3) keywords. Thus, it is a tough work to find those terms describing the wanted content in a manner such that the search delivers the intended results already on the first result pages. In the iterative process of obtaining the desired web pages, pictures with their tremendous context information may be a big help. This contribution presents an approach to include picture processing by humans as a means for context search selection and determination in a locally working search control.

Keywords: search engine, context, keyword, metadata, evaluation, picture information.

1 Introduction

The today's World Wide Web (WWW) consists of approx. 785 million websites [1] that offer 80% of their information in textual form [2]. Since there is no connection between information and the location where it is stored, big search engines like Google are needed to support users in finding the needed documents. Therefore, the requested contents must be described in a few search words, which are used to find matching web documents within the huge indexes of the search engine databases. Usually, this method results in a plenty of documents presented to the users, who mostly review only the first 10-30 results. Algorithms like PageRank [3] try to ensure that the most relevant search results appear among those first 30 result entries.

Thus, the search for non-trivial information in the web becomes an iterative process. In a first step, the user formulates a query and sends their search words to a search engine, which mostly serve as contextual information. A few returned results are read afterwards in order to specify the wanted information more precisely and find out more specific keywords from the target topic. This is an especially hard and tedious process, if fully new topics are searched as their description with well-known terms is likely not available. Then, the newly found terms are used for another request and the described process continues until the desired documents are found. Depending on the experience of the user, this may take more or less time.

S. Boonkrong et al. (eds.), *Recent Advances in Information and Communication Technology*, Advances in Intelligent Systems and Computing 265,
DOI: 10.1007/978-3-319-06538-0_25, © Springer International Publishing Switzerland 2014

Consequently, in several research projects, methods have been developed to identify additional search words/search word alternatives and to present them to the user in this process. Google itself uses a statistical approach [4] and offers frequently co-occurring terms from its users' queries as suggestions to refine queries. This method of course fails if completely new or very seldom needed content is searched. However, obtained information about the user is also often used for other purposes than just for the improvement of web search, e.g. to place targeted commercial advertisement based on the area of his or her search activities.

Therefore, other (and mostly scientific) approaches apply locally working methods. They analyse local documents in order to provide a more precise description of the user's information needs in form of appropriate keywords for the current search context and do not transfer any user related information to the (centralised) search machine. One example is the FXResearcher [5]. This tool carries out a text analysis of a set of whole documents kept in specific directories of the user in order to extract additional keywords for the next search iterations. In addition, recently downloaded and evaluated documents may be also used for improvements in the next steps. The DocAnalyser [6], as another example, uses the locally displayed contents of currently visited webpages as the source of search context information and additional, selectable keywords. In the connected WebNavigator [7] project, webpages linked from the currently viewed one are analysed in order to give the user valuable directions for a suitable selection of the next webpage to visit.

Additionally, most search engines today allow a search for pictures with amazing results. Either the content of pictures must be described by so called metadata (which are again keywords describing contents and context) [8] or a search for similar pictures (using the color, contrast and other hard image values) [9] is provided. In fact, visual information plays an important role in humans' life. The human brain is able to process complex picture information with a very high speed due to a massive parallelisation [10]. A lot of details in a picture can be used to distinguish even similar contexts in a precise manner. This is a quite strange but a long known fact. Already in 1911, the expression "Use a picture. It's worth a thousand words." appears in a newspaper article by Arthur Brisbane discussing journalism and publicity [11]. The roots of that phrase are even older and have been expressed by earlier writers. For example, the Russian novelist Ivan Turgenev wrote (in *Fathers and Sons* in 1862), "A picture shows me at a glance what it takes dozens of pages of a book to expound.".

Nevertheless, up til now, there was no tool supporting a combined search of web documents using pictures so far. Therefore, the tool PDSearch has been designed and developed to specifically address this problem. In the next two sections, the concept of PDSearch as well as its implementation and use cases will be explained in detail. Afterwards, section four discusses the validity of this new approach to web search and outlines possible options to improve it. Section five summarises the article and points out the advantages of the presented search paradigm and the realised solutions.

2 PDSearch: The Concept

The above said has been the idea for a combined, still iterative search methodology, which might become a part of the concept presented in [6]. There, the hypothesis was stated that centralised search engines will hardly be able to satisfy all special user needs. Although it is consensus that search engines work well today and will (probably) always be a part of the Internet, a dedicated interaction with the respective user and an access to protected, local information may increase the quality and efficiency of search in most cases. In order to reduce internet traffic and support security, this user-oriented processing should be carried out by a local agent on the user's machine. The developed architecture is depicted in Fig. 1.

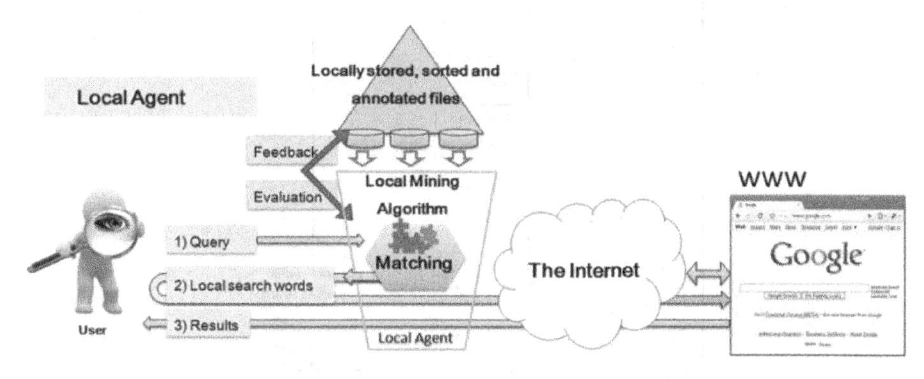

Fig. 1. Cooperation of the local agent and web search engines

The main idea for this improvement is that the used search words seldom can describe the environment of thoughts, which motivates the user to submit a given search request. Hence, the search words should be expanded with other terms that can disambiguate the initial query and specify the search context more precisely. In [5], this has been successfully done using local text documents. However, this approach makes the support of fully new search topics impossible.

It was found, that pictures may fulfil this task in a perfect manner as users can instantly recognise their contents and easily decide whether they fit in the search context of their current information need. Therefore, matching pictures (obtained from a local repository of web documents or from a web search engine) should be presented as results in the first iterations of a complex search in order to narrow down the search area/context. The textual information of the (not presented) web documents, which those pictures appear on, can then be used to extract additional keywords from. Afterwards, consecutive queries consisting of these keywords (now they become search words) can be automatically constructed and sent to web search engines, whereby the user has to decide whether to retrieve more similar images and other web documents. In Fig. 2, this iterative search process is shown in form of a flow chart using Google as a means to retrieve the pictures and their contextual information. The approach has been prototypically implemented in the search application PDSearch whose implementation details and user guidance will be explained in the next section. Although Google is suggested to retrieve the images, this step can be easily carried out by local search applications that indexes e.g. previously downloaded web documents with their textual contents and images. The open source search engine library Lucene, as an example, can be used for this purpose. This way, the algorithm can be integrated into local search agents, too.

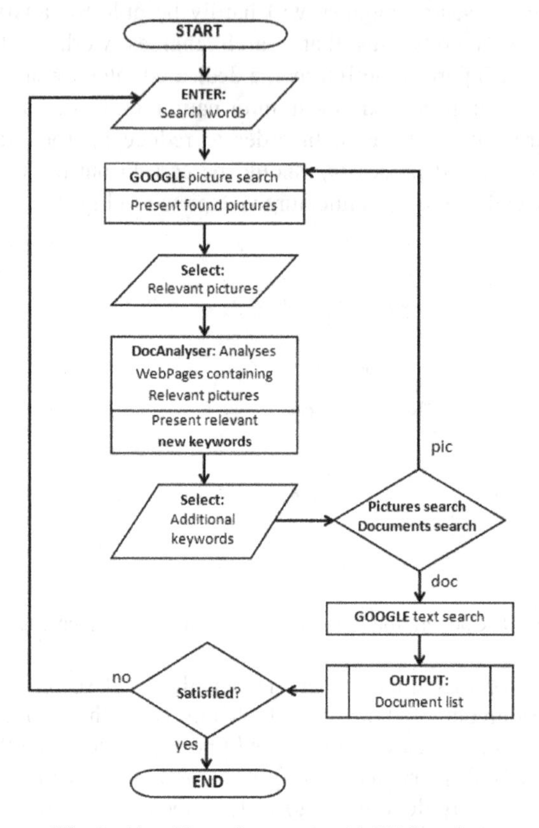

Fig. 2. Algorithm of a search with PDSearch

It has been found that search topics with a high potential for disambiguation (words with two or more meanings, called homonyms, e.g. mouse, names e.g. "Thomas Böhme") get high benefits from our new approach.

3 Implementation and Use

The prototypical search application PDSearch consists of a multi-threaded text analysis server written in Java and a client that has been conceived with easy and intuitive usability in mind. Therefore, the interactive graphical user interface (GUI) is realised as a dynamically created webpage in HTML and JavaScript with a query input field, a table showing the images matching the current query as pictograms and an option to search for similar images based on the selected images. The GUI can run in all modern web browsers and is generated by the text analysis server. Also, this way, it is easily possible to integrate the GUI into future web-based mobile applications, called Apps, that offer options for fast and convenient visual search.

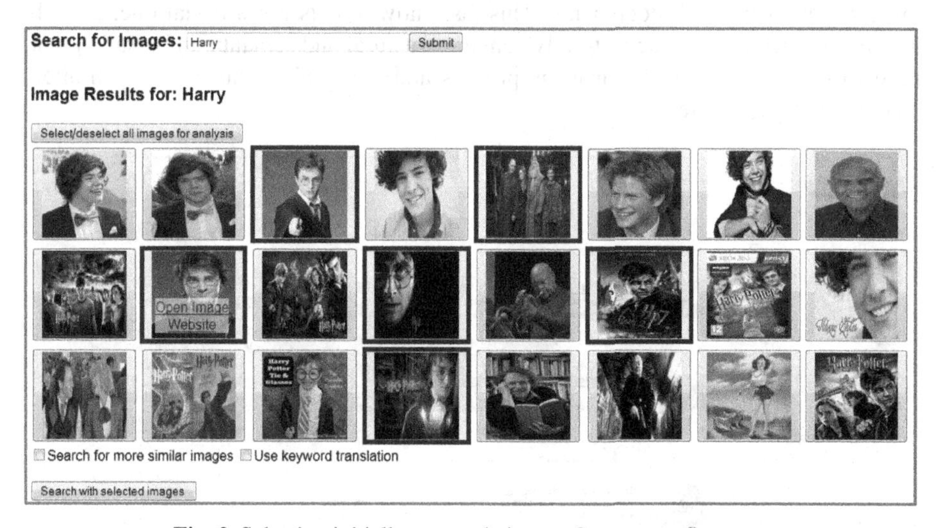

Fig. 3. Selecting initially returned pictures for query refinement

When the user selects one or more images for further analysis (see Fig. 3), the URLs of the web documents they are embedded in are sent to the analysis server that extracts their most prominent keywords and returns either another set of images that match these keywords for further query refinement as it can be seen in Fig.4 or will return documents containing these keywords that additionally can be interactively selected and deselected as shown in Fig. 5. Furthermore, the user has the possibility to directly open the respective web documents when pointing at a specific image.

Fig. 4. Similar pictures to the selected ones returned by PDSearch

In order to receive relevant documents and images that are similar to the selected images, the text analysis server needs to first determine high-quality keywords of the images' contextual web documents. This task, however, is not a trivial one, as web documents such as web pages usually contain structural and semantic information that need to be separated in the analysis process and may only contain a very limited amount of textual content.

Fig. 5. Similar documents to the selected pictures returned by PDSearch

Therefore, algorithms are needed that can even identify the topics of short texts in satisfying quality. Here, classical algorithms for frequency-based keyword extraction like TF-IDF [12] and difference analysis [13] will yield unsatisfying results. However, the graph-based algorithm presented in [14] extends the HITS algorithm [15] by taking into account the strength of the term associations provided by the edge weights of the (directed) co-occurrence graphs of the texts to be analysed and can determine their keywords with a high quality even when the textual basis is sparse. The consideration of the semantic relations between the terms in texts therefore plays a crucial part when it comes to identifying characterising and discriminating terms. Second, this algorithm can also detect the source topics in texts. These terms strongly influence the main topics in texts but are not necessarily important keywords themselves. Using these terms as queries, it is possible to find related (not necessarily similar) documents or images to the previously selected ones. The PDSearch analysis server employs this technique in order to extract the most important phrases and terms and uses them as search words in automatically generated queries to find similar and related images and documents. However, in case that no keywords can be extracted this way from the selected images' contextual web documents, the PDSearch suggests related terms to the images' captions (if they are available) or to the original query from Google Autocomplete [4] in order to help the user to find the right search words.

4 Discussion

The new search methodology discussed herein has been conceived to facilitate the search for web documents using relevant pictures as queries. However, not the images themselves are being analysed in order to identify the objects they show, but the textual contents of the web documents they are embedded in. These contents will usually describe the respective images with their entities properly. As mentioned in section two, this search approach is well suited to define an initial search context more precisely without the need for users to enter additional search words manually, which is a tedious and often error-prone task. This approach is additionally extended by offering selectable keywords of the analysed web documents to be used as query terms. That means, during the (first) search iterations using pictures as queries, a high recall of the search results is aimed at. When, however, the user is looking for web documents, as seen in Fig. 5, based on the previous selection of pictures, he or she can easily get to know new technical terms relevant for an in-depth research on a topic of interest. In this search step, the precision of the search results will be increased by the fine-grained selection of proper keywords as search words/query terms. Here again, the user does not need to enter search words manually. A simple click on the keywords is sufficient to change the current query, which is a major assistance during the search process. This is especially true when the user needs to delve into a new topic.

In first user tests, the validity and flexibility of this search paradigm could be proven. However, as the text analysis process is one of the most important steps (it is responsible to identify the web document's keywords and for the automatic query construction containing them) during the search sessions with PDSearch, it is sensible to think about options to increase the relevance of both kinds of search results (images and web documents). One way would be to realise a topically depending term weighting and clustering in conjunction with a named-entity recognition in order to improve the identification of relevant keywords and phrases. Another way would be to extend the already included cross-language search for English and German queries with the automatic translation of the initial and subsequent search queries into further languages. This way, relevant images and web documents in other languages could be found, too. Future search applications implementing this or a similar methodology can then offer their users a drastically enhanced usability when searching the WWW.

5 Conclusion

A new paradigm for web search is introduced based on the combined use of keywords and pictures. On the one hand, users may choose a set of similar and related pictures very fast to easily narrow down the search context. On the other hand, the search process benefits from taking into account the textual information of the web documents, which those pictures are embedded in, using an automatic extraction of additional search words. This method has been prototypically implemented in the search application PDSearch with a locally working frontend for the web search and an interface to the Google search engine. A couple of users already tried the new tool and found it very helpful in their daily application. Its advantages and possible ways for its improvement have been discussed, too.

References

1. November 2013 Web Server Survey 2013/11/01/november-2013-web-server-survey.html (2013), http://news.netcraft.com/archives/ (last retrieved on November 29, 2013)
2. Grimes, S.: Unstructured Data and the 80 Percent Rule (2008), http://breakthroughanalysis.com/2008/08/01/unstructured-data-and-the-80-percent-rule/ (last retrieved on November 29, 2013)
3. Page, L., Brin, S., Motwani, R., Winograd, T.: The pagerank citation ranking: Bringing order to the web. Technical report, Stanford Digital Library Technologies Project (1998)
4. Website of Google Autocomplete, Web Search Help (2013), http://support.google.com/websearch/bin/answer.py?hl=en&answer=106230 (last retrieved on November 29, 2013)
5. Kubek, M., Witschel, H.F.: Searching the Web by Using the Knowledge in Local Text Documents. In: Proceedings of Mallorca Workshop 2010 Autonomous Systems. Shaker Verlag, Aachen (2010)
6. Website of DocAnalyser (2013), http://www.docanalyser.de (last retrieved on November 29, 2013)
7. Website of WebNavigator (2013), http://www.docanalyser.de/webnavigator (last retrieved on November 29, 2013)
8. Yee, K., Swearingen, K., Li, K., Hearst, M.: Faceted metadata for image search and browsing. In: CHI 2003 Proceedings of the SIGCHI Conference on Human Factors in Computing Systems, pp. 401–408 (2003)
9. Tushabe, F., Wilkinson, M.H.F.: Content-based Image Retrieval Using Combined 2D Attribute Pattern Spectra. In: Peters, C., Jijkoun, V., Mandl, T., Müller, H., Oard, D.W., Peñas, A., Petras, V., Santos, D. (eds.) CLEF 2007. LNCS, vol. 5152, pp. 554–561. Springer, Heidelberg (2008)
10. Hawkins, J., Blakeslee, S.: On Intelligence: How a New Understanding of the Brain will Lead to the Creation of Truly Intelligent Machines. Times Books (2004)
11. Brisbane, A.: Speakers Give Sound Advice. Syracuse Post Standard, 18 (March 28, 1911)
12. Salton, G., Wong, A., Yang, C.S.: A vector space model for automatic indexing. Communications of the ACM 18(11), 613–620 (1975)
13. Heyer, G., Quasthoff, U., Wittig, T.: Text Mining: Wissensrohstoff Text: Konzepte, Algorithmen, Ergebnisse. W3L-Verlag, Dortmund (2006)
14. Kubek, M., Unger, H., Loauschasai, T.: A Quality- and Security-improved Web Search using Local Agents. Intl. Journal of Research in Engineering and Technology (IJRET) 1(6) (2012)
15. Kleinberg, J.M.: Authoritative sources in a hyperlinked environment. Journal of the ACM 46(5), 604–632 (1999)

Reducing Effects of Class Imbalance Distribution in Multi-class Text Categorization

Part Pramokchon and Punpiti Piamsa-nga

Department of Computer Engineering,
Faculty of Engineering, Kasetsart University, Thailand
{g4785036,pp}@ku.ac.th

Abstract. In multi-class text classification, when number of entities in each class is highly imbalanced, performance of feature ranking methods is usually low because the larger class has much dominant influence to the classifier and the smaller one seems to be ignored. This research attempts to solve this problem by separating the larger classes into several smaller subclasses according to their proximities, by k-mean clustering then all subclasses are considered for feature scoring measure instead of the main classes. This cluster-based feature scoring method is proposed to reduce the influence of skewed class distributions. Compared to performance of feature sets selected from main classes and ground-truth subclasses, the experimental results show that performance of a feature set selected by the proposed method achieves significant improvement on classifying imbalanced corpora, the RCV1v2 dataset.

Keywords: feature selection, ranking method, text categorization, class imbalance distribution.

1 Introduction

Multi-class text categorization is an automatic process for classifying text documents contents into predefined categories by assessment of their [1]. Most text categorization methods typically employ feature vector models for representing text documents, such as Bag-of-Words (BOW) [1-3] that generally uses all words in data collection as features. However, most words in documents are usually redundant and irrelevant to the classification. Hence, classification in this data space is not only computationally expensive but it also degrades the learning performance [1-5].

To reduce computation, a small set of the most relevant words to targeted classes is selected as features for classification process in order, while accuracy is still acceptable [1-5]. Filter-based methods, one of the most frequently-used methods, are to select features by statistical "usefulness" of each individual feature for classification. The usefulness score is determined by correlation measurement of feature characteristics to its class [1-5]. A feature that has the higher score is more relevant and features that have higher scores than a cut-off threshold are selected.

S. Boonkrong et al. (eds.), *Recent Advances in Information and Communication Technology*, Advances in Intelligent Systems and Computing 265,
DOI: 10.1007/978-3-319-06538-0_26, © Springer International Publishing Switzerland 2014

Usually, a cut-off threshold could be either a predefined maximum number of features or a predefined minimum score [6].

There are many feature scoring measures in text classification, such as Information Gain and Chi-square etc. [2-4, 7]; however, these measures often fails to produce good performance on multi-class text categorization, especially recall rate, due to the nature of skewed class distribution problem. The number of features and number of documents in each category is drastically different in real life [8-10]. The feature ranking methods [11] tend to rank features in large classes (majority), while those in small classes (minority), which are difficult to be learned, are rarely considered [9-10]. This not only leads to a low recall rate for the smaller classes but also reduces the overall performance of the multi-class text categorization.

We hypothesize that if each class has comparably equal number of features to the small main classes, rare but important features of the small main classes would have more appropriate chances to be selected as feature set. In this research, we proposed a technique to reduce the effect of imbalanced class distribution by creating intermediate subclasses by clustering features in each large class and then use these subclasses to be analyzed along with small main classes. Technically, large classes of training dataset are clustered by k-mean clustering [12] into k subclasses and those k subclasses are used to be replaced with their main class for computing feature scores. Unlike most of the re-sampling paradigm (under-sampling, over-sampling, and hybrids [9]), our proposed method avoids the problem of important concept missing, overfitting, and overgeneralization [9] by considering both within-class and between-class relationships based on the clustering method.

The performance of the proposed algorithm is evaluated on the RCV1v2 dataset [13]. Using accuracy and macro-averaged F1 in classification of the dataset on supported-vector machine (SVM), experiments showed that a feature set selected by proposed algorithms is significantly better than the one selected from only main classes. At 4000 features, macro-averaged F1 is increased from 0.813 to 0.941; accuracy is increased from 87.86 to 95.4. The feature selected by the proposed method not only increases overall accuracy, but also improves F1-measues of the smallest class from 0.547 to 0.881.

The rest of the paper is organized as follows. Background of text categorization, feature selection and problem of class distribution imbalance is summarized in section 2. Section 3 presents our proposed feature selection algorithm and describes its characteristics. Experiments are explained and the results are discussed in section 4. Section 5 concludes this work.

2 Related Works

2.1 Multi-class Text Categorization

Let $C = \{c_1, \dots, c_{|C|}\}$ be a set of categories of interest and $D = \{d_1, \dots, d_{|D|}\}$ be a set of documents. A decision matrix of text categorization is $K = D \times C$. Each element $\langle d_l, c_k \rangle$ is a Boolean value, a value of true indicates a decision to label d_j under c_k, while a value of false indicates a decision not to label d_j under c_k. A classifier is a function $\Phi := K \rightarrow \{true, false\}$ [1]. Let $F = \{f_1, \ldots, f_{|F|}\}$ be a set of original distinct features that occurs at least once in at least one document in the data collection; $Tr = \{d_1, \ldots, d_{|Tr|}\}$ be a training set of text documents; and $w(f_i, d_j)$ be the feature weight of feature f_i in document d_j. A document d_j is represented as a feature weight vector $\bar{d}_j = [w(f_1, d_j), \ldots, w(f_{|F|}, d_j)]$ [1,13-14]. This feature weight $w(f_i, d_j)$ quantifies the importance of the feature f_i for describing semantic content of document d_j. The documents in the training set are represented by these feature weight vectors which are used in the learning classifier steps. The classifier is then built by a machine-learning algorithm.

2.2 Feature Scoring Measure

Feature scoring measure is an approach to evaluate the usefulness of an individual feature by analyzing general characteristics of the training examples such as information, dependency, distance, consistency, etc [5]. In this research, we used document frequency and information gain in our experiments.

Document Frequency ($DF(f_i)$) [1,10] of a feature is the number of documents containing that feature defined as:

$$DF(f_i) = |\{d_j | f_i \text{occur in } d_j, d_j \in Tr\}| \tag{1}$$

Information Gain (IG) [1-2,4,7,10-11] is used to measure the amount of information a feature can contribute to the classification result. The Information Gain of feature t_i is defined as:

$$IG(f_i) = \sum_{k=1}^{|C|} \sum_{f \in \{f_i, \bar{f}_i\}} P(c_k, f) \log \frac{P(c_k, f)}{P(c_k) \cdot P(f)}, \tag{2}$$

where $P(c_k|f_i)$ is the conditional probability that the feature f_i occurs in category c_k, and $P(c_k|\bar{f}_i)$ is the conditional probability that f_i does not occur in c_k.

2.3 Class Imbalance and Its Effect on Feature Selection

Class imbalance of dataset is a situation that number of entities in each class is drastically different. Typically, classifiers are optimized for overall accuracy without taking distribution of class sizes into account. Thus, the larger classes are dominant by the higher frequencies of support data; and the smaller classes become statistically unimportant. Feature ranking methods also fail when applied to multiple classes with non-uniform class distributions (class skew) as well. The method pays more attention to the large classes to compensate for their weakness that tends to ignore small classes. Thus, the relevant features for large classes can be selected but features for small classes are ignored. In [10], it has been showed that Information Gain assigns higher scores to the terms (features) correlated with large classes and such that small classes are under-represented in top-rank features. Their experimental result also showed the overall accuracy is improved even though there is a low recall rate for small classes.

2.4 *k*-Means Clustering

The *k*-means clustering is an automatic and unsupervised approach to separate a dataset into *k* clusters. A data entity belongs to a cluster if that entity is the nearest to that cluster centroid. Given a set of document data (d_1, d_2, …, $d_{|D|}$), where each data is a $|F|$-dimensional vector, *k*-means clustering aims to partition the |D| observations into *k* clusters ($k \leq |D|$) $S = \{s_1, s_2, \ldots, s_k\}$ so as to minimize the within-cluster sum of squares (WCSS): $arg\ min_S \sum_{j=1}^{k} \sum_{x_i \in s_j} \|x_i - \mu_j\|^2$, where μ_j is the mean of the data in S_j. The most common algorithm uses an iterative refinement technique [12]. The process of algorithm starts by selecting *k* initial cluster centers. Then, the centers are iteratively updated by finding the means of the points closest to teach center. The process is done when there is no more updating needed on the centroids.

3 Proposed Method

We proposed to separate data in large classes into several subclasses according to their proximity, then all the subclasses are used for feature scoring measure instead of the main classes. The proposed algorithm is showed in Algorithm 1 (Cluster-based feature scoring measure algorithm).

Algorithm 1 Cluster-based feature scoring measure algorithm

Input: Original feature set (F)
  Training data ($\mathrm{Tr} = \{\langle d_1, c_i\rangle, \ldots, \langle d_{|Tr|}, c_{|Tr|}\rangle\}$)
  Main class ($C = \{c_1, \ldots, c_{|C|}\}$)
  Predefined number of cluster ($K = \{k_1, \ldots, k_{|C|}\}$)
Output: Feature set ranked by score ($FScore$)
1: for all classes in C do
2:   cluster all data $\{\langle d, c_j\rangle\}$ within the class c_j into $\{\langle d, c_j^t\rangle\}\ t = 1, \ldots, k_j$ using k-Means clustering
3: $\mathrm{Tr}' = \{\langle d_1, c_1^t\rangle, \ldots, \langle d_{|Tr|}, c_{|Tr|}^t\rangle\}$ //new training set
4: **for each** feature f_i **in** F **do**
5:   $s_i \leftarrow ComputeScore(f_i, Tr')$
6: $FScore \leftarrow SortedFeatureByScore(s_i)$
7: Return $FScore$

In Algorithm 1, given a set of training data with known main classes, $Tr = \{\langle d_i, c_j\rangle\}\ i = 1, \ldots, |Tr|, j = 1, \ldots, |C|$, the proposed method starts by separately clustering each class cj into kj clusters using the k-means clustering algorithm (lines 1-2). Then, in line 3, we have a new training set $Tr' = \{\langle d_i, c_j^t\rangle\}, i = 1, \ldots, |Tr|, j = 1, \ldots, |C|, t = 1, \ldots, |K|$ where t is the label of a specific cluster (subclass) of main class C and $|K| = \sum_{j=1}^{|C|} k_j$. Then in lines 4-5, feature score s_i of Tr' is measured with associated subclasses instead of the main classes. Feature results are sorted by feature score (line 6) and then feature set which is ranked by feature score is returned in line 7.

Separating the main classes into the subclass before computing feature score not only reduces the influence of large classes but also increases chance that important but rare features of small classes can be selected. In order to avoid the impact of class separating with small classes, we assign a small kj for small classes and large kj for large classes. The kj value is obtained by following formula:

$$k_j = round\left(\frac{|Tr|_j}{|Tr|} \times |K|\right) \tag{3}$$

where $|K|$ is an empirical parameter of the proposed algorithm, which requires the experiment to find out the best value, represented the summation of resulted subclass and $|Tr|_j$ is a number of text document in the class c_j.

4 Experiments

4.1 Dataset

In the experiment, the RCV1v2 dataset is used to evaluate our algorithm [13]. Table 1 shows details of dataset characteristics. It is organized in four "Topic Codes" hierarchical groups: CCAT, ECAT, GCAT, and MCAT. Each class also has different defined subclasses, which we used it as ground truth. Totally, dataset contains 19,806 training documents and 16,886 testing documents. Table 1 shows sizes of main classes and number of their ground truth subclasses. The largest class (CCAT) is almost five times as large as the smallest class (ECAT). Fig. 1 depicts distribution of ground truth subclasses. Note that size distributions of different subclasses are likely similar.

Table 1. Dataset characteristics

Class	Number of subclasses	Number of documents	
		training	testing
CCAT	18	8410	6993
ECAT	10	1690	1467
GCAT	23	4959	4580
MCAT	4	4747	3846

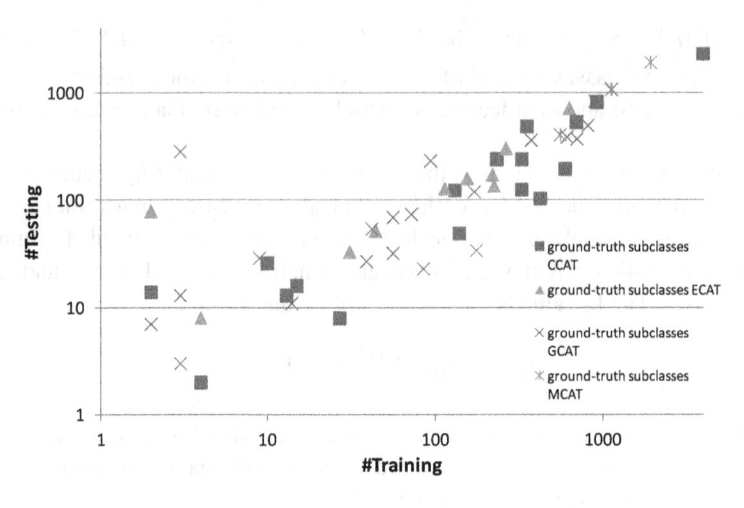

Fig. 1. Distribution of ground-truth subclasses

4.2 Performance Measure

To evaluate the overall performance of the proposed method in the multi-class text categorization, we employ the Accuracy defined as follows:

$$Accuracy = \frac{\sum_{j=1}^{|C|} TP_j}{\sum_{j=1}^{|C|} (TP_j + FN_j)} \times 100 \tag{4}$$

True positive TP_j is numbers of test documents that are classified into class Cj correctly, and false negative FN_j is the number of test documents incorrectly classified under other categories.

We also use F1 ($F1_j$) as the local performance measure, which is a combination of precision P_j and recall R_j. The definition of this measure is $F1_j = \frac{2 \times P_j \times R_j}{(P_j + R_j)}$, where $P_j = \frac{TP_j}{TP_j + FP_j}$ and $R_j = \frac{TP_j}{TP_j + FN_j}$, which False positive FP_j is numbers of test documents that are classified into class cj incorrectly. To exhibit the performance of the proposed method on the smaller class, the F1 measures over the multiple classes are summarized using the macro-averaged F1 [1] defined as follows:

$$MF1 = \frac{\sum_{j=1}^{|C|} F1_j}{|C|} \tag{5}$$

4.3 Results

In the experiment, we assumed that reducing size of large classes by clustering to smaller subclasses will make features in the small classes statistically more visible to the learning machine than results without clustering. We determined effectiveness of feature scoring measure among 1) the main classes, 2) the ground truth subclasses from the dataset, and 3) resulted K subclasses from the Algorithm 1. We applied Information Gain (IG) to measure feature scores on training data. Feature ranking method is executed with numbers of selected features |F|, ranged from 100 to 2000, stepped by 100 and from 2000 to 4000, stepped by 500. According to Algorithm 1, we generated a vector model for training data by using only top-ranked features instead of all features in order to reduce training time. Features are ranked by Document Frequency (DF). By empirical experiments, 4000 features are sufficient for acceptable clustering result. Then, we used k-means clustering onto the vector model. We selected several arbitrary values of |K| in the evaluation. First, we selected |K| = 55 since it is the same number as of ground-truth subclasses and then other lower values (22, 33, and 44) and one higher value (66) are arbitrarily selected in order to investigate effects of subclass sizes on the classification performance. The result, which is a set of clusters, was used as a subclass for measuring feature scores. Next, IG score for each feature was computed and the traditional feature ranking method proceeded.

We adopt Support Vector Machine (SVM) [15-16] with linear kernel which is a favorite and robust learning algorithm for highly dimensional data as the classifier in the experiments [16]. The experiment was performed using the LibSVM package of WEKA, with the default values of parameters [17]. We investigated the effectiveness of the features for multi-class text categorization. Using traditional feature ranking method, the selected feature subsets were used to train the classifier; then the classification performance was evaluated on training documents.

Fig. 2 shows the ranking accuracy of feature ranking method using clustering resulted subclass, ground truth subclass and main class on the RCV1v2 dataset. We observe that 1) the ground truth subclasses outperformed the main class when the number of feature further increases; 2) compared with main class, using ground truth subclasses for feature scoring measure can improve classification accuracy substantially; and 3) with |K| = 22, 33, 44, 55 and 66, our algorithm achieved higher accuracy than the main classes at number of feature from 100 to 4000 and case |K| = 44 performs the best. The accuracy values of both methods are approximately equal when number of features is more than 1800 with all |K| values. Moreover, the slope of ground-truth subclasses line is zero at 1800 features. Basically, the feature ranking method will select 1800 top-ranked features as the optimal feature subset.

Next, classifiers from training process were used to classify testing documents and the classification performance was then evaluated. Fig. 3 and Fig. 4 illustrate the performance comparison at selected numbers of features (1800 and 4000 features) of the main class, ground truth subclass and resulted class where K = 44. Fig. 3 shows that micro-averaged F1 of resulted class is slightly lower than of the ground truth subclasses. Since clustering technique is unsupervised learning method which has no prior knowledge, thus, a number of errors are normally occurred; 2) F1-measures in the smallest class (ECAT) are increased from 0.392 to 0.857 at 1800 features and from 0.547 to 0.881 at 4000 features; and 3) macro-averaged F1 are increased from 0.703 to 0.931 at 1800 features and 0.813 to 0.941 at 4000 features. Fig. 4 shows that overall accuracy measures are increased from 79.4 to 93.9 at 1800 features and 87.86 to 95.4 at 4000 features.

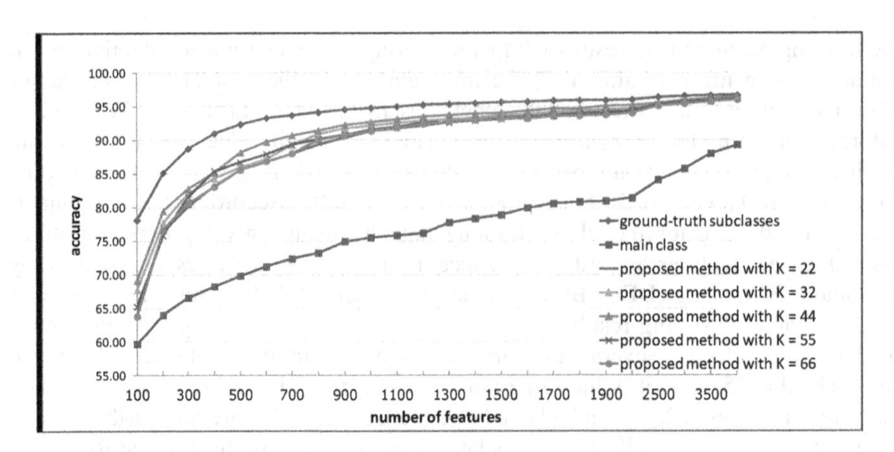

Fig. 2. The ranking accuracy of feature ranking method using clustering resulted subclass (|K| = 22, 33, 44, 55, 66), ground truth subclass and main class on the RCV1v2 dataset

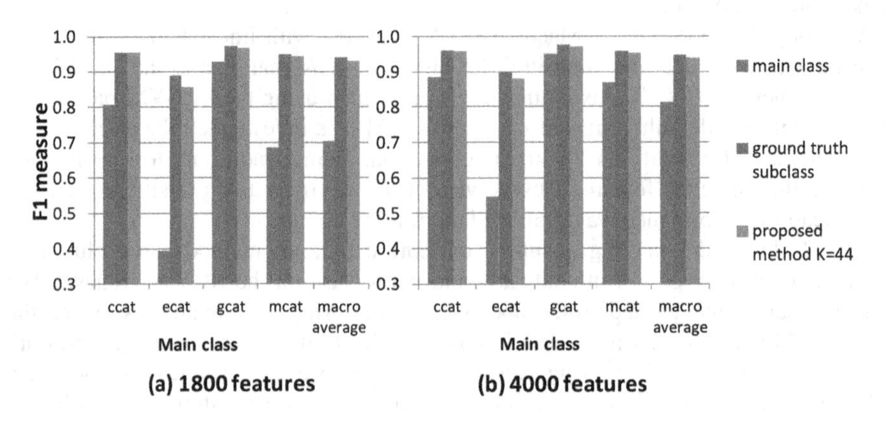

Fig. 3. The F1 measures for 1800 features and 4000 features for class variants

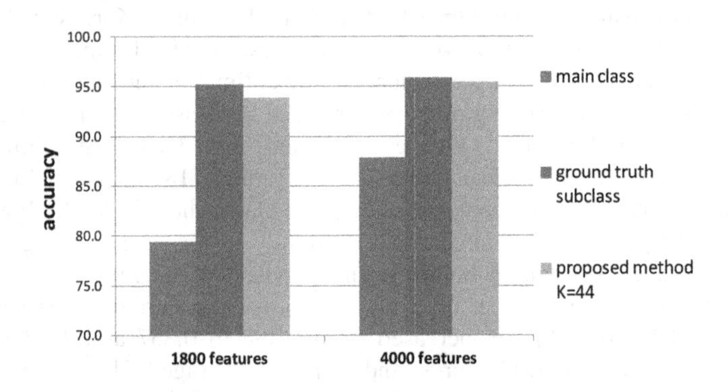

Fig. 4. The accuracy for 1800 features and 4000 features for class variants

By the experimental results, we can summarize that using only 1800 features performances (F1 and accuracy) of our proposed with K=44 are slightly lower than ground truth; however, it is much better than using main class only. More importantly, performances on the smaller classes are much improved. Therefore, cluster based method can obtain resulted cluster which can be used instead of main class for feature scoring measure on multi-class dataset with highly imbalance category sizes. Particularly, if the ground truth subclasses are unavailable, the proposed algorithm based on clustering technique can still process and yields the sufficient results. The suitable parameter |K| may be tuned by empirical process but we argue that parameter |K| should be defined as equal as or lower than number of ground truth subclasses.

5 Conclusion

In this paper, we proposed a method that improves the effectiveness of feature ranking method in the multi-class text categorization in the case that class distribution is imbalanced. The proposed method is about applying k-Means clustering to separate data from large classes into several subclasses according to their proximities, then all subclasses are considered for feature scoring measure instead of main classes. The proposed method aim to reduce influence of the large classes to feature ranking method, and increase chance of the small classes to be selected its rare and important feature. Our algorithm is developed and tested on the RCV1v2 dataset. We found that considering information between features and sub-hierarchy classes, instead of main classes for feature scoring measure improves the performance of multi-class classification with imbalance dataset. Experiments showed that proposed method works well and can separate main class to resulted subclass which balances the effect of size of the main class between large classes and small classes. We conclude that our proposed method is suitable for the data sets with high imbalance class distribution and the sub-hierarchies class is unavailable.

In future work, a quantitative analysis with the state-of-the-art methods still has to be done to justify the advantage of the proposed approach. Other topics still remain in the research field of this paper. For example making a method for determining appropriate K value or applying other clustering techniques that can efficiently and effectively separate the imbalance dataset are interesting for multi-class categorization research.

References

1. Sebastiani, F.: Machine learning in automated text categorization. ACM Comput. Surv. 34, 1–47 (2002)
2. Forman, G.: An extensive empirical study of feature selection metrics for text classification. J. Mach. Learn. Res. 3, 1289–1305 (2003)
3. Forman, G.: Feature Selection for Text Classification. Computational Methods of Feature Selection. Chapman and Hall/CRC Press (2007)

4. Yang, Y., Pedersen, J.O.: A Comparative Study on Feature Selection in Text Categorization. In: 14th International Conference on Machine Learning, pp. 412–420. Morgan Kaufmann Publishers Inc., 657137 (1997)
5. Liu, H., Yu, L.: Toward integrating feature selection algorithms for classification and clustering. IEEE Transactions on Knowledge and Data Engineering 17, 491–502 (2005)
6. Soucy, P., Mineau, G.W.: Feature Selection Strategies for Text Categorization. In: Xiang, Y., Chaib-draa, B. (eds.) AI 2003. LNCS (LNAI), vol. 2671, pp. 505–509. Springer, Heidelberg (2003)
7. Uchyigit, G., Clark, K.: A new feature selection method for text classification. International Journal of Pattern Recognition and Artificial Intelligence 21, 423–438 (2007)
8. Zheng, Z., Wu, X., Srihari, R.: Feature selection for text categorization on imbalanced data. SIGKDD Explor. Newsl. 6, 80–89 (2004)
9. He, H., Garcia, E.A.: Learning from Imbalanced Data. IEEE Trans. on Knowl. and Data Eng. 21, 1263–1284 (2009)
10. Makrehchi, M., Kamel, M.S.: Impact of Term Dependency and Class Imbalance on The Performance of Feature Ranking Methods. International Journal of Pattern Recognition and Artificial Intelligence 25, 953–983 (2011)
11. Makrehchi, M., Kamel, M.S.: Combining feature ranking for text classification. In: IEEE International Conference on Systems, Man and Cybernetics, ISIC 2007, pp. 510–515 (2007)
12. MacQueen, J.B.: Some Methods for Classification and Analysis of MultiVariate Observations. In: Cam, L.M.L., Neyman, J. (eds.) Proc. of the Fifth Berkeley Symposium on Mathematical Statistics and Probability, vol. 1, pp. 281–297. University of California Press (1967)
13. Lewis, D.D., Yang, Y., Rose, T., Li, F.: RCV1: A New Benchmark Collection for Text Categorization Research. Journal of Machine Learning Research 5, 361–397 (2004)
14. Lee, L.-W., Chen, S.-M.: New Methods for Text Categorization Based on a New Feature Selection Method and a New Similarity Measure Between Documents. In: Ali, M., Dapoigny, R. (eds.) IEA/AIE 2006. LNCS (LNAI), vol. 4031, pp. 1280–1289. Springer, Heidelberg (2006)
15. Drucker, H., Wu, D., Vapnik, V.N.: Support vector machines for spam categorization. IEEE Transactions on Neural Networks 10, 1048–1054 (1999)
16. Elias, F.C., Elena, M., Irene, D.A., Jos, R., Ricardo, M.: Introducing a Family of Linear Measures for Feature Selection in Text Categorization. IEEE Transactions on Knowledge and Data Engineering 17, 1223–1232 (2005)
17. Witten, I.H., Frank, E.: Data Mining: Practical Machine Learning Tools and Techniques, 2nd edn. Morgan Kaufmann, San Francisco (2005)

Towards Automatic Semantic Annotation of Thai Official Correspondence: Leave of Absence Case Study

Siraya Sitthisarn* and Bukhoree Bahoh

Division of Computer and Information Technology,
Faculty of Science, Thaksin University, Thailand
{ssitthisarn,ibook}@gmail.com

Abstract. The realization of semantic web depended on the availability of web of data associated with knowledge and information in the real world. The first stage for web of data preparation is semantic annotation. However framing such manual semantic annotation is inappropriate for inexperienced users because they require specialist knowledge of ontology and syntax. To address this problem, this paper proposes an approach of automatic semantic annotation based on the integration between natural language processing techniques and semantic web technology. A case study on leave of absence correspondence in Thai language is chosen as the domain of interest. Our study shows that the proposed approach verified the effectiveness of semantic annotation.

Keywords: semantic annotation, ontology, NLP, RDF.

1 Introduction

Nowadays, there is a significant number of information within Thai government organizations. Much of the information can be found in text repository and is represented in unstructured formats. Therefore, information systems could not understand the meaning of information content. As a result, the systems are not able to process the information in intelligent ways. For example, the systems could not provide an effective search because search engines are not understood either the keywords within user request or content of information. Consequently the systems just retrieve information that has the users' keyword terms included. Such the search may have high recall but has limited precision. This results in much irrelevant information being identified. In addition, without understanding information content is difficult for automatic interpreting into new knowledge. This will be useful for executive decision making.

Semantic Web technology matters for information management. It enables solution to address the above limitations [1-2]. Semantic web is intended for including meaning to documents on the web and representing it in RDF format that can be understood by applications or agents. It also allows systems to understand the content of the information. Consequently, systems enable high precision of search and intelligent document generation.

* Corresponding author.

S. Boonkrong et al. (eds.), *Recent Advances in Information and Communication Technology*, Advances in Intelligent Systems and Computing 265,
DOI: 10.1007/978-3-319-06538-0_27,

Semantic annotation is the process of annotating unstructured documents with semantic metadata. Most of the current technology is based on human-centered annotations. However, manual semantic metadata is difficult and time consuming. In addition, it also requires specialists who have knowledge on ontology and RDF syntax, to avoid incorrect and incomplete annotation [3-4]. In order to address human limitations, the automatic approach is required.

This paper proposes an automatic semantic annotation approach for official correspondence in Thai language. This is the initial stage of intelligent interpreting the correspondence into new knowledge for supporting executive decision making. The approach is based on integration between natural language processing techniques and semantic web technology. Ontology is for identifying vocabulary used for the correspondence annotation. The linguistic tool takes an important role for word segmentation. The longest string matching and pre-defined entity patterns are used as main techniques of named entity identification. The evaluation results provide high effectiveness of annotation for the case study.

The paper is structured in the following way. The next section introduces the background knowledge and related work. Section 3 presents the automatic semantic annotation architecture and the outlines of the main components are explained. Section 4 explains the design of the case study ontology. Then the details of implementation of each component are described in Section 5. The evaluation study is provided in section 6 followed by the conclusions and recommendations for future work.

2 Background Knowledge and Related Work

2.1 RDF

RDF (Resource Description Framework) is a formalized language for representing information in the web [5]. It is aimed at representing information which needs to be processed by an application rather than only being displayed to people. RDF provides a common set of assertions, know as statements, for expressing information. Each statement consists of three elements: subject, predicate and object [5].

RDF is a graphical representation in which statements form a directed graph. Subject and object are represented as nodes and the predicates are represented as edges. There are two types of nodes: resources and literals. Literals represent a constant value, such as a number or a string. In contrast, resources representing everything else by using URI and resources can be either subjects or objects. Predicates represent the connection between resources, or between resource and literal. RDF also provides an XML-based syntax for recording and exchanging such graphs.

RDF uses statements for expressing the information which combines data and metadata based on semantic data model inside. As a result the information exchange can be performed without the problem of syntax and loss of meaning. In addition, representing information using statement is also a powerful tool for information/knowledge integrations.

2.2 Related Work

Semantic annotation is the process of inserting tags into documents. The tags are specific metadata modeled in the domain ontology. The goal of semantic annotation is to provide documents with metadata represented in machine readable formats such as XML and RDF. Having examined the semantic annotation, which could be implemented in different approaches. Unfortunately, approaches can be classified using automation degree: manual and automatic annotations.

CREAM [6] is a manual semantic annotation framework. It allows users to type or drag and drop tags associated to instances of concepts in the ontology. OntoMat annitizer and OntoAnnotate [6] are the semi-automatic tools based on CREAM framework. These tools exploit HTML document structures and then provide user friendly functions for manual annotating web pages in OWL, SQL and Xpoint formats. Torres and his colleagues [7] proposed the tool named Semdrops to enrich web resources with semantic data. Semdrops is based on the social semantic tagging approach that allows users to collaboratively annotate the web resources. The annotated resources are represented in RDF format.

For the automatic annotation, KIM [8] proposed information extraction techniques for creating a semantic knowledge base. The annotation in KIM is metadata of people, place and so on. The metadata is vocabulary modeled in the KIMO ontology. KIM can support named entities appeared in documents in broad usage, such as television news. Some of automatic semantic annotation systems are specific to domains. For example, Smine and colleagues [9] presented the automatic annotation system for learning object annotation. In any text file, named entities related to the pre-defined learning object category are identified. Then the relationships between named entities are detected before generating metadata statements in XML format. Albukhitan and Helmy [10] provided an automatic annotation system for the domain of the Arab food recipe. Their approach is similar to Smine and colleagues but word segmentation and named entity identification techniques are specific for Arab language and the annotated documents are represented in RDF format. Similarly, we also propose an automatic semantic annotation but in different domains (i.e. Thai government official correspondence). Our approach is based on NLP techniques used in the Thai language. The Thai linguistic tool is used for supplementing NLP capability of our system. Metadata related to ontology will enrich the correspondence with the semantic RDF format. The architecture and detail of ontology are explained in the next section.

3 System Architecture

The automatic semantic annotation architecture used for Thai government official correspondence is shown in Fig. 1. At a high level, there are 2 main parts (i.e. ontology construction and semantic annotation processing).

With regard to the domain of interest, the leave of absence correspondence ontology (in OWL) is the core of semantic annotation. Its concepts and relationships are used as metadata for the correspondence annotation. The design of the ontology is discussed in Section 4.

The semantic annotation processing is based on NLP techniques. The architecture shares the same generic components which also have been used by other work [8-10]. As depicted in Fig.1, the semantic annotation processing consists of 4 components: (i) Thai word tokenizer; (ii) named entity recognition; (iii) relationship extraction; (iii) RDF generation.

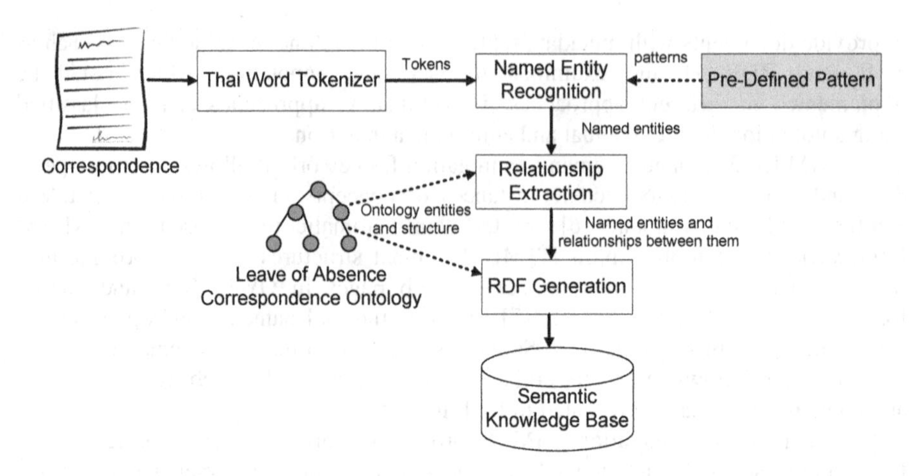

Fig. 1. The automatic semantic annotation architecture

The correspondence in the text file is the input of semantic annotation processing. Firstly the word tokenizer component separates the token entities in the document with unlabeled definition. The named entity recognition component will then identify type/name of token entities. Afterward the relationship extraction will detect relationships between named entities. The detected relationships associated to object/data properties in the ontology. Finally, the RDF generation component will create the RDF statements corresponded to ontology entities as well as named entities and their relationships which are extracted from the documents. These semantic documents are stored in the knowledge base repository. Each component will be discussed in detail to illustrate the extension made by this paper.

4 Leave of Absence Correspondence Ontology

Leave of absence correspondence is the government official correspondence sent between staffs in the organization. The purpose of the correspondence is to ask the commander in chief's permission for a short time leaving with causes of: unwell condition, personal business, academic conference, vacation and so on. Similar to other types of official correspondence, the leave of absence correspondence is used as evidence for personal performance assessment.

This section explains the design of leave of absence correspondence ontology. The ontology is to provide vocabulary for annotating the correspondence. The ontology consists of four main classes: (i) Correspondence, (ii) Sender, (iii) Receiver, (iv) Leave. As depicted in Fig. 2 the class "Correspondence" relates to the class "Sender" by an object property *sentFrom*. Similarly, "Correspondence" and "Receiver" are related by the object property *sentTo*. The class "Sender" and "Leave" are associated with the object property *involve*. Both "Sender" and "Receiver" are subclasses of "Staff". In addition, the class "Staff" is also connected to the class "Position" and "Sub_Organization" by the object properties *havePosition* and *workIn* respectively.

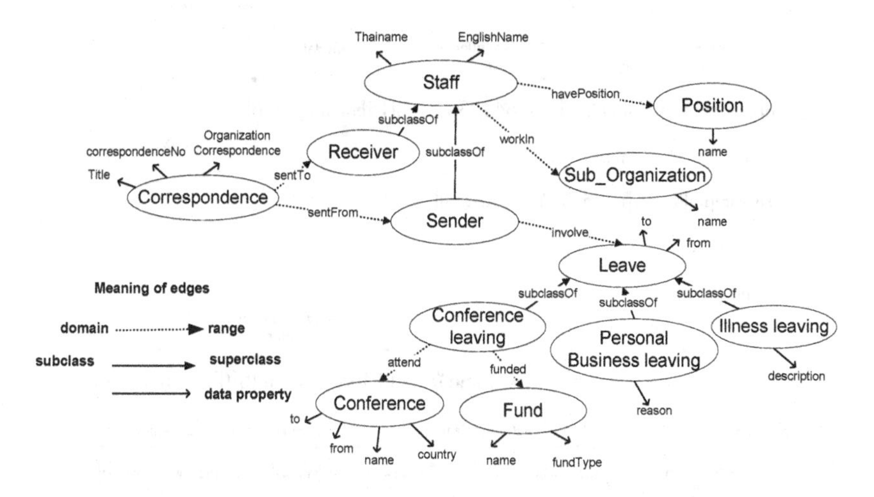

Fig. 2. The leave of absence correspondence ontology diagram

In our case study the class "ConferenceLeaving", "PersonalBusinessLeaving" and "IllnessLeaving" are subclasses of the "Leave". As can be seen in Fig. 2 the subclass "ConferenceLeaving" relates to the class "Conference" by an object property *attend*. Besides, the subclass "ConferenceLeaving" also associate to the class "Fund" by an object property *funded*.

5 Semantic Annotation Processing

This section explains the implementation of semantic annotation processing. Detail of each component will be described as below.

5.1 Thai Word Tokenizer

Thai word tokenizer is the first stage of semantic annotation processing. It is used to identify and extract tokens that appear in Thai documents. However, Thai language is different from others. It consists of 44 consonants and 21 vowel symbols. The direction of writing system is left to right without spaces between words. Thai word is mixed between consonants and one vowel symbols. Besides, there are no symbols like a full stop or space to identify the end of a sentence. Example of correspondence in Thai language is depicted in Fig.3.

To segment words, we used a Thai tokenizer tool named LongLexto. It was launched by Thai National Electronic and Computer Technology Center (NECTEC). The LongLexto is tokenizer based on the longest matching approach [11]. The segmentation performance is based on pattern matching using dictionaries. However, if a token in document does not exist in the dictionary. The longest string matching is used.

Apart from Thai dictionary, our work provides additional dictionary to gather lists of sub organizations in Thaksin University, positions and name of staffs. The additional dictionary will supplement the specific name matching. Example of output from this stage is also illustrated in Fig.3.

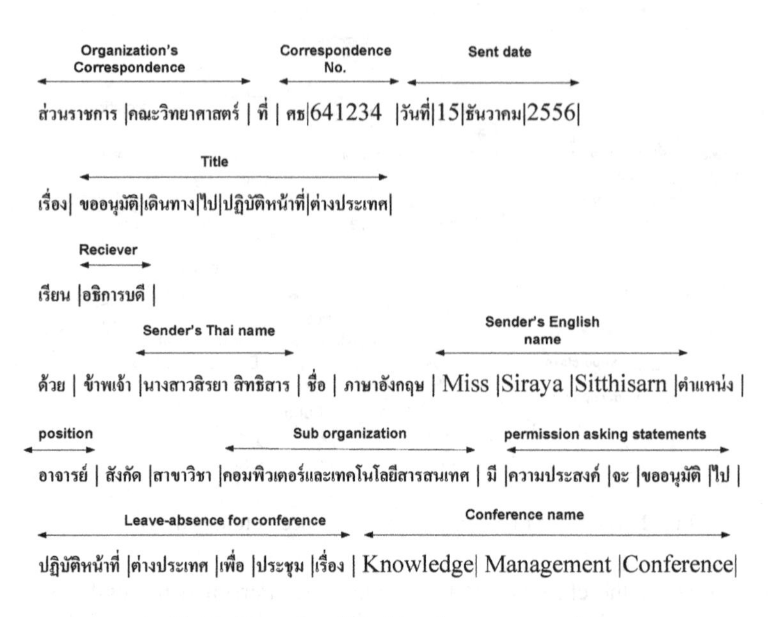

Fig. 3. Example of Thai text in a correspondence

5.2 Named Entity Recognition

Basically, named entity (NE) recognition is a process of identifying name of: people, organizations, locations and so. This paper, we used 2 main techniques for recognizing name entities: string matching and predefined pattern.

As depicted in Fig.4 the longest string matching technique is used for matching tokens with concepts (comments/labels of classes) and instances in the ontology. For example, tokens: | Knowledge | management |conference|, we can match |conference| with Class: Conference. In addition, |Knowledge | management| are mapped to instances of Class: Conference. Therefore we can identify that a group of word: Knowledge management conference is referred to name of the conference.

However, there are some ambiguities on named entity recognition. Sometimes, it is difficult to identify that named entities are relevant to which ontology concepts. For example, there are 2 named entities referred to positions: (i) university president and (ii) lecturer. The question is "which one is the position of receiver". Therefore pre-defined patterns are required for supplementing this process. As can be seen, the patterns of a receiver can be recognized because it usually follows word | เรียน | (To).

Receiver ::= 'เรียน' (Position name)

Sender ::= 'ข้าพเจ้า' (Gender Pronoun) (Sender name) 'ตำแหน่ง'(position name)

Gender Pronoun ::= 'ดิฉัน' | 'ผม'

The results of named entity recognition are stored in the table. These will use as input for next stage (i.e. relationship extraction).

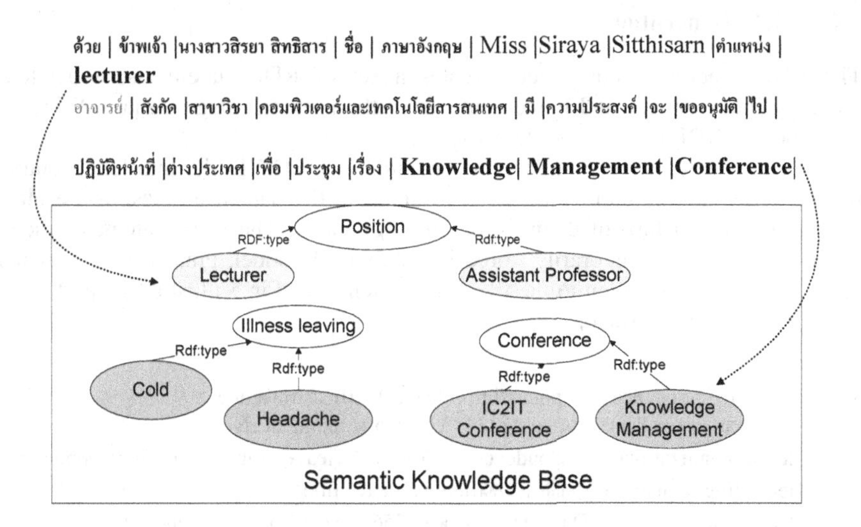

Fig. 4. String matching between tokens and instances/classes in the ontology

5.3 Relationship Extraction

The relationship extraction is to define the relationships of named entities. In order to do that, we need the ontology index for containing the ontology structure. The ontology index is used for mapping relationships between entities.

The ontology index contains entries of object/data properties in the ontology. Each entry has 3 fields: (i) domain, (ii) object property/data property, (iii) range/literal. As depicted in Table 1, one of the entries contains object property "sentFrom" which has domain as "Correspondence" and range as "Sender". The class of "Correspondence" is also the domain of data property "OrganizationCorrespondence". Hence, the entries of the index include data property "OrganizationCorrespondence" which has domain as "Correspondence" and range is literal (i.e. named entity referred to OrganizationCorrespondence). For "Sender" the entries of the index include data property "Thai name" which has domain as "Sender" and range is literal (i.e. named entity referred to Sender's ThaiName). Each ontology index entry will be mapped to named entities. As a result we have a set of statements that consists of subjects, predicates and objects. This set is input for RDF generation.

Table 1. Examples of entries in the ontology index

Domain	Object/Data Property	Range
Correspondence	*sentFrom*	Sender
Correspondence	*organization Correspondence*	Named entity referred to *organizationcorrespondence*
Correspondence	*correspondenceNo*	Named entity referred to *correspondenceNo*
Sender	*ThaiName*	Named entity referred to Sender's *ThaiName*

5.4 RDF Generation

The RDF generation component creates a set of RDF statements associated to ontology index entries and named entities in the previous stage. Jena API [12] is a main tool for RDF information construction.

The construction starts with loading the ontology (in OWL file) into Jena's ontology model. The ontology entities are used to annotate the named entities, as we already define the relationships of them in the ontology index. The RDF statements then are constructed and are temporarily stored in Jena's RDF model. Finally all statements in the RDF model are written to the file (RDF extension). Fig.5 illustrates a part of RDF information from this stage.

```
<leave:Correspondence rdf:about="http://leaveab/2013/siraya.doc">
  <leave:correspondenceNo>ศธ641234</eave:correspondenceNo>
  <leave:organizationCorrespondence>คณะวิทยาศาสตร์</leave:organizationCorrespondence>
  <leave:title>ขออนุมัติเดินทางไปปฏิบัติหน้าที่ต่างประเทศ</leave:title>
  <leave:CorrespondenceDate>15ธันวาคม พ.ศ.2556</leave:CorrespondenceDate>
  <leave:sentTo>
     <leave:Receiver rdf:about="http://www.tsu.ac.th/presidentInfo.html">
     <leave:havePosition>
        <leave:Position rdf:about="http://tsu.ac.th/position/president">
          <leave:name>อธิการบดี</leave:name>
        </leave:Position>
     </leave:havePosition>
   </leave:Receiver>
  </leave:sentTo>
```

Fig. 5. RDF information of the leave of absence correspondence

6 Evaluation

The objective was to evaluate the capability of the named entity component. If the system reaches the high effectiveness named entity identification so that it also enables high correctness of annotation. The used metrics were precision and recall. Precision = $\frac{|Ra|}{|A|}$ and Recall = $\frac{|Ra|}{|R|}$ where |R| is the number of entities corresponded to each ontology entity; |A| is the entities identified; and | Ra | is the number identified entities that are relevant to meaning of content.

The experiment was conducted on 3 groups of leave of absence correspondence (i.e. conference, personal business and illness). Each group contains 30 text files of correspondence. The results of named entity recognition are stored in the tables. Each table involves an ontology class and its data properties. Then the average of precision and recall of each ontology entity is calculated.

The results concerning the average precision and recall of each class and its data properties are summarized in Table 2.

Table 2. Experiment results

Test sets of entities	Precision (%)	Recall (%)
Correspondence No and Title	100.00	100.00
Receiver: (position)	100.00	100.00
Sender :(Thai name, English name)	95.00	100.00
Position	98.24	93.33
Organization	89.28	83.33
Business reason	95.00	63.33
Illness symptom	88.00	73.33
Conference	74.07	66.66
Fund	84.61	73.33

The results show that there is very high effectiveness of named entity identification on the entities of "Correspondence", "Receiver" and "Sender". This is because the set of entities appears in the correspondence with the fixed positions. While, entities of "Position" and "Organization" have a lower level of precision and recall because our named entity techniques could not handle the problem of the misspelled word. Similarly, the remainder set of entities has the lowest level recall because: (i) there are some tokens that couldn't be exactly mapped onto any entity in ontology as well as any pre-defined pattern; (ii) there are misspelled tokens in the correspondence.

7 Conclusion and Future Work

This paper, we have proposed the automatic semantic annotation approach based on integration between natural language processing techniques and semantic web technology. The domain of interest is official leave of absence correspondence. The initial results are promising and show that the component of named entity recognition provides the satisfaction level of precision and recall.

Unfortunately, we are planning to improve the performance of the named entity extraction. In particular, the investigation of algorithms/approaches for addressing the problem of the misspelled word that appears in the correspondence.

References

1. Strasunskas, D., Tomassen, S.L.: A Role of Ontology in Enhancing Semantic Search: the EvOQS Framework and its Initial Validation. International Journal of Knowledge and Learning 4, 398–414 (2009)
2. Qu, Y., Cheng, G.: Falcons Concept Search: A Practical Search Engine for Web Ontologies. IEEE Transaction on Systems, Man, and Cybernetic—Part A: Systems and Humans 41 (2011)
3. Kiyavitskayaa, N., Zenic, N., Cordy, J.R., Michc, L., Mylopoulosa, J.: Cerno: Light-weight tool support for semantic annotation of textual documents. Data & Knowledge Engineering 68, 1470–1492 (2009)

4. Urena, V., Cimianob, P., Iriac, J., Handschuhd, S., Vargas-Veraa, M., Mottaa, E., Ciravegnac, F.: Semantic annotation for knowledge management: Requirements and a survey of the state of the art. Web Semantics: Science, Services and Agents on the World Wide Web 4, 14–28 (2006)
5. Manola, F., Miller, E.: RDF Primer, http://www.w3.org/TR/2004/REC-rdf-concepts-20040210/
6. Handschuh, S., Staab, S., Studer, R.: Leveraging metadata creation for the Semantic Web with CREAM. In: Günter, A., Kruse, R., Neumann, B. (eds.) KI 2003. LNCS (LNAI), vol. 2821, pp. 19–33. Springer, Heidelberg (2003)
7. Torres, D., Diaz, A., Skaf-Molli, H., Molli, P.: Semdrops: A Social Semantic Tagging Approach for Emerging Semantic Data. In: IEEE/WIC/ACM International Conference on Web Intelligence, WI 2011 (2011)
8. Kiryakov, A., Popov, B., Terziev, I., Manov, D., Ognyanoff, D.: Semantic annotation, indexing, and retrieval. Web Semantics: Science, Services and Agents on the World Wide Web 2, 49–79 (2004)
9. Smine, B., Faiz, R., Descles, J.P.: Extracting Relevant Learning Objects Using a Semantic Annotation Method. In: International Conference on Education and e-Learning Innovation (2012)
10. Albukhitan, S., Helmy, T.: Automatic Ontology-Based Annotation of Food, Nutrition and Health Arabic Web Content. In: The 4th International Conference on Ambient Systems, Networks and Technologies, ANT 2013 (2013)
11. Haruechaiyasak, C.: Longlexto: Tokenizing Thai texts using longest matching approach (2006)
12. Mccarthy, P.: Introduction to Jena: Use RDF Models in Your Java Applications with the Jena Semantic Web Framework, http://www.ibm.com/developerworks/xml/library/j-jena/

Semantic Search Using Computer Science Ontology Based on Edge Counting and N-Grams

Thanyaporn Boonyoung[1] and Anirach Mingkhwan[2]

[1] Information Technology, King Mongkut's University of Technology
North Bangkok, Bangkok, Thailand
thanya.nb@gmail.com
[2] Industrial and Technology Management, King Mongkut's University of Technology
North Bangkok, Bangkok, Thailand
anirach@ieee.org

Abstract. Traditional Information Retrieval systems (keyword-based search) suffer several problems. For instance, synonyms or hyponym are not taken into consideration when retrieving documents that are important for a user's query. This study adopts an ontology of computer science and proposes an ontology indexing weight based on Wu and Palmer's edge counting measure for solving this problem. This paper used the N-grams method for computing a family of word similarity. The study also compares the subsumption weight between Hliaoutakis and Nicola's weight and query keywords (Decision Making, Genetic Algorithm, Machine Learning, Heuristic). A probability value (p-values) from the t-test (p = 0.105) is higher 0.05 and indicates no significant evidence, of not differences between both weights methods. The experimental results show that the document similarity score between a user's query and the paper suggests that the new measures were effectively ranked.

Keywords: Computer Science Ontology, Semantic Search, Ontology Indexing.

1 Introduction

The advancement of Information Technology and Information Society has increased the amount of information available and, as a result, the development of information storage and retrieval systems has become a significant challenge. The goal of Information Retrieval discipline is to search and retrieve the most relevant documents to a user's query. Therefore a good IR system should retrieve only those documents that satisfy the user needs, not a lot of unnecessary data. Traditional Information Retrieval systems (keyword-based search) suffer several problems such as synonyms or hyponym are not taken into consideration to document retrieval so users must input several similar keywords to retrieve relevant documents. However, the search system teats all keywords with the same importance and retrieves all the relevant documents. The result is information overload that makes it difficult for users to find really useful information from a large amount of search results[1]. Aiming to solve the limitation of a keyword-based search is a semantic similarity measure [2-5] and an ontology-based approach [6-7] with which researchers try to understand a user's query.

S. Boonkrong et al. (eds.), *Recent Advances in Information and Communication Technology*, Advances in Intelligent Systems and Computing 265,
DOI: 10.1007/978-3-319-06538-0_28, © Springer International Publishing Switzerland 2014

Semantic search systems consider various points including the context of the search, location, intent, variation of words, synonyms, generalized and specialized queries, concept matching and natural language queries to provide relevant search results. For example, consider an article on general health topics and another article specifically on Diabetes. If someone searches for health information, both articles could match even though the article on Diabetes does not talk specifically about "health".[8] Many semantic similarity searches use domain ontologies to consider the hierarchical structure and compute the relationship between terms. Most semantic similarity researches compute the similarity between concept using Wordnet[1], which is an online lexical and can also be seen as an ontology. It contains terms, organized into taxonomic hierarchies. Although Wordnet is widely used, it is still limited, and does not offer specific domain, the results are not in an hierarchical structure and researchers can't compute the relationship between the terms. Consequently, specific domains such as the United States National Library of Medicine (NLM)[2] offering a hierarchical categorization of medical and biological terms called Medical Subject Headings (MeSH)[3] [2,9] facilitates searching. Computer Science field are specific domain as the medical and biological domain that some term and relation not offered in Wordnet taxonomic hierarchies. For example, when users are interested topic about "decision making", their hierarchical structure does not propose in Wordnet, in which decision making is related to supporting decision making, decision trees and many others in the knowledge of Computer Science.

The difference of document keyword and Ontology keyword (Computer Science Ontology) is keyword determining the document (academic articles), often result in the terminology (specific domain), the target group, and the methodologies mixing together. For example "Medical decision making", "Heuristic evaluation" and "Genetics based machine learning" are keywords not found and not related in the ontology. Thus the results cannot measure the path length on Computer Science Ontology of a query's keyword and document keywords. So this study used the N-grams method to solve this problem because it is an efficient method for computing a family of word similarity.

In order to be able to improve the accuracy of semantic similarity measure between words in Computer Science field, the study adopts a Computer Science Ontology, which can also be represented as a hierarchical structure to measure the path length of the words in the taxonomy. This paper conducts a semantic search using Computer Science ontology based on edge counting and N-grams.

The rest of this paper in organized as follows: A review related work on semantic similarity measure Section 2. Computer Science Ontology, taxonomic hierarchy used in this work, as well as a comparison of semantic similarity methods with standard weight on the ontology in Section 3. Calculating a document similarity score is presented in Section 4. The experiment results of our approach are presented in Section 5, followed by Conclusion and future studies.

[1] `http://wordnet/princeton.edu`
[2] `http://www.nlm.nih.gov`
[3] `http://www.nlm.nih.gov/mesh`

2 Related Work

In an Ontology based similarity measure, searching is performed by interpreting the meanings of keywords (i.e. semantics). The comparison of two terms in a document using ontology usually has properties in the form a attributes, level of generality or specificity, and relationships with other concepts. The method used in this current study was a Computer Science ontology based similarity measure based on edge counting methods. This is a semantic similarity measure that computes the similarity between two concepts based on edge counting method. Wu and Palmer Measure [10] proposes a similarity measure for finding the most specific common concept that subsumes both of the concepts being measured. The length from the most specific shared concept is scaled by the sum of IS-A links from it to the compared two concepts.

$$ConSim(C_1, C_2) = \frac{2N}{N1+N2+2N} \quad (1)$$

Hliaoutakis et al.[2] investigated approaches to computing the semantic similarity between medical terms (using the Mesh ontology of medical and biomedical terms). Wu and Palmer Measure values are normalized and have a high correlation with other approaches. N-Grams Similarity and Distance is a contiguous sequence of n items from a given sequence of text or speech. Grzegorz Kondrak [11] formulates a family of word similarity measures based on N-grams, computes word similarity based on N-grams, Sembok and Bakar [12] and proposes N-grams string similarity and stemming on Malay documents.

$$\frac{2x|n\text{−}Keyword(QW) \cap n\text{−}Keyword(OW)|}{|n\text{−}Keyword(QW)|+|n\text{−}Keyword(OW)|} \quad (2)$$

In this study's semantic similarity measure, the edge counting method (Wu and Palmer Measure) was used to compute the hierarchy relation on Computer Science Ontology. The study found keywords string matching with the tri-grams method that is often used in natural language processing for doing statistical analysis of texts.

3 Computer Science Ontology

This section describes taxonomic hierarchy (Computer Science ontology) and ontology indexing weight.

3.1 Computer Science Taxonomy

The present study adopts an ontology for knowledge of computer science, reference from computer science curricula 2013 draft report[4] that has been endorsed by the Association for Computing Machinery (ACM) and IEEE Computer Society. Computer Science Curricula 2013 (CS2013), represents a comprehensive revision. 11

[4] http://ai.stanford.edu/users/sahami/CS2013

CS2013 redefines the knowledge units in Computer Science(CS). The last complete Computer Science curricular volume was 9 released in 2001 (CC2001) , and an interim review effort concluded in 2008 (CS2008)[5]. The CS2013 Body of Knowledge is organized into a set of 18 Knowledge Areas (KAs) in Table 1, 70 corresponding to topical areas of study in computing.

The taxonomic hierarchy (ontology) model of computer science keywords(terms) is organized in Is-A relationships (Hyponym/Hypernym) with more general terms (e.g "Operating System and Digital Forensic", "Information Management and Database System") higher in Information Management taxonomy than more specific terms (e.g "Object-oriented model" ,"Indexing"). A keyword (term) may appear in more than one taxonomy, such as "Information Retrieval" is term of Information Management and Intelligent Systems show that in Fig. 1. There are four levels, eighteen taxonomies and more than 200 terms.

Table 1. Body of Computer Science Knowledge

Knowledge Areas	Abbr.	Knowledge Areas	Abbr.
Algorithms and Complexity	AL	Networking and Communication	NC
Architecture and Organization	AR	Operating Systems	OS
Computational Science	CN	Platform-based Development	PBD
Discrete Structures	DS	Parallel and Distributed Computing	PD
Graphics and Visual Computing	GV	Programming Languages	PL
Human-Computer Interaction	HC	Software Development Fundamentals	SDF
Security and Information Assurance	IAS	Software Engineering	SE
Information Management	IM	Systems Fundamentals	SF
Intelligent Systems	IS	Social and Professional Issues	SP

3.2 Ontology Indexing

This study computes the weight of subsumption (hypernym/ hyponym or meronym/holonym hierarchy) in Computer Science ontology using Wu and Palmer measure [10]. This weights are shown in Table 2. and document similarity scores shown in Table 3. The study compares the our Subsumption weight with Hliaoutakis et al.[2] and Stokes's weight [13] with four query keywords (Decision Making, Genetic Algorithm, Machine Learning,Heuristic. A probability value (p-values) from the t-test (p = 0.105) is higher 0.05 and indicates no significant evidence that our weight based on Wu and Palmer Measure and their methods were not different.

For example, if query keyword(QK) and ontology keyword (OK) are the same word or synonymous, the weight is assigned as 1. For example, 'Database' and 'DB', 'Information

[5] http://www.acm.org/education/curricula/ComputerScience2008.pdf

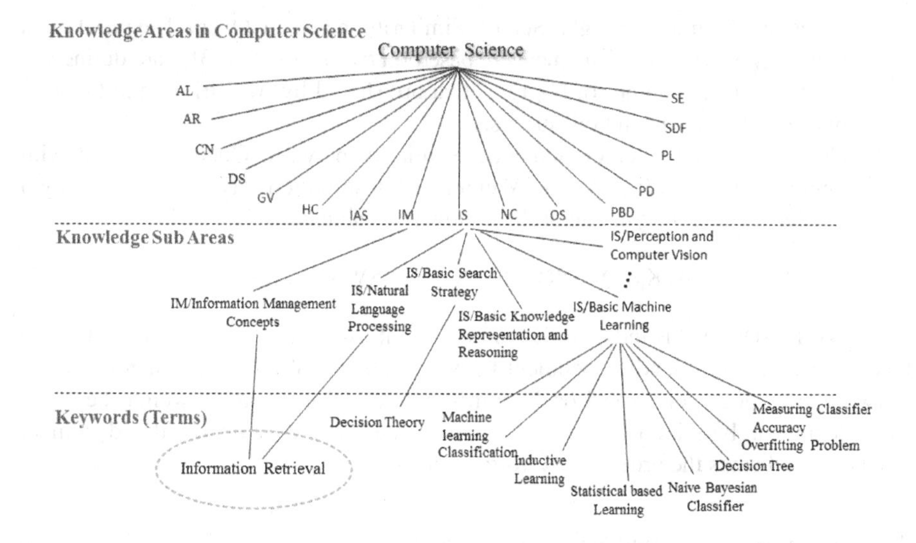

Fig. 1. Illustration of computer science ontology hierarchy

Retrieval' and 'IR', 'Entity-Relationship' and 'E-R' in different documents express the same meaning. The weight for 'Decision Making' with 'Team organization' is 0.75 Both keywords are members in Software Engineering/Software Project Management.

Table 2. Computer Science ontology Weight based on Wu and Palmer Measure

Relationship Type	Our Weights
Repetition /Synonymy	1
query keyword and document keyword have same sub area	0.75
query keyword and document keyword have same area	0.5
query keyword and document keyword have different area	0.25
query keyword not found In CS ontology	0

4 Calculating Document Similarity Score

This section describes the document similarity score between a document and the query. In the document keyword set, a document is represented by a keyword vector, i.e., document = (keyword1, . . . , keywordi, . . . , keywordn) ($1 \leq i \leq n$). We first compute keyword weight of document.

4.1 Calculating Document Keywords Score

We first compute a total of semantic keywords weight of document (3) with N-grams (2) and hierarchy relation in Computer Science ontology. This study proposes two method for calculating keywords weight set of document:

(a) Calculate N-grams Weight Set of Similarity between Query Keyword with Ontology Keyword : The measure based on *tri-grams* (1) [13], are defined as the ratio of the number of N-grams that are shared by two strings and the total number of N-grams in both strings.
(b) Calculate Weight Set of Semantic Similarity between Query Keyword with Ontology Keyword: Semantic Weight of Subsumption (Hypernym/ Hyponym or Meronym/Holonym Hierarchy), shown in Table 2.

$$SWK_{d,i}(QK,DK) = SWK_{n\text{-}gram} + SWK_{cs\text{-}onto} \qquad (3)$$

where $SWK_{d,i}(QK,DK)$ is the total Semantic Weight Keyword (SWK) document of the Query Keyword (QK) and a document keyword (DK); $SWK_{n\text{-}gram}$ is semantic weight keyword each document based on N-grams; $SWK_{cs\text{-}onto}$ is semantic weight keyword each document based on Computer Science ontology; d indicates the document number ; i indicates the order of keyword on document;

4.2 Calculate Document Similarity Score

The similarity between a document and the query is computed as

$$\mathbf{DSS}\,(D_d, \mathbf{q}) = \begin{cases} 1 + \dfrac{\sum_{i=1}^{N} \mathrm{SWK}_{d,i}}{N-1} & \text{; One of the query and Keywords document are exactly alike.} \\ \dfrac{\sum_{i=1}^{N} \mathrm{SWK}_{d,i}}{N} & \text{; otherwise} \end{cases}$$

Where $DSS(d_i,q)$ is the document similarity score (DSS) of a document (d) and the query (q) ; N indicates the total number of keywords in $document_i$;

5 Experiment Results

This section we summarizes the main experiments and the results obtained in the study. To test the proposed system, this present study used Computer Science Ontology and Computer Science documents, which consist of 1769 documents. As an example, 'Decision Making' is the query for the experiment. First, the research found that the knowledge area of 'Decision Making' in Computer Science ontology and its Computer Science area are Software Engineering and Social and Professional Issues. As a result, the study computed both the distance area of the query and the document keywords (see Fig.2).

This paper also computes the string similarity with N-grams and used Wu and Palmer measure to find the distance in Computer Science Ontology, described in section 3 and 4. The result in Table 3. shows that the document similarity score of documents and the query ("Decision Making").

Table 3. Accuracy Comparison Between Keyword-based with Our Method

Decision Making								
PaperID	**Sim_{SE}**	**Sim_{SP}**	**PaperID**	**Sim_{SE}**	**Sim_{SP}**	**PaperID**	**Sim_{SE}**	**Sim_{SP}**
115	1.493	1.118	**553**	0.558	0.558	**750**	0.290	0.290
1055	1.250	1.250	**180**	0.443	0.318	**857**	0.290	0.290
847	1.208	1.083	**1537**	0.403	0.403	**882**	0.264	0.250
1700	1.125	1.125	**758**	0.365	0.428	**584**	0.250	0.063
1506	1.100	1.100	**859**	0.365	0.428	**155**	0.230	0.230
852	0.930	0.680	**135**	0.353	0.253	**829**	0.207	0.207
779	0.730	0.417	**372**	0.334	0.334	**40**	0.195	0.195
1455	0.652	0.589	**1046**	0.332	0.282	**1026**	0.186	0.186
810	0.615	0.428	**775**	0.316	0.441	**726**	0.164	0.129
69	0.604	0.438	**544**	0.297	0.197	**1681**	0.163	0.413
1706	0.588	0.438	**1410**	0.291	0.191	**59**	0.050	0.050

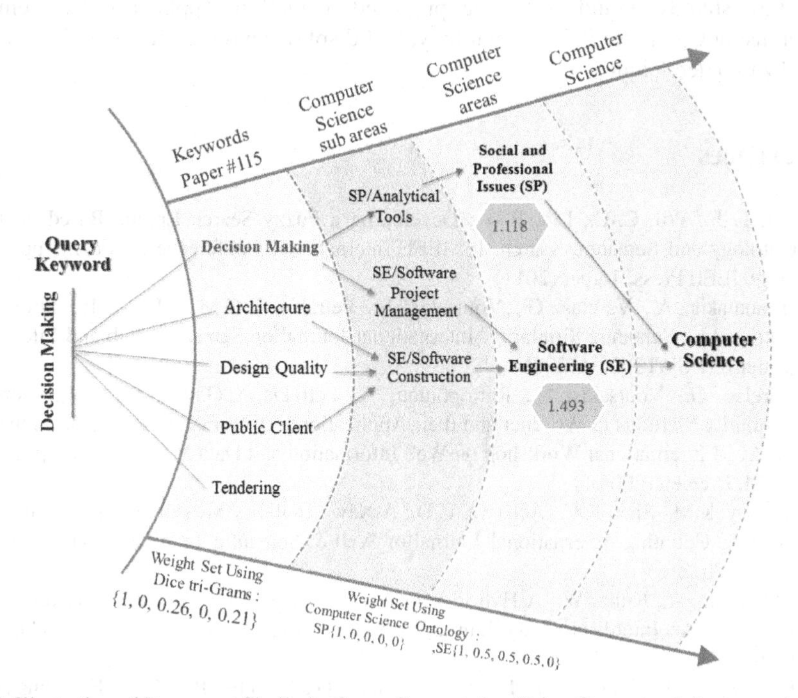

Fig. 2. Illustration of Document Similarity Score Computation Using Computer Science ontology

For example, document keywords on paperID#115 consist of "Decision Making", "Architecture", "Design Quality", "Public Client" and "Tendering". The current study first Computed the set score of string matching(tri-grams) followed by formula (2) between query's keyword and document keywords is {1,0,0.26,0,0.21}. The distance set score of query's keyword and document keywords based on edge counting (Wu and Palmer Measure) is shown in Table 3. The set of weight score is {1,0.5,0.5,0.5,0}. The keywords are keywords of Software Engineering on Computer Science Ontology. "Tendering" was not found in Computer Science Ontology. so it has a score of 0. The semantic similarity score was computed using the formula (3) and is 1.493, as shown in Fig 2.

6 Conclusion

This present study proposes a new method for document similarity scores between a query's keyword and a document. The paper constructs a Computer Science Ontology to distance hierarchy computation based on Wu and Palmer measure and tri-grams to find the similarity of keywords string matching that it can keep a comprehensive keyword and discard ponderous keywords. The experimental results present document similarity scores that suggest that the new measures are effectively ranked.

Furthermore, if document similarity scores are more than one, that mean user's query keyword also appeared in documents. Such as paperID#115,1055,847,1700 and 1506 in Table 3.

Future studies should apply the proposed method to applications of semantic search using Computer Science ontology, and display the results using an information visualization technique.

References

1. Lai, L.-F., Wu, C.-C., Lin, P.-Y.: Developing a Fuzzy Search Engine Based on Fuzzy Ontology and Semantic Search. In: IEEE International Conference on Fuzzy, pp. 2684–2689. IEEE Press, Taipei (2011)
2. Hliaoutakis, A., Varelas, G., Voutsakis, E., Petrakis, E.G.M., Milios, E.: Information Retrieval by Semantic Similarity. International Journal on Semantic Web and Information Systems (IJSWIS) 2(3) (2006)
3. Varelas, G., Voutsakis, E., Raftopoulou, P., Petrakis, E.G.M., Milios, E.: Semantic Similarity Methods in Wordnet and their Application to Information Retrieval on the web. In: ACM International Workshop on Web Information and Data Management, pp. 10–130. ACM, Bremen (2005)
4. Shenoy, K.M., Shet, K.C., Acharya, U.D.: A New Similarity Measure for Taxonomy based on Edge Counting. International Journal of Web & Semantic Technology (JJWesT) 3(4), 23–30 (2012)
5. Schwering, A., Kuhn, W.: A Hybrid Semantic Similarity Measure for Spatial Information Retrieval. An Interdisciplinary Journal of Spatial Cognition & Computation 9(1), 30–63 (2009)
6. Fernandez, M., Cantador, I., Lopez, V., Vallet, D., Castells, P., Motta, E.: Semantically enhanced Information Retrieval: An ontology-based approach. Journal of Web Semantics: Science, Services and Agents on the World Wide Web 9, 434–452 (2010)
7. Weng, S.-S., Tsai, H.-J., Hsu, C.-H.: Ontology construction for information classification. Journal of Expert Systems with Applications 31(1), 1–12 (2006)

8. John, T.: What is Semantic Search and how it works with Google search, http://www.techulator.com/resources/5933-What-Semantic-Search.aspx
9. Batet, M., Sanchez, D., Valls, A.: An Ontology-based measure to compute semantic similarity in biomedicine. Journal of Biomedical Informatics 44, 118–125 (2011)
10. Wu, Z., Palmer, M.: Verb semantics and lexical selection. In: Proceeding of the 32nd Annual Meeting of the Association for Computational Linguistics, Las Cruces, New Mexico, vol. 13, pp. 133–138 (1994)
11. Kondrak, G.: N-Gram Similarity and Distance. In: Consens, M.P., Navarro, G. (eds.) SPIRE 2005. LNCS, vol. 3772, pp. 115–126. Springer, Heidelberg (2005)
12. Sembok, T.M., Bakar, Z.A.: Effectiveness of Stemming and N-grams String Similarity Matching on Malay Documents. International Journal of Applied Mathematics and Informatics 5(3), 208–215 (2011)
13. Stoke, N.: Applications of Lexical Cohesion Analysis in the Topic Detection and Tracking Domain. A thesis submitted for the degree of Doctor of Philosophy in Computer Science Department of Computer Science Faculty of Science National University of Ireland, Dublin (2004)
14. Watthananon, J., Mingkhwan, A.: A Comparative Efficiency of Correlation Plot Data Classification. The Journal of KMUTNB 22(1) (2012)
15. Lertmahakrit, W., Mingkhoan, A.: The Innovation of Multiple Relations Information Retrieval. The Journal of KMUTNB 20(3) (2010)

LORecommendNet: An Ontology-Based Representation of Learning Object Recommendation

Noppamas Pukkhem

Department of Computer and Information Technology, Faculty of Science,
Thaksin University, Thailand
noppamas@tsu.ac.th

Abstract. One of the most problems facing learners in e-learning system is to find the most suitable course materials or learning objects for their personalized learning space. The main focus of this paper is to extend our previous rule-based representation recommendation system [1] by applying an ontology-based approach for creating a semantic learning object recommendation named "LORecommendNet". The "LORecommendNet" ontology represents the knowledge about learning objects, learner model, semantic mapping rules and their relationship are proposed. In the proposed framework, we demonstrated how the "LORecommendNet" can be used to enable machines to interpret and process learning object in recommendation system. We also explain how ontological representations play a role in mapping learner to personalized learning object. The structure of "LORecommendNet" extends the semantic web technology, which the representation of each based on an OWL ontology and then on the inference layer, based on SWRL language, making a clarify separation of the program components and connected explicit modules.

Keywords: learning object, ontology, recommendation, semantic web.

1 Introduction

Online learning resources are commonly referred to as learning objects in e-learning environment. They will be a fundamental change way of thinking about digital learning resource. Actually, learning objects can be learning components presented in any format and stored in learning object repositories which facilitate various functions, such as learning object metadata, learning object creation, search, rating, review, etc. Rapidly evolving internet and web technologies have allowed a using of learning objects in Learning Management System, but the problem is that it does not offer personalized services and dues to the non-personalized problem. All learners being given access to the same set of learning objects without taking into consider the difference in learning style, prior knowledge, motivation and interest. This gives result in lack of learner information to perform accurate recommending of the most suitable learning objects. This work provides prior knowledge about learners and learning objects in semantic web approach that can be used in our semantic-based recommendation model.

S. Boonkrong et al. (eds.), *Recent Advances in Information and Communication Technology*, Advances in Intelligent Systems and Computing 265,
DOI: 10.1007/978-3-319-06538-0_29,

An ontology is an important tool in representing knowledge of any resources in WWW. Until now, knowledge bases are still built with little sharing or reusing in related domains. Our focus is on developing the ontology-based representation called "LORecommendNet" in order to present the knowledge about the learner, learning object and propose an effective process for enhancing learning object selection of learners through our semantic-based recommendation model. In reasoning process, we proposed a set of personalization rules that will allow reasoning on the instances of "LORecommendNet".

The remainder of this paper is organized as follows. Section 2 gives background and previous work. An overview of the learning object concept, learning style and related works is also included. Section 3 presents the analysis and design of learning objects and the learner model ontology of LORecommendNet. Then, section 4, we propose our designing of an inference layer by using SWRL and Jess Rule. Finally, section 5 concludes this paper, giving a summary of its main contribution and pointing towards future research directions.

2 Background Knowledge and Previous Work

2.1 Learning Objects and Learning Object Metadata

Learning objects are a new way of thinking about learning content design, development and reuse. Instead of providing all of the material for an entire course or lecture, a learning object only seeks to provide material for a single lesson or lesson-topic within a larger course. Examples of learning objects include simulations, and adaptive learning component. In general, the learning objects must have the following characteristics; self-contained, can be aggregated, reusable, can be aggregated, tagged with metadata, just enough, just in time and just for you [2].

International efforts have been made on developing standards and specifications about learning objects since late 1990's. IEEE Learning Technology Standards Committee, IMS Global Learning Consortium, Inc., and CanCore Initiative [3] are organizations active in this area. IEEE LOM Standard is composed of Standard for Learning Object Metadata Data Model, Standard for XML Binding and Standard for RDF Binding which is a multipart standard. The first part of the standard, IEEE 1484.12.1 LOM Data Model standard [4], has been accredited and released. The LOM Data Model is the core of existing metadata specifications. It defines a hierarchical structure for describing a learning resource by data elements that are grouped into nine categories; *General, Lifecycle, Technical, Meta-metadata, Educational, Relation, Rights, Classification and Annotation.*

2.2 Learning Styles and Preferences

Learning style is an important criterion towards providing personalization, since they have a significant influence on the learning process. Attempting to represent the learners' learning styles and adapting the learning object so as the most suit them is a challenging research goal. Learning style designates everything that is characteristic for an individual when the learner is learning, i.e. a specific manner of approaching a learning activity, the learning strategies activated in order to fulfill the task.

Felder-Silverman learning style model [5] is the one of the most widely used learning style in adaptive hypermedia system. Another important reason noted by Sangineto [6],

Felder-Silverman learning styles was widely experimented and validated on an engineering and science student population. Furthermore, this model contains useful pragmatic recommendations to customize teaching according to the students' profiles.

2.3 Semantic-Based Tools

There are several semantic-based tools for information extraction and transform into meaningful which can be used through the process of building ontology-based model in our work:

- *XML Schema* is a structure of blocks of XML document formatting, similar as DTD (Document Type Definition). It can be used to store content and document's structure, but not all knowledge about content can be represented in the tree structure, and it is time consuming to maintain the order of presentation of knowledge. So, only XML Schema is not reasonable to represent documents.
- *RDF (Resource Definition Framework)* is a set of recommendations for well-formed XML documents. It is a more data aspect framework than human aspect. So, it is designed for enabling machine processing rather than for ease of human understanding.
- *OWL (Web Ontology Language)* is a language for definition of web ontologies which explicitly represents the semantics of terms in vocabularies and its relationships between these terms [7]. This is an accepted standard, language and platform independent and well- formed XML- markup language.
- *SWRL (Semantic Web Rule Language)* is a proposal in submission by the W3C, aiming at combining OWL and inference rules language [8]. SWRL rules reason about OWL individuals in terms of classes and properties. SWRL provides seven types of atoms: *Class, Individual, Data valued property, Different individual, Same individual, Built-in* and *Data range atoms.*

2.4 Previous Works

In our previous work [1,9], we developed the method for generating the course concept map called "Course Concept Map Combination Model: CMCM. The course concept map is the domain model that represents all possible sequences of learning concept for a specific course. The domain model stores the knowledge about the course preferences, instructor's characteristics and experiences. The main concept map was implemented by using the Cmaptool [10]. In recommendation model, we proposed three recommendation algorithms: i) preferred feature-based, ii) neighbor-based collaborative filtering and iii) non-personalized. The result is the preferred feature-base algorithm having more accuracy prediction than others. Previous work of recommendation system was tested and several experiments were proposed in order to show the suitability of using recommendation algorithm for recommending learning object to learners based on their learning style.

For improving the semantic recommendation in our previous work, we extended the LORecommend ontology for reasoning rules by using SWRL. The idea behind the semantic recommendation is, as the name suggests, to add a level of meaning to the Web. As this reason, it can be more easily manipulated by computer programs, and more effectively by humans. The proposed ontology-based representation model was presented in section 3. With this improvement, supported our recommendation system higher scalability and easier maintenance of the approach are expected.

3 LORecommendNet: Ontology-Based Representation Model

Ontologies in our system are written in OWL. To support the development of the ontologies and rendering in OWL, we use the open source tool Protégé 4.3[11]. Next subsection, concrete examples of the LORecomendNet within our recommendation system will be presented.

3.1 Learning Object Ontology

The main part of LORecommedNet is learning object ontology. Properties of learning objects as well as relationships to other learning objects are defined by the learning object ontology. Based on IEEE LOM standard, there are many kinds of learning object feature. The summarized results that rating from 15 experts in feature selection process is presented in Table 1. and their description is presented in Table. 2.

Table 1. The result summary of learning object feature analysis and selection

LO Feature		Feature Rating	Normalized Score
LOM category	**Element**	**Score***	**(α = 0.7)**
General	Language	140	0.9032
	Description	119	0.7677
Technical	Format	129	0.8323
Educational	Interactivity Type	111	0.7161
	Learning Resource Type	143	0.9226
	Interactivity Level	109	0.7032
	Semantic Density	112	0.7226

*Number of experts = 15

Table 2. The description of the selected features for LORecommendNet

ID	Class Name	Element Path	Instance
F1	Format	LOM/Technical/Format	Video, Image,Text, Audio, Animation
F2	Interactivity Type	LOM/Educational/ Interactivity_Type	Active, Expositive, Mixed
F3	Interactivity Level	LOM/Educational/ Interactivity_Level	Very low (0), Low (1), Medium (2), High (3), Very high (4)
F4	Semantic Density	LOM/Educational/ Semantic_Density	Very low (5), Low (6), Medium (7) High (8), Very high (9)
F5	Learning Resource Type	LOM/Educational/ Learning_Resource_Type	Exercise, Algorithm, Experiment, Example, Definition, Slide, Index

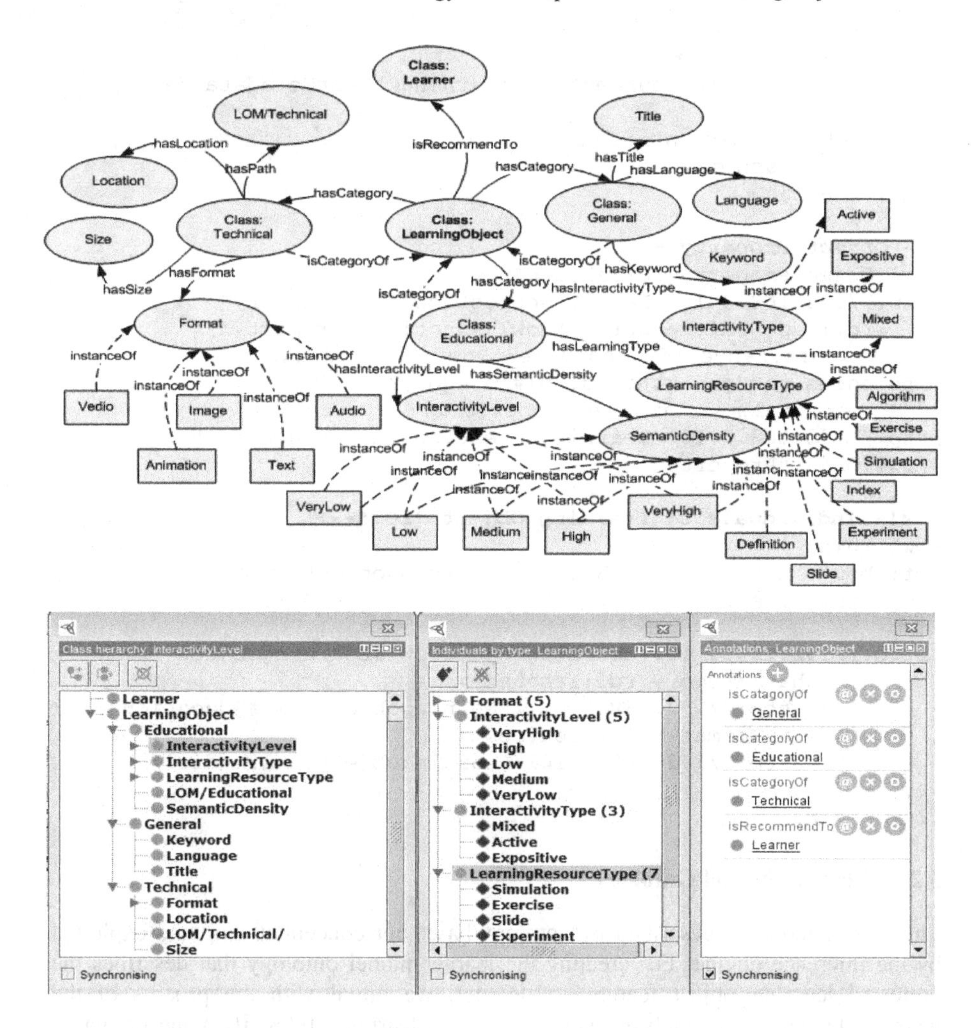

Fig. 1. A learning object ontology and the part of representation with Protégé

In our work, the learning object ontology describes about the properties of learning objects with five LOM features; format, interactivity type, interactivity level, semantic density and learning resource type. The learning object class ontology and the part of representation implementing with Protégé are presented in Fig.1. The OWL rendering format that describes about learning object class, object properties and individuals of them are shown as follows.

```
<!--The Learning Object Class-->
<owl:Class rdf:about ="#LearningObject">
<rdfs:subClassOf rdf:resource="&LO;LORecommend"/>
   <owl:Restriction >
      <owl:maxCardinality
            rdf:datatype="&xsd;nonNegativeInteger">1</owl>
```

```
        <owl:onProperty
                rdf:resource="&ls;learningObjectMetadata"/>
                ...
    </owl:Restriction>
</rdfs:subClassOf>
</owl:Class>
...
<!--Object Property-->
<owl:ObjectProperty rdf:about="
http://192.168.0.101/Learningobject.owl#hasCategory">
    <rdf:type rdf:resource="&owl;FunctionalProperty"/>
        <rdfs:range rdf:resource="&owl: http://www.owl-
ontologies.com/generations.owl#Category"/>
        <rdfs:domain rdf:resource="&owl:
http://192.168.0.101/Learningobject.owl#Learningobject"/>
</owl:ObjectProperty>
  ...
  <!--Individuals of Learning Object "LO1"-->
<owl:Thing
rdf:about="http://192.168.0.101/Learningobject.owl#LO1">
     <rdf:type rdf:resource=

"http://192.168.0.101/Learningobject.owl#LearningObject"/>
         <hasCategory rdf:resource=
         "http://192.168.0.101/Learningobject.owl#Education"/>
         <hasFormat rdf:resource=
         "http://192.168.0.101/Learningobject.owl#Audio"/>
...
</owl:Thing>
```

3.2 Learner Model Ontology

The learner model is described by ontology-based for conceptualizing and exploited by the inference engine. For creating the learner model ontology that describes the preferred learning object features of learner, we initial with the process of the learners' learning style analysis using an index of learning styles (ILS) questionnaire. The ILS is a 44-question instrument designed to assess preference on the four dimensions of the Felder-Silverman learning style model. Each dimension of the ILS has a 2-pan scale which represents one of the two categories (eg. Visual/Verbal).

The valid rule is the rule that is the member of the intersection set of word meaning between semantic group (SG) and LO features. Table 3 shows the detail of semantic mapping, the semantic groups (SG) within the dimensions provide relevant information in order to identify learning styles. If a learner has a preference for tends to be more impersonal oriented and trying things out learner would have a balanced learning style on the active/reflective dimension. However, a learner has also a balanced learning style if they tend to be more socially oriented and prefer to think about the material. Although both learners have different preferences and therefore different behavior in an online course, both are considered according to the result of ILS.

Table 3. Example of semantic groups associated with the ILS

Dimension	Set of Questions	Symbol	Semantic Group (SG)
Dimension 1	1, 5, 9, 13,17, 21, 25, 29,33, 37,41	**A**-**A**ctive	Trying something out(SG1) Social oriented (SG2)
		R-**R**eflective	Think about material(SG3) Independent (SG4)
Dimension 2	2, 6, 10, 14, 18, 22, 30, 34, 38, 42	**S**-**S**ensing	Existing way(SG5) Concrete material(SG6) Careful with details (SG7)
		I-**I**ntuitive	New ways(SG8) Abstract material (SG9) Not careful with detail (SG4)
Dimension 3	3, 7, 11, 15, 19, 23, 31, 35, 39, 43	**U**-Vis**U**al	Picture (SG11)
		B-ver**B**al	Spoken word (SG12) Written word (SG13) Difficulty visual style (SG14)
Dimension 4	4, 8, 12, 16, 20, 24, 32, 36, 40, 44	**Q**-Se**Q**uential	Detail oriented (SG15) Sequential progress (SG16) From part to the whole (SG17)
		G-**G**lobal	Overall picture (SG18) Non-sequential (SG19) Relation (SG20)

We adopt OWL (Web Ontology Language) to express ontology enabling expressive knowledge description and information interoperability of knowledge. According to the learner model ontology, the following OWL based markup segment describes the user contexts (learner) about "Learner1". The learner ontology show in Fig. 2 depicts contexts about a learner that corresponds to Table 3. We are collecting learning style scores of a learner in four learning style dimensions (A/R, S/I, U/B and Q/G) which are their weight have an interval 0-1. The relation between a learner and their learning styles certifies by hasDimension, hasSemanticGroup and hasLearning Style.

For example, the class Learner is described by the class learning style. The class learning style is built from four components: dimension1, 2, 3 and 4. Each dimension explains about learner learning styles. Moreover, we can describe the relationship between learning style and learner preference (TryingSometingOut, SocialOriented, ThinkAboutMaterial etc.).

4 SWRL as an Inference Layer

SWRL as an inference Layer, is used to establish individual relationships and adaptation rule. In this paper, we extend the mapping rule in our previous work into reasoning rules. Various relationships are captured in the body of SWRL rules that described about the relation among learning object, learner model and its environment.

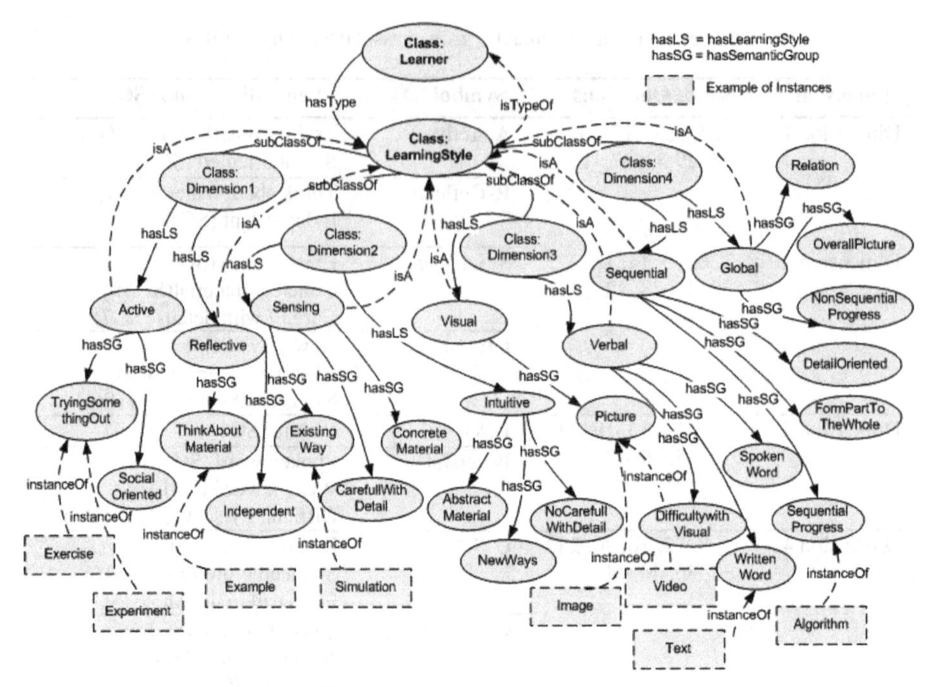

Fig. 2. Learner model ontology

The association between each learning style and the learning object features is represented by a rule-based representation. We demonstrate the example of validated mapping rule selection from all possible mapping rules as follows.

Example of Mapping Rules

```
Mapping Rule:  Reflective Learner
Propose:  Recommend Learning Object for  "R-Reflective" Learner
Rule-based Representation:
If "R" ∈ Learner(L) Then  LOM.educational.interactivity_type =
"expositive" and LOM.educational.LearningResourceType =
"definition" or  "algorithm" or "example"
Reflective := {think about it, try to understand, listen}
Map to: Interactivity Type:= "expositive" := {audio}
    Semantic density := "medium":={audio}
    Semantic density := "high":= {video}
    Learning resource type:="definition":={explanation,give
meaning }
    Learning resource type := "algorithm"= { step for action}
    Learning resource type := "example"= { show how to act}
```

In this section we show rules are employed to reason over ontology-based model (learner model, learning object model). The communication between reasoning rules and the other resource information will take place by exchanging RDF annotations. Several rules is can be derived. The example of rules is presented as follows.

```
1.  Person := Learner
2.  Resource := LearningObject
3.  LearningStyle:= Active U Reflective  U Sensing  U Intuitive
    U Visual U Verbal U Sequential U Global
4.  ActiveLearner := Person ∩ Active
5.  ReflectiveLearner := Person ∩ Reflective
6.  LOFeature:= format U interactivityType U interactivityLevel
    U SemanticDensity U LearningResourceType
7.  LOforActive :=(active U mixed)∩(exerciseU simulations U
    experiment)
8.  InteractivityType:= active U mixed U expositive
9.  InteractivityLevel:= verylow U low U medium U high U
    veryhigh
10. SemanticDensity:=  verylow U low U medium U high U veryhigh
11. LearningResourceType := exercise U experiment U definition
    U algorithm U example U slide U index
12. LOforVisual := (video U image U animation) ∩ (medium U high
    U veryhigh)∩ simulation
```

From the ontology extracting above, the initial relations are identified to be a member of the class. Finally, adding rule mapping ontology knowledge to them, like SWRL Rule1, 2, 3 and Rule 4 etc.

Example of SWRL Reasoning Rules

```
Rule1: LearingObject(?LO) ∧hasInteractiveType(?LO, expostive) ∧
       LearningResourceType(?LO, algorithm) → LOforReflective(?LO)
Rule2: LearingObject(?LO) ∧hasFormat(?LO, video) ∧
       hasInteractiveLevel(?LO, high) ∧
       LearningResourceType(?LO, simulation)→ LOforVisual(?LO)
Rule3: Learner(?L) ∧hasSemanticGroup(?L, picture) ∧
       hasLearningStyle(?L, visual) → VisualLearner(?L)
Rule4: VisualLearner(?L)  ∧LOforVisual(?LO) → recommend(?L,?LO)
```

From the reasoning rules, we can infer the suitable learning object to the specific learner by using the relationship which presented in LOReccommendNet.

Representing SWRL Rules as Jess Rules

The SWRL rules can be represented in Jess using their facts is relatively straightforward. Once the OWL concepts and SWRL rules have been represented in Jess (using Jess Tab in Protégé 3.5), the Jess engine can perform inference process. For example, take the example SWRL Rule as:

```
LearingObject(?LO) ∧hasInteractiveType(?LO, expostive) ∧
LearningResourceType(?LO, algorithm) → LOforReflective(?LO)

VisualLearner(?L)  ∧LOforReflective(?LO) → recommend(?L,?LO)
```

These rules can be represented in Jess by following Jess Rule.

```
(defrule aRule (LearningObject (name ?LO))
               (hasInteractiveType ?LO expositive)
               (LearingResourceType ?LO algorithm)
 => (assert (LOforReflective (name ?LO))

(defrule aRule (VisualLearner (name ?L))
               (LOforReflective (name ?LO))
 => (assert (recommend (?L ?LO))
```

When the inference process completes, these individuals will be transformed into OWL knowledge.

5 Conclusion

This paper has described "LORecommendNet", which is an ontology-based modeling strategy that has been employed by a personalized e-learning system, recommendation system and learning style-based adaptation. The ontology-based model constructed from the potential knowledge and competence state of the learner and the relationships between concepts in our domain. We designed the models based on an OWL ontology representation, SWRL rules and Jess Rules to infer from the ontology content. The learners can be provided with the intelligent and personalized support that recommend the most suitable learning object to them. Moreover, syntactically different but semantically similar learning objects can more easily be located. The idea behind the semantic-based representation is, as the name suggests, to add a level of meaning to the Web. As this reason, it can be more easily manipulated by computer programs, and more effectively by humans. Judging from current research directions, the future work will hold greater shareability, interoperability and reusability among existing learning management systems or e-learning application via a semantic web approach, rather than an all-encompassing knowledge model.

References

1. Pukkhem, N., Vatanawood, W.: Personalised learning object based on multi-agent model and learners' learning styles. Maejo International Journal of Science and Technology 5(3), 292–311 (2011)
2. Alderman, F.L., Barritt, C.: Creating a Reusable Learning Objects Strategy: Leveraging Information and Learning in a Knowledge Economy. Pfeifer Publishing (2002)
3. CanCore, Educational guidelines, http://www.cancore.ca/guidelines/1.9/CanCore_guideline_Educational_1.9.pdf (accessed: February 25, 2012)
4. IEEE, Draft Standard for Learning Object Metadata (2002), http://ltsc.ieee.org/wg12/files/LOM_1484_12_1_v1_Final_Draft.pdf (accessed: January 20, 2012)

5. Felder, R.M., Spurlin, J.: Application, reliability and validity of the index of learning styles. Int. J. Engineering Educational 21, 103–112 (2005)
6. Sangineto, E., Capuano, N., Gaeta, M., Micarelli, A.: Adaptive course generation through Learning styles representation. Universal Access in the Information Society 7, 1–23 (2008)
7. OWL Web Ontolgoy Language Reference, http://www.w3.org/TR/owr-ref, Horrocks, I.: SWRL: A Semantic Web Rule Language-Combining OWL and RuleML, http://www.w3.org/Submission/SWRL_2004
8. Pukkhem, N., Vatanawood, W.: Multi-expert guiding based learning object recommendation. In: Proceedings of International Conference on Computer Engineering and Application, Manila, Philippines, pp. 73–77 (2009)
9. Novak, J., Caňas, A.: The theory underlying concept maps and how to construct and use them. Technical Report IHMC Cmap Tools 2006-0I, the Florida Institute for Human and Machine Cognition, http://cmap.ihmc.us/publications/researchpapers/TheoryUnderlyingConceptMaps.pdf (accessed: January 30, 2013)
10. Protégé Ontology and Knowledge Base Framework, http://protege.standford.edu (accessed: May 25, 2013)

Secure Cloud Computing

Wolfgang A. Halang[1], Maytiyanin Komkhao[2], and Sunantha Sodsee[3]

[1] Chair of Computer Engineering, Fernuniversität in Hagen, Germany
wolfgang.halang@fernuni-hagen.de
[2] Faculty of Science and Technology,
Rajamangala University of Technology Phra Nakhon, Bangkok, Thailand
maytiyanin.k@rmutp.ac.th
[3] Faculty of Information Technology,
King Mongkut's University of Technology North Bangkok, Thailand
sunanthas@kmutnb.ac.th

Abstract. The security risks of cloud computing include loss of control over data and programs stored in the cloud, spying out these data and unnoticed changing of user software by the cloud provider, malware intrusion into the server, eavesdropping during data transmission as well as sabotage by attackers able to fake authorised users. It will be shown here how these security risks can effectively be coped with. Only for preventing the cloud provider from wrong-doing no technical solution is available. The intrusion of malware into cloud servers and its malicious effects can be rendered impossible by hardware-supported architectural features. Eavesdropping and gaining unauthorised access to clouds can be prevented by information-theoretically secure data encryption with one-time keys. A cryptosystem is presented, which does not only work with one-time keys, but allows any plaintext to be encrypted by a randomly selected element out of a large set of possible ciphertexts. By obliterating the boundaries between data items encrypted together, this system removes another toehold for cryptanalysis.

Keywords: Cloud computing, malware prevention, hardware-based security, security by design, eavesdropping, unbreakable encryption.

1 Introduction

Cloud computing means providing services such as data storage, data processing and file repository by a central server to remote users via the Internet. Since already in the 1960ies mainframe computers with so-called time-sharing operating systems were connected via telecommunication lines to user terminals, cloud computing is half a century older than its name and the present hype about it. At that time, the security risks associated with this kind of centralised computing were not so severe, however, as they are today. These risks include loss of control over data and programs stored in the cloud, spying out these data and unnoticed changing of user software by the cloud provider, malware intrusion into the server, eavesdropping during data transmission as well as sabotage by attackers able to fake authorised users.

S. Boonkrong et al. (eds.), *Recent Advances in Information and Communication Technology*, Advances in Intelligent Systems and Computing 265,
DOI: 10.1007/978-3-319-06538-0_30,

In this paper, it will be shown how these security risks can effectively be coped with. Only for preventing the cloud provider from wrong-doing no technical solution is available. The intrusion of malware and its malicious effects are made impossible by hardware-supported architectural features. Eavesdropping and gaining unauthorised access to a cloud is prevented by information-theoretically secure data encryption with one-time keys.

The technical security measures outlined herein are all not new, but there are consistently disregarded for decades by mainstream academic computer science as well as by the IT industry. The reasons for this disregard are up to speculation. Maybe, it is just ignorance or the inability to take notice of published results. But for academic circles it could also be the unwillingness to recognise certain problems as solved preventing further research with its possibility to publish more papers. The branch of the IT industry supplying malware detection software and related tools would simply lose its complete business. The suppliers of computers want to sell new computers before the old ones cease functioning, and the customers of operating systems, which as software never wear out, need to be convinced by errors and new features to buy new operating system versions. Hence, of each generation of new hardware and operating systems not more than only slight improvements with respect to security can be expected to allow for many more generations to be marketed. And, finally, we should not forget those who are interested in spying out personal data such as secret services, governments and big corporations.

2 Spying and Sabotage by the Cloud Provider

If one uses a cloud as a data depository, only, and encrypts the data in an unbreakable way as, for instance, outlined in Section 4, then the cloud provider can neither spy out these data nor modify them without being noticed. The provider could just destroy the data which, however, does not make any sense.

There is on-going research, e.g. [1], on the possibility to enable processing of encrypted data without the need to decrypt them first. To carry out a dyadic operation f on two arguments x and y encrypted with the function e this requires, for instance, the existence of an operator g, such that

$$g(e(x), e(y)) = e(f(x, y))$$

holds. This constraint is extremely restrictive. For most combinations of e and f no g will exist. If for a given f a g could exist, then the encryptions allowing this will most likely be rather weak and, hence, easily breakable. Furthermore, to be able to apply an operator g, it must always be known where in a ciphertext the encryptions of two arguments can be found. This offers another point of departure for cryptanalytic attack. In Section 4 it will be shown that the locations of encrypted values should and can be blurred as well.

As a result, one has no other means than to trust the cloud provider if one wants to process data in a cloud. This should not be a problem if the cloud is operated by the own organisation. A trustable cloud provider could also be a co-operative society in which one is a member. An example for this is DATEV e.G. (www.datev.de) in Nuremberg, which was founded in 1966 by tax accountants, lawyers, chartered accountants and their clients and which has presently some 40,000 members.

3 Hardware-Supported Malware Prevention

The easiest approach to prevent malware intrusion into cloud servers is just to run the server software on classical mainframe computers, because up to now malware and hackers have never succeeded in entering such a machine [13]. Thus, security updates have almost never been required for their operating systems and their most important subsystems.

An analysis of the various intruders, particularly in form of programs and executable Internet content with malicious intentions, which are making their way into differently designed computers, reveals that they are based on some common principles of operation. If these operation principles are thwarted by appropriate measures, malware is prevented from spreading and from launching its destructive effects. There is a single measure, which already makes any form of malware intrusion impossible, viz. separation of program and data memory. The now for two thirds of a century predominant and thoughtlessly perpetuated Von Neumann architecture with its minimalistic principles is totally inadequate for systems that need to be secure, as it does not separate data from instructions and, thus, does not permit to protect both kinds of information properly. The Harvard architecture, on the other hand, provides this separation throughout and, therefore, represents an adequate construction principle. It is interesting to note that the Harvard architecture is even older than Von Neumann's, and that it actually dates back to Konrad Zuse's first working computer Z1 of 1936. In [2] it was shown that the Harvard architecture can be emulated on Von Neumann computers. It is remarkable to note that this emulation could be implemented with just a single logic gate with two inputs, as shown in Fig. 1.

The constructive security measures discussed in the sequel are less restrictive than the Harvard architecture. Nevertheless, they disable the operation principles of all known malevolent programs in effective ways. In devising them, great importance was attached to the presented solutions being simple and easy to duplicate, in order to be understood and applied without any problems by the users of computers, as unnecessary complexity is the enemy of any effort towards enhancing security.

Software with malicious intentions often interferes with application programs or even with operating system routines in order to manipulate them for its destructive purpose or to deactivate software-implemented security functions. Here a memory segmentation measure as developed in [3] takes effect. It reliably prevents unauthorised accesses to the storage areas of operating system and application programs. To this end, a hardware-supervised segmentation of memory is employed, which protects program code against modifications not permitted by a hardware-implemented write-protection.

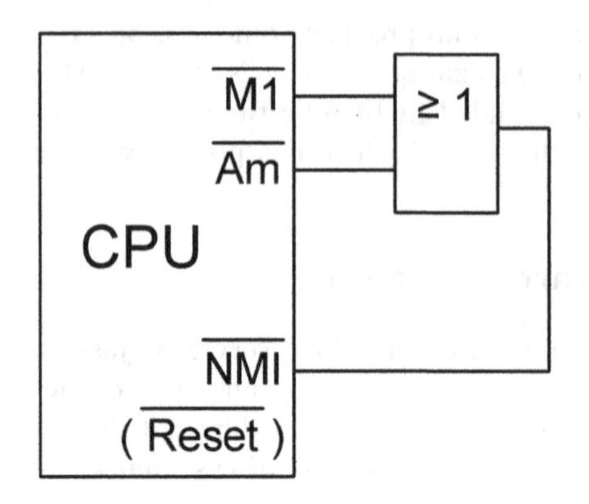

Let the main storage be divided into the lower half for program storage and the upper half for data storage. If the processor tries to fetch an instruction (indicated by its output line $\overline{M1}$) from data storage (indicated by an address with the most significant address line $\overline{Am}$ active), then the circuitry shown causes a non-maskable interrupt or a processor reset.

Fig. 1. Emulating the Harvard architecture on a Von Neumann computer

In contrast to programs, data are subject to frequent modifications. Therefore, a hardware-implemented write-protection as above is not feasible for handling reasons. Data can be protected against programs for spying out and modification, however, by a context-sensitive memory allocation scheme [4]. Applying this measure, any unauthorised access to data is precluded. To this end, a system's mass storage, in particular the data area, is further subdivided by a partitioning into context-dependent segments. In an installation mode it is precisely specified which accesses to these segments are permitted to the programs. This is oriented at the data to be protected and not the programs, i.e. in general to each program there exist several data segments separated from one another. In other words, this method permits memory references to any application program and operating system service only by means of using access functions write-protected by hardware, which release the storage areas required for the respective application case for writing and reading or just for reading accesses.

In order not to endanger the advantages of memory areas write-protected by hardware measures during the installation phases of programs, it is necessary to accommodate service programs and their databases also in areas write-protected by hardware and separated from the program area. During installation phases, a hardware device according to [5] constructively excludes attackers from gaining administrator rights by authenticating an authorised person by biometrical or other means.

Destructive programs and software-based aggression from the Internet often use components of digital systems, which they would not need for their feigned nominal functions. Here the hardware-supported security measure detailed in [6] takes effect. It lets any program, any interpretable file and any executable Internet content first disclose which resources are required for execution. The disclosure of a program's nominal functions enables to install boundary values for systems and to supervise their operation. By this and, at any instant, by locking all resources not needed at that time by means of hardware, it is warranted that the desired nominal functionality is observed.

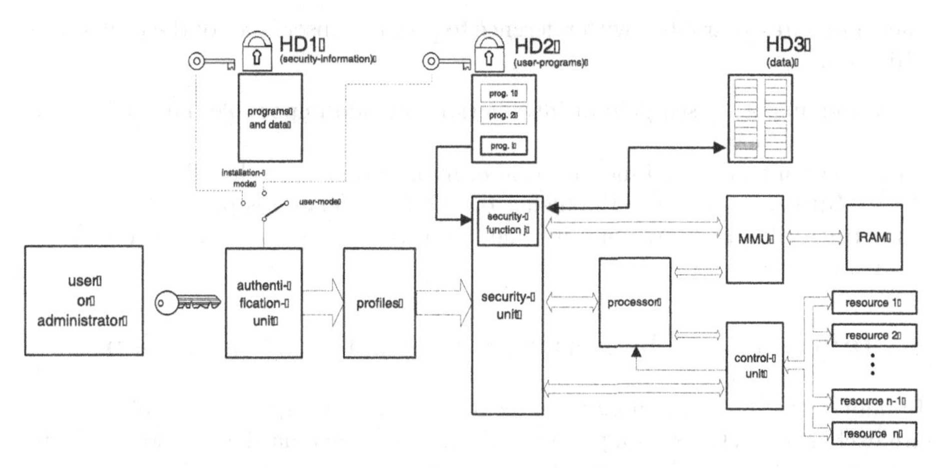

Fig. 2. Hardware-supported security measures

Utilising the measures outlined above, computers are effectively protected against inadmissible accesses, and become immune to intruders and espionage. This holds in particular for still unknown attack patterns or malware, too, because there is no more need for databases of malicious code or attack prototypes, which become obsolete within hours anyway due to the swift spreading of current malware via the Internet. In addition, separation and structuring considerably facilitates the maintainability of computers. Systems protected by the mentioned measures exhibit, on the basis of disclosing their nominal functions, of the permanent supervision against set bounds, of the context-sensitive allocation of data and of the impossibility to attack operating systems and application programs, a degree of robustness which allows them to maintain their functionality despite some failing application programs. In essence, with their features

- Data and instructions are separated throughout.
- Authentications are not influenceable by software.
- Protection systems are not attackable themselves; their implementation is proven correct and safely protected against modifications not permitted.
- The protection of systems cannot be put out of effect during the installation phases of application programs or of operating system components as well.
- All storage levels (main memory, mass storage etc.) are protected throughout against unauthorised accesses by means of authentication-dependent virtual address spaces.
- Constraints and nominal functionalities of programs are defined in installation phases, and permanently supervised during operation. Their observance is guaranteed under real-time conditions.
- To protect data against effects of software errors or malicious interpretable files, and to enable context-sensitive memory allocation, programs can be instantiated with access functions.
- Requests arriving from the outside are always placed first in separate and enclosed data areas to be preprocessed there.

these measures guarantee, with reference to [11], the observance of the protection objectives

Privacy: unauthorised gain of information is made impossible, i.e. spying out of data is obviated,
Integrity: unauthorised modification of information is precluded,
Availability: unauthorised influence on the functionality is precluded and
Attributability: at any point in time the responsible persons can be identified with certainty.

4 Information-Theoretically Secure Data Encryption

In information and communication technology, increasingly datasets of any size are exchanged between computers in form of streams via data networks. This holds particularly for the communication between cloud servers and cloud clients. To guarantee the confidentiality of such messages' contents, a plentitude of methods for the encryption of the data streams was developed [10]. Currently used encryption methods usually employ the same keys during longer periods of time, lending themselves to cryptanalytic attacks. It was shown, for instance, that the rather widespread asymmetrical RSA-cipher with 768 bits long keys has at least theoretically been broken. The symmetrical cryptosystem DES is already regarded as unsafe, too. Other ciphers such as 3DES or AES are still being considered safe, but only because the presently available computing power is insufficient to carry out simple brute-force attacks. In some countries law requires to deposit the keys used with certain agencies. Thus, these countries' secret services do not need any cryptanalysis whatsoever to spy out encrypted data.

In consequence, only perfectly secure one-time encryption is feasible in the long run. Perfect security is achieved, if encryption of a plaintext yields with equal probability any possible chiphertext, and if it is absolutely impossible to conclude from the ciphertext to the plaintext. According to the theorem of Shannon [12] fundamental for information theory, a cryptosystem is only then regarded as perfectly safe, if the number of possible keys is at least as large as the number of possible messages. Hence, also the number of keys is at least as large as the one of possible ciphertexts which, in turn, must be at least as large as the number of possible plaintexts.

Based on these considerations, in the sequel a cryptosystem is presented, which does not only work with one-time keys. It allows any plaintext to be encrypted by a randomly selected element out of a large set of possible ciphertexts, and it obliterates the boundaries between data items encrypted together. Thus, it is impossible to conclude from boundaries between data items in ciphertexts to the boundaries of data items in the plaintexts, removing another toehold for cryptanalysis.

The general operational principle of cryptosystems can be described mathematically as follows. Let sequences of message symbols (plaintext) s_0, s_1, ... from an arbitrary alphabet S to be transmitted. Messages are encrypted by the sender with an encryption function and, after transmission, decrypted by the receiver with the inverse decryption function. Generally, both functions are publically known according to Kerckhoffs' principle [8], but parameterised with a secret key K, which is agreed upon by the the communicating units via a confidential and authentic channel. In dependence upon this key a sequence of states

$$\sigma_{t+1} = f(\sigma_t, K) \tag{1}$$

is generated both in the sender and the receiver at discrete points in time $t \geq 0$ with the state transfer function f, and a key stream

$$z_t = g(\sigma_t, K)$$

is generated with the key stream generation function g. The initial state σ_0 may be known publically, or may also be derived from the key K. With the key stream, plaintext is then transformed in a state-dependent way by the invertible mapping

$$c_t = h(z_t, s_t) \tag{2}$$

to ciphertext, which is decryptable by applying the inverse mapping

$$s_t = h^{-1}(z_t, c_t).$$

The key stream sequence must be as similar to a genuine random sequence as possible. In case of self-synchronising stream ciphers, the determination of the state σ_{t+1} additionally depends on the last-generated cipher symbols $c_t, \ldots, c_{t-l+1}$ with fixed l, $l \geq 1$.

It is a common feature of all known cryptosystems that they subject the data elements to be transmitted, may that be bits, alphanumerical characters or bytes containing binary data, may they be single or in groups, always as unchanged entities to encryption. The information-theoretical model of cryptosystems according to Shannon [12] is founded on this restrictive basic assumption as well. Consequently, information such as the boundaries between data elements and their number perpetuates observably and not encrypted into the ciphertext: as a rule, to any plaintext symbol there corresponds exactly one ciphertext symbol. Since even block ciphers seldom work with data entities exceeding 256 bits, the symbols in plaintexts and in ciphertexts are ordered in the same sequences or, at least, their positions lie very close together. Thus, corresponding symbols in plain- and ciphertext can rather easily be associated with one another, which considerably facilitates code breaking.

A solution of this problem [7] is based on the observation that, ultimately, in the technical realisation of all cryptosystems the symbols of the plaintext alphabets (or modifications thereof) and of the ciphertext alphabets are all represented by binary encodings. Correspondingly, for encryption the most general among all possible forms of replacing one bit pattern by another one is utilised. In course of this encryption, particularly the boundaries between the plaintext symbols are blurred, the binary positions of several plaintext symbols are functionally interrelated with each other, and for any bit pattern to be encrypted an encryption is randomly selected out of a corresponding set.

The encryption according to [7] differs from the state of the art outlined above as follows. A state sequence is generated similar to Eq. (1). In every state σ_t, however, a number m_t of bit positions to be encrypted is determined according to an arbitrarily selectable method. The parameter m_t can – and should – be different from the number of bit positions, with which the plaintext alphabet is encoded, whereby the boundaries between the plaintext symbols are annihilated. Then, for m_t bit positions each, an encryption with $n > m_t$ bit positions is determined by means of a state-dependent relation

$$R_t \subset \{0,1\}^{m_t} \times \{0,1\}^n.$$

Here, the parameter n may not be smaller than m_t, as information would get lost otherwise, and it should not be equal to m_t either, in order to prevent the disadvantages mentioned above. Contrary to the function h of Eq. (2), the relation R_t does not need to be a mapping: it is even desirable that with every element in $\{0,1\}^{m_t}$ as many elements of $\{0,1\}^n$ as possible are related by R_t, allowing to randomly select among them one as encryption. Moreover, every element in $\{0,1\}^n$ should be a valid enciphering of an element in $\{0,1\}^{m_t}$, to completely exhaust the encryption possibilities available. Unique decryptability is given, when the inverse relation is a surjective (onto) mapping:

$$R_t^{-1} : \{0,1\}^n \longrightarrow \{0,1\}^{m_t}.$$

Contrary to Kerckhoffs' principle, this decryption function is not known publically – and the relation R_t used for encryption is not only publically unknown, but no function either. Publically known is only, that R_t^{-1} is a totally arbitrary mapping among all possible ones mapping the finite set $\{0,1\}^n$ onto another finite set $\{0,1\}^{m_t}$.

The conditions mentioned above already hamper to a very large extent breaking a data encryption performed according to [7]. A decryption would only be possible, if an attacker had such an amount of ciphertexts available as required by pertaining analyses – totally disregarding the necessary computational power. The following further measures prevent, however, that sufficiently long encipherings, generated with a certain choice of a parameter set and an encryption relation, arise in the first place.

During operation, an encryption unit can – at randomly selected points in time – sufficiently often vary the parameter m_t between 1 and the length of an input register and, thus, modify the encryption relation correspondingly. The unit as well as an inversely constructed and working decryption unit can co-ordinate sporadic modifications of the encryption using a protocol. In order to transmit between the units as few details on the encryption relation as possible for confidentiality reasons, just the instants of re-parametrisations ought to be co-ordinated. The respective new parameter values and identifiers of the en- and decryption relations to be applied should, however, be generated in both units with synchronously running algorithms. Random number generators based on chaos-theoretical principles [9], for instance, are very suitable for this purpose.

By selecting the parameters $m_t \neq k$ and $n > m_t$ it is inherently achieved, that it cannot be concluded in an easy way anymore from the boundaries between the symbols in a ciphertext to the boundaries of the symbols in the plaintext data stream. For $n > m_t$, the set of possible encryption elements is embedded in a considerably larger image set, significantly impeding code analysis for an attacker. The number of all possible relations $R_t \subset \{0,1\}^{m_t} \times \{0,1\}^n$, for which R_t^{-1} is a surjective mapping, amounts to $\frac{2^n!}{(2^n-2^{m_t})!}$. For the choice $m_t = 17$ and $n = 24$, considered to be practically feasible, this is in the order of magnitude of $10^{946,701}$ different relations – an extremely big number. The set of these relations comprises, among others, all possibilities to permutate bits in their respective positions, to insert $n - m_t$ redundant bits, which may each have an arbitrary of both possible values 0 or 1, at $\binom{2^n}{2^{n-m_t}}$ positions in the output bit patterns as well as to functionally interrelate the values of the bit positions of the encryption elements in a fully general way.

5 Conclusion

The security risks brought about by cloud computing have been identified. If clouds are just used as data depositories and unbreakable encryption is employed, then cloud providers can neither spy out these data nor modify them without being noticed. The on-going research on enabling the processing of encrypted data without having to decrypt them is doomed to failure since the conditions, under which this might be possible, are too restrictive by far. Therefore, in this respect the only choices are not to use cloud computing, to trust the provider or to own the cloud.

Currently applied cryptosystems to secure confidential data transmission to or from clouds have either already been broken or are expected to be broken soon. Also, their keys need to be deposited with government agencies in some countries. Therefore, to prevent eavesdropping and gaining unauthorised access to clouds, information-theoretically secure one-time encryption is the method of choice. It was shown that encryption with one-time keys can even be enhanced by encrypting any plaintext by a randomly selected element out of a large set of possible ciphertexts. Blurring the boundaries between data items encrypted together, it is made impossible to conclude from boundaries between data items in ciphertexts to the boundaries of data items in the plaintexts. Thus, a toehold for cryptanalysis left open by a silent assumption in Shannon's communication theory is eliminated. The operation principles of malware – present and future – intruding both cloud servers and client computers can easily thwarted if one resorts to hardware-supported measures, whereas software-based ones can inherently not meet the expectations. The simple measure of separating program and data memories, i.e. the Harvard architecture, is fully effective. Perpetuating the inadequate Von Neumann architecture any longer is dangerous and does not make sense. If the Harvard architecture should be considered too restrictive, it is also possible to efficiently harden the Von Neumann architecture with several hardware-based security features. For the majority of these features, here it was referred to patent applications of the year 2000 [3,4,6], which were not granted later. The patent office showed the applicants that they had re-invented the wheel, and could prove that similar ideas had been published already decades before. Fourteen years have passed since then, but the ignorance of the professional circles persists.

References

1. Fahrnberger, G.: SecureString 2.0 – A Cryptosystem for Computing on Encrypted Character Strings in Clouds. In: Eichler, G., Gumzej, R. (eds.) Networked Information Systems. Fortschr.-Ber. 10, 826, pp. 226–240. VDI Verlag, Düsseldorf (2013)
2. Halang, W.A., Witte, M.: A Virus-Resistent Network Interface. In: Górski, J. (ed.) SAFECOMP 1993, pp. 349–357. Springer, Heidelberg (1993)
3. Halang, W.A., Fitz, R.: Speichersegmentierung in Datenverarbeitungsanlagen zum Schutz vor unbefugtem Eindringen. German patent application DE 100 31 212 A1 (2000)
4. Halang, W.A., Fitz, R.: Kontextsensitive Speicherzuordnung in Datenverarbeitungsanlagen zum Schutz vor unbefugtem Ausspähen und Manipulieren von Daten. German patent application DE 100 31 209 A1 (2000)
5. Halang, W.A., Fitz, R.: Gerätetechnische Schreibschutzkopplung zum Schutz digitaler Datenverarbeitungsanlagen vor Eindringlingen während der Installationsphase von Programmen. German patent 10051941 since 20 October (2000)
6. Halang, W.A., Fitz, R.: Offenbarendes Verfahren zur Überwachung ausführbarer oder interpretierbarer Daten in digitalen Datenverarbeitungsanlagen mittels gerätetechnischer Einrichtungen. German patent application DE 100 55 118 A1 (2000)
7. Halang, W.A., Komkhao, M., Sodsee, S.: A Stream Cipher Obliterating Data Element Boundaries. Thai Patent Registration (2014)
8. Kerckhoffs, A.: La cryptographie militaire. Journal des Sciences Militaires. 9. Serie (1883)
9. Li, P.: Spatiotemporal Chaos-based Multimedia Cryptosystems. Fortschr.-Ber. 10, 777. VDI-Verlag, Düsseldorf (2007)
10. Menezes, A.J., van Oorschot, P.C., Vanstone, S.A.: Handbook of Applied Cryptography. CRC Press, Boca Raton (1997)
11. Rannenberg, K., Pfitzmann, A., Müller, G.: Sicherheit, insbesondere mehrseitige IT-Sicherheit. In: Mehrseitige Sicherheit in der Kommunikationstechnik, pp. 21–29. Addison-Wesley, Bonn (1997)
12. Shannon, C.E.: Communication Theory of Secrecy Systems. Bell System Technical Journal 28, 656–715 (1949)
13. Spruthm, W.G., Rosenstiel, W.: Revitalisierung der akademischen Großrechnerausbildung. Informatik Spektrum 34(3), 295–303 (2011)

Enhanced Web Log Cleaning Algorithm for Web Intrusion Detection

Yew Chuan Ong and Zuraini Ismail

Advanced Informatics School, Universiti Teknologi Malaysia, Malaysia
ycong4@live.utm.my,
zurainiismail.kl@utm.my

Abstract. Web logs play the crucial role in detecting web attack. However, analyzing web logs become a challenge due to the huge log volume issue. The objective of this research is to create a web log cleaning algorithm for web intrusion detection. Studies on previous works showed that there are five major web log attributes needed in web log cleaning algorithm for intrusion detection, namely multimedia files, web robots request, HTTP status code, HTTP method and other files. The enhanced algorithm is based on these five major web log attributes along with a set of rules and conditions. Our experiment shows that the proposed algorithm is able to clean noisy data effectively with a percentage of reduction of 40.41 and at the same time maintain the readiness for web intrusion detection at a low false negative rate (0.00531). Future works may address the web intrusion detection mechanism.

Keywords: web log, data cleaning, preprocessing, intrusion detection.

1 Introduction

With the increasing use of World Wide Web as a platform for online services such as banking and shopping, protecting the web site security becomes essential. From a defensive point of view, web log analysis plays a vital role in detecting web attacks. Web log files allow in-depth analysis as it contains fruitful information which included Internet Protocol (IP) address, requested Uniform Resource Locator (URL) and the date and time of the request.

One of the challenges of web log analysis is the rapid growth of web log files size. As mentioned by [1], the biggest challenge for intrusion detection is the "Big-Data" which caused by the large amount of data collected during the intrusion detection process. Thus, web log mining come into place to ease the extraction of information from huge volume of web log files. However, log files contain undesirable and irrelevant data which may affect the mining result. To ease mining process and improve mining result, web log files should be preprocessed. The methods of web log files preprocessing may vary for web usage mining and web intrusion detection [2]. To the best of our knowledge, only few researches [2], [3] focus purely on web log preprocessing for intrusion detection.

This study focuses on one of the steps in log file preprocessing which is data cleaning. A web log cleaning algorithm specifically for web intrusion detection is

S. Boonkrong et al. (eds.), *Recent Advances in Information and Communication Technology*, Advances in Intelligent Systems and Computing 265,
DOI: 10.1007/978-3-319-06538-0_31, © Springer International Publishing Switzerland 2014

proposed in this study. The proposed algorithm aimed to reduce the size of the web log files and at the same time prepare the web log file for intrusion detection purpose.

Our work will identify the attributes involved in web log cleaning for intrusion detection. The readiness of the web log cleaning algorithm to prepare the web log file for web intrusion detection will also be tested. These are the two main areas which are not addressed by the previous studies [2], [3].

The rest of the paper is organized as follows: Section 2 discussed the existing web log cleaning methods for intrusion detection and the web log attributes. Section 3 presents the proposed algorithm while the implementation is described in Section 4. The experimental results and analysis are shown in Section 5. Subsequently, Section 6 concludes this paper as well as future works.

2 Literature Review

2.1 Web Log Cleaning for Intrusion Detection

Due to the escalation of computer activities, especially Internet browsing, the volume of intrusion detection dataset from real world became huge. This dataset which consists of various attributes is complex and highly susceptible to inconsistent data and noise. Thus, more and more data preprocessing algorithms have been applied to tackle the huge volumes of intrusion detection dataset [4].

There are few researches [2], [3], [5], [6], [7] which used web logs as the dataset for intrusion detection. A detailed data cleaning process had been presented by [2] to remove unnecessary and noisy data. This data cleaning algorithm will remove logs which contained multimedia files extension such as jpg, gif and css. It will also remove logs with HTTP status code 200 and "-" in uri-query node. It is claimed that the probability for such request to contain web intrusion is almost zero. [5] implemented the same cleaning algorithm proposed by [2] in a similar study. On the other hand, the web log preprocessing algorithm designed by [3] will eliminate logs with multimedia files request and logs with HTTP status code 200. The data cleaning method remains to be similar to [8] study, which is used for web usage mining. [6] proposed a system which can differentiate web logs triggered by Denial of Service (DoS) attacks from legitimate web request. Data cleaning is included in the study. But, it does not focus on the web log attributes as the study is focus on the web user session. [7] presented the analysis of web logs for intrusion detection. Unfortunately, the work does not clearly outline the details of data cleaning in the log file preprocessing stage.

2.2 Attributes Involved in Web Log Cleaning

As web logs preprocessing served different purposes, the method of data cleaning may vary [2]. The following comprehensively describe the attributes involved in web log cleaning.

Multimedia File: Remove multimedia file is common in web logs data cleaning. Multimedia files included images, videos and audio. Multimedia files are categorized as useless files in web log preprocessing for both web usage mining and web intrusion detection. Web log files size can be reduced to less than 50% of its original sizes by eliminating the image request [9].

Web Robots Request: Robots request is another attribute in web logs data cleaning. Web robots act as automated tool which will periodically scan web sites for content extraction [10]. Log files generated by web robots will dramatically affect the web sites traffic statistics [9]. In their research, [11] demonstrated how the web logs file size reduced to 43.81% of its original size after the robots request was removed.

HTTP Status Codes: HTTP status code is also the attribute involved in web log cleaning. Log files with unsuccessful HTTP status code are usually eliminated during the web log cleaning process. The widely acceptable definition for unsuccessful HTTP status codes is a code under 200 and over 299. Such selection criteria only applicable in web log preprocessing for web usage mining. Algorithm proposed by [2] suggested that log files with unsuccessful HTTP status codes (400 series and 500 series) should be kept because it may be an indicator for anomaly actions. A data cleaning algorithm was developed by [3] which removed all log files with status code 200. However, [2] argued that log files with status 200 series should be remained as these log files may include web attacks like Structured Query Language (SQL) injection and Cross Site Scripting (XSS) which have been executed successfully.

HTTP Methods: A few researches [9], [10], [12] have included HTTP method as an attribute in web log cleaning. In the LODAP Data Cleaning Module, [12] recommended that all log files with HTTP request method other than GET should be removed as these are non-significant in web usage mining. Besides request with GET method, [9] added that HTTP request with POST method should be kept. [10] keep the log files with HTTP GET and HEAD request to obtain more accurate referrer information.

Other Files: There are some other files which are included as attributes in web log cleaning. For instance, log files with page style files request [10], [13]. The example of page style file is .css file. Log files with request to accessorial resources embedded in HyperText Markup Language (HTML) file should be removed too as these accessorial resources like scripts have no relationship with the specific page requested by user [14].

3 The Proposed Algorithm

The ultimate goal of the proposed algorithm is to clean web log files for web intrusion detection. The algorithm included five major attributes namely multimedia files, web robots request, HTTP status code, HTTP method and other files.

3.1 Algorithm Design

There is a total of six cleaning conditions in the design of this algorithm. First, all web logs with HTTP status code 200 and "-" in uri-query node will be removed. The probability for web logs with such criteria to contain malicious web attacks is almost zero [2].

Second, the algorithm will remove web logs with multimedia file extensions in uri-stem. The same approach was proposed by [2] and [3]. However, there are two additional conditions. The web logs with multimedia file extensions in uri-stem will be removed if (1) The HTTP request in the web log is not HTTP POST and (2) The HTTP status code is not 400 series and 500 series. Web logs with status code 400 series and 500 series should be kept as these may consider as malicious attempt. Users who triggered many web logs with HTTP error status code are subject to suspect [2]. Besides, web log with POST method should be kept as violating content can be transmitted via HTTP POST [15]. In common case, to launch web defacement attack, attacker will use HTTP POST method to replace part or all of the web interface components.

Third, legitimate web robots requests will be removed. The intention of web robots detection is to detect malignant robots which involved unethical content usage [16]. In this algorithm, specific IP address will be included in the web robot IP whitelist. If there are web logs with web robots request from these IP addresses, the web log will be removed. Example of legitimate robots request will be the Googlebot. Googlebot is the Google's web crawling bot.

Fourth, the proposed algorithm will remove web log with legitimate file extension if the web logs contain no HTTP status codes with 400 series and 500 series and the HTTP method is not HTTP POST. The most common legitimate file extension is .css which indicates the page style file. Some others file extension are .pdf, .txt and .doc [12]. Similar with multimedia files extension, HTTP errors code may reflect malicious attempt while HTTP POST allow attacker to upload file to a website for malicious purpose. The list of file extension included in the proposed algorithm is based on the experiment result by [12].

Fifth, web log with HTTP HEAD method and legitimate IP will be removed. HTTP HEAD method is used in a web monitoring system. The monitoring system will send HTTP request with HEAD method to the monitored website to ensure the availability of the website. However, a large number of HTTP HEAD requests may indicate malicious web robots activities [17]. Thus, web log with HTTP HEAD method can only be removed if the legitimacy of the IP address is confirmed. For example, the IP which belonged to the web monitoring system can be considered as legitimate.

Lastly, web log with HTTP POST method will be removed if the file posted are legitimate. For instance, it is legitimate if there is web log with .svc file extension in uri-stem and with HTTP POST method. .svc file is a special content file which represents the Windows Communication Foundation (WCF) service hosted in IIS. Thus, posting .svc file can be considered as legitimate web server activity.

4 Implementation

There are three machines involved in the implementation. The first machine is the e-commerce site web server which generates web log files. The second machine is the log collector server which receives logs from the e-commerce site web server. The third machine is a workstation which used for simulation attack. The web log format is Internet Information Services (IIS) Log Format (Microsoft) while the web log files collected are in text file format. This text file will be preprocessed by the written code to obtain the clean web log file. Figure 1 showed the architectural diagram for simulation attack and web log collection.

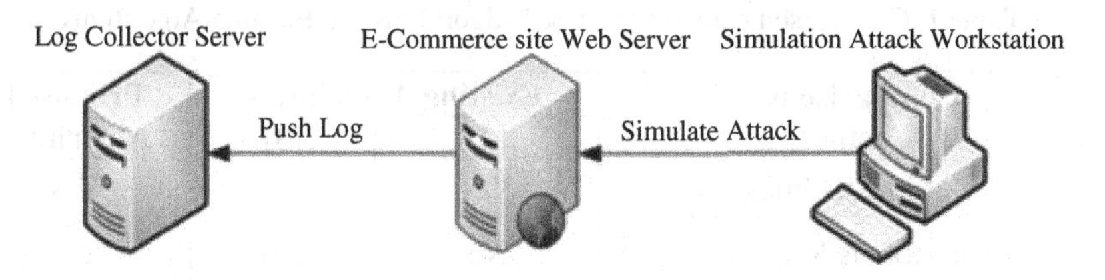

Fig. 1. Architectural Diagram for Simulation Attack and Web Log Collection

4.1 Simulation Attacks, Web Log Collection and Web Log Cleaning

To produce the web log intrusion dataset for experimental purpose, simulation attack is carried out by using three web vulnerability assessment tools. These tools are Acunetix, Nikto and w3af. The simulation attack is carried out in a workstation with Microsoft Windows installed and there is another operating system, BackTrack GNOME running in a virtual environment. Acunetix is running on Microsoft Windows while Nikto and w3af are running on BackTrack GNOME. For log collection, the analysis target, which is an e-commerce site web server is configured to send web logs to the log collector server via User Datagram Protocol (UDP).

The web logs are collected in between 6:00 a.m. to 6:00 p.m. while all three simulation attacks are performed in between 10.00 a.m. to 5:00 p.m.. Once the complete set of web logs had been collected, codes are executed to identify the intrusion dataset. Identification is made based on the IP address as all three simulation attacks are carried out via three different IP address.

The proposed algorithm which had been coded into Python program is executed on the web log files. These included the original web log files and the original intrusion dataset. The output are the clean web logs file and the clean intrusion dataset. Furthermore, the program for analysis is executed on both original web log files and clean web log files to obtain the web log file size and the number of log entries.

5 Experimental Results and Analysis

This section presents the experimental result and analysis of the proposed algorithm. The effectiveness of the proposed algorithm is measured via two different factors which are the percentage of reduction and the false negative rate.

5.1 Algorithm Comparison

The existing algorithms which are selected for comparison purpose are the algorithms presented by [2] and [3]. Both of these algorithms are focusing on web log preprocessing for web intrusion detection. The comparison is done to evaluate if the five major web log attributes are included in the web log cleaning algorithms. It is important as all five major web log attributes should be considered during the algorithm design. Comparison is also done to check if the concept of rules and conditions is applied in the web log cleaning algorithms.

Table 1. Comparison between Proposed Algorithms and Existing Algorithms

Comparisons Factor	Existing Algorithms		Proposed Algorithm
	[2]	[3]	
Multimedia Files	Yes	Yes	Yes
Web Robots Request	No	No	Yes
HTTP Status Code	200, 400 series, 500 series	200	200, 400 series, 500 series
HTTP Method	No	GET	GET, POST, HEAD
Others Files	Yes	Yes	Yes
Number of Rules and Conditions	2	1	6

Table 1 clearly showed that the proposed algorithm includes all five major web log attributes. Algorithm presented by [2] and [3] ignored one of the web log attributes which is the web robot request. Apart from this, they ignored the HTTP HEAD method. The proposed algorithm clearly state that it will remove web logs with HTTP HEAD and web robot request provided that the IP address are legitimate. Besides, there are only three examples of files extensions provided by these algorithms [2], [3], namely .jpg, .gif and .css. However, the proposed algorithm defined a total of sixteen multimedia file extensions and four other files extension.

The proposed algorithm contained six rules and conditions as discussed in the algorithm design. Even though [3] introduced the concept of rules and conditions in the cleaning algorithm, there is only one example provided, which is removing web log with HTTP status code 200 and HTTP GET. Obviously, such condition is not suitable for intrusion detection as web log with HTTP status code 200 may indicate successful attack [2].

5.2 Measuring Factors of Web Log Cleaning Capability

Two measuring techniques are used to evaluate the cleaning capability of the proposed algorithm. These are the web log file size and the total number of web log entries. The web log file size is measured in bytes while total number of web log entries is calculated based on the total number of lines in the web log file. Based on the data obtained, the percentage of reduction will be calculated. The percentage of reduction is calculated based on the total number of web log entries which have been cleaned by the web log cleaning algorithm. It can be computed by using the following equation:

Percentage of reduction = (total number of web log entries removed / total number of web log entries) × 100%

The higher is the percentage of reduction, the better is the cleaning capability. Note here that the number of web log entries is used for the calculation of percentage as it yields more accurate result. For instance, two text file with same number of web log entries may have different file size as the length of the web log entries may differ.

5.3 Measuring Factors of Intrusion Detection Readiness

The false negative rate is introduced to measure the readiness of the web log cleaning algorithm to prepare the web log file for web intrusion detection. To calculate the false negative rate, assumption was made that all web log entries are considered as non-malicious unless the web log entries contained the source IP address which belonged to the simulation attack workstation's IP address. Since this work does not focus on the intrusion detection mechanism, this is the best way available to identify the intrusion dataset. Lower false negative rate reflects better intrusion detection readiness.

The false negative rate is computed as:

False negative rate = Total number of malicious request removed / total number of malicious request

5.4 Results and Analysis

The initial size of the web log file is 56730361 bytes with 379536 entries. The proposed algorithm managed to clean up 153372 entries which carried percentage of reduction 40.41. It is the second highest in percentage compared to algorithm presented by [3] which yielded the percentage of reduction 56.81. Algorithm by [2] carries the lowest percentage of reduction, which is 13.94. This indicates that the proposed algorithm has good capability in term of web log cleaning.

Apart from this, the total number of malicious request and the total number of malicious request removed are obtained to compute the false negative rate. The malicious request is extracted from the original web log file based on the source IP addresses used for simulation attack. The malicious request obtained is further categorized into three different intrusion dataset based on the types of vulnerability assessment tools used. The number of entries for Acunetix intrusion dataset, Nikto intrusion dataset and w3af intrusion dataset are 382, 1207 and 482 respectively.

Table 2. Comparison of Percentage of Reduction between Proposed Algorithms and Existing Algorithms

Measuring Factors	Existing Algorithms		Proposed Algorithm
	[2]	[3]	
File Size Reduced (bytes)	6945603	32423581	18957149
Number of Entries Removed	52916	215616	153372
Percentage of Reduction (%)	13.94	56.81	40.41

Table 2 showed the number of entries which has been reduced after the cleaning process by using three different algorithms. Note here that the web log entries which have been reduced from the intrusion dataset are considered as false negative. Based on the formula of false negative rate which had been defined earlier, the false negative rate is computed. Result in Table 3 showed that the proposed algorithm has a false negative rate of 0.00531. Algorithm proposed by [2] has the best false negative rate which is 0.00144 while algorithm by [3] has a false negative rate of 0.15789.

Compared to [2], the proposed algorithm removed eight extra web requests which caused a higher false negative rate. These web logs are manually reviewed. All the web requests contained HTTP GET method with request to .txt file. According to our algorithm, it is considered as legitimate. It also indicates that vulnerability assessment tools may request for legitimate files during the reconnaissance process. [2] does not clean this request as the .txt file extension is not defined in their algorithm.

Table 3. Comparison of False Negative Rate between Proposed Algorithms and Existing Algorithms

Measuring Factors	Tools	Existing Algorithms		Proposed Algorithm
		[2]	[3]	
Number of Entries Reduced	Acunetix	0	57	6
	Nikto	0	5	0
	w3af	3	265	5
False Negative Rate	Acunetix	0.0000	0.1492	0.0157
	Nikto	0.0000	0.0041	0.0000
	w3af	0.0062	0.5497	0.0131
	Overall	0.00144	0.15789	0.00531

Even though algorithm proposed by [3] has the highest percentage of reduction, it has the highest false negative rate (0.15789) too. Thus, it indicates that this algorithm is not suitable for intrusion detection. Meanwhile, algorithm presented by [2] has the lowest false negative rate (0.00144). But, the percentage of reduction is much lower than the proposed algorithm, which is 13.94%.

6 Conclusion

A web log cleaning algorithm for intrusion detection had been presented. The enhanced algorithm had taken five major web log attributes into consideration. Moreover, it is designed with the concept of rules and conditions in mind. Comparison revealed that the proposed algorithm includes all five major web log attributes. Besides, statistical analysis recorded that the algorithm has the second highest percentage of reduction and second lowest false negative rate compared to the existing web algorithms. Thus, it can be concluded that the proposed web log cleaning algorithm served better cleaning capability and at the same time provided better readiness for the clean web log files to be used for web intrusion detection. In future, we planned to extend the web logs collection time as a larger web log file may yield different results. We also aimed to develop a web intrusion detection mechanism algorithm which can be combined with the enhanced web log cleaning algorithm.

References

1. Suthaharan, S., Panchagnula, T.: Relevance feature selection with data cleaning for intrusion detection system. In: Proceedings of the IEEE SoutheastCon, pp. 1–6. IEEE (2012)
2. Salama, S.E., Marie, M.I., El-Fangary, L.M., Helmy, Y.K.: Web Server Logs Preprocessing for Web Intrusion Detection. Computer and Information Science 4, 123–133 (2011)
3. Patil, P., Patil, U.: Preprocessing of web server log file for web mining. World Journal of Science and Technology 2, 14–18 (2012)
4. Farid, D.M., Rahman, M.Z., Rahman, C.M.: Adaptive Intrusion Detection based on Boosting and Naive Bayesian Classifier. International Journal of Computer Applications 24, 12–19 (2011)
5. Eshaghi, M., Gawali, S.Z.: Web Usage Mining Based on Complex Structure of XML for Web IDS. IJITEE International Journal of Innovative Technology and Exploring Engineering 2, 323–326 (2013)
6. Suen, H.Y., Lau, W.C., Yue, O.: Detecting Anomalous Web Browsing via Diffusion Wavelets. In: International Conference on Communications, pp. 1–6. IEEE (2010)
7. Chauhan, P., Singh, N., Chandra, N.: Deportment of Logs for Securing the Host System. In: 5th International Conference on Computational Intelligence and Communication Networks, pp. 355–359. IEEE (2013)
8. Aye, T.T.: Web log cleaning for mining of web usage patterns. In: 3rd International Conference on Computer Research and Development, pp. 490–494. IEEE (2011)
9. Raju, G., Satyanarayana, P.: Knowledge Discovery from Web Usage Data: Complete Preprocessing Methodology. IJCSNS International Journal of Computer Science and Network Security 8, 179–186 (2008)
10. Vellingiri, J., Pandian, S.C.: A Novel Technique for Web Log mining with Better Data Cleaning and Transaction Identification. Journal of Computer Science 7, 683–689 (2011)
11. Reddy, K.S., Varma, G., Babu, I.R.: Preprocessing the web server logs: an illustrative approach for effective usage mining. ACM SIGSOFT Software Engineering Notes 37, 1–5 (2012)
12. Castellano, G., Fanelli, A., Torsello, M.: Log data preparation for mining web usage patterns. In: Proceedings of IADIS International Conference Applied Computing, pp. 371–378 (2007)
13. Suneetha, K., Krishnamoorthi, R.: Identifying user behavior by analyzing web server access log file. IJCSNS International Journal of Computer Science and Network Security 9, 327–332 (2009)

14. Anand, S., Aggarwal, R.R.: An Efficient Algorithm for Data Cleaning of Log File using File Extensions. International Journal of Computer Applications 48, 13–18 (2012)
15. Stamm, S., Stern, B., Markham, G.: Reining in the web with content security policy. In: Proceedings of the 19th International Conference on World Wide Web, pp. 921–930. ACM (2010)
16. Bomhardt, C., Gaul, W., Schmidt-Thieme, L.: Web robot detection-preprocessing web logfiles for robot detection. In: New Developments in Classification and Data Analysis, pp. 113–124 (2005)
17. Doran, D., Gokhale, S.S.: Web robot detection techniques: overview and limitations. Data Mining and Knowledge Discovery 22, 183–210 (2011)

Possible Prime Modified Fermat Factorization: New Improved Integer Factorization to Decrease Computation Time for Breaking RSA

Kritsanapong Somsuk[1] and Sumonta Kasemvilas[2]

[1] Department of Electronics Engineering, Faculty of Technology, Udon Thani Rajabhat University, UDRU, Udon Thani, Thailand
kritsanapong@udru.ac.th

[2] Department of Computer Science, Faculty of Science, Khon Kaen University, KKU, Khon Kaen, Thailand
sumkas@kku.ac.th

Abstract. The aim of this research is to propose a new modified integer factorization algorithm, called Possible Prime Modified Fermat Factorization (P^2MFF), for breaking RSA which the security is based upon integer factorization. P^2MFF is improved from Modified Fermat Factorization (MFF) and Modified Fermat Factorization Version 2 (MFFV2). The key concept of this algorithm is to reduce iterations of computation. The value of larger number in P^2MFF is increased more than one in each iteration of the computation, it is usually increased by only one in MFF and MFFV2. Moreover, this method can decrease the number of times in order to compute the square root of some integers whenever we can strongly confirm that square root of these integers is not an integer by using number theory. The experimental results show that P^2MFF can factor the modulus faster than MFF and MFFV2.

Keywords: RSA Scheme, Modified Fermat Factorization (MFF), Modified Fermat Factorization Version 2 (MFFV2), Computation time, Integer Factorization.

1 Introduction

Communicating Information over Computer Network, such as the Internet is prevalent. However, computer network is the insecure channel. Thus, the third party can trap the contents which are sent through this channel. From this problem, many methodologies were proposed to protect the information in this channel. Cryptography is one of many methodologies used to secure the information by encryption and decryption. Symmetric key cryptography is the first proposed cryptography used the same key to encrypt and decrypt the messages. However, the secure channel is needed for exchanging the same key between the sender and the receiver. In 1976, asymmetric key cryptosystem [1], also known as public key cryptosystem, was proposed to solve the key-exchange problem in symmetric key cryptosystem. This technique uses a pair of key for encryption and decryption that is

S. Boonkrong et al. (eds.), *Recent Advances in Information and Communication Technology*, Advances in Intelligent Systems and Computing 265,
DOI: 10.1007/978-3-319-06538-0_32,

public key which is used to encrypt the message and private key which is used to decrypt the encrypted message. RSA [1] is a type of public key cryptosystem proposed by Ron Rivest, Adi Shamir and Leonard Adleman. It is the most well-known algorithm that computing the product of two large prime numbers is uncomplicated, but computing its inverse which is integer factorization is very difficult. Nevertheless, RSA is broken whenever the third party can factor the modulus which is the product of two large prime numbers.

Many integer factorization algorithms were introduced continuously such as Trial Division algorithm (TDA) [2], Quadratic Sieve (QS) [3, 4], Pollard's p-1 algorithm [5, 6], Monte Carlo Factorization algorithm [2, 6], VFactor [7, 8], Modified VFactor (MVFactor) [8] improved from VFactor, Fermat's Factorization algorithm (FFA) [2, 5] and Modified Fermat Factorization (MFF) [3, 5] improved from FFA. Both of FFA and MFF are integer factorization algorithms that the modulus must be rewritten as the difference of squares. Finding two large prime factors of the modulus, we have to find two perfect square integers that their subtraction is equal to the modulus. In addition, we recently proposed Modified Fermat Factorization Version 2 (MFFV2) [3] to reduce number of times to compute integer square root when compared with MFF.

This research is to propose Possible Prime Modified Fermat Factorization (P^2MFF) which is a new modified integer factorization algorithm to decrease computation time in comparison to MFF and MFFV2. The key concept to increase computation speed of P^2MFF is from three ideas that are, 1) reducing iterations of computation, the value of larger number in this algorithm is increased more than one in each iteration of the computation; 2) reducing number of times to compute integer square root whenever its least significant digit is *2, 3, 7* or *8* [3]; and 3) reducing number of times to compute integer square root whenever the result of this integer modulo *4* is equal to *2* or *3* [9]. The experimental results show that P^2MFF can factor the modulus faster than MFF and MFFV2 for all value of the modulus.

The rest of this paper is organized into 6 parts: Section 2 describes the algorithm of RSA. Section 3 explains related works i.e. FFA, MFF and MFFV2. The proposed method is presented in Section 4. Section 5 demonstrates the experimental results. The last section concludes this research.

2 RSA Scheme

RSA is a type of public key cryptosystem used for data encryption and digital signature. The algorithm of RSA for data encryption is as the following; Firstly, two large primes, p and q, must be generated to find the modulus, $n = p*q$ and the Euler' s totient function, $\Phi(n) = (p - 1)*(q - 1)$ respectively. Then the public key, e, and private key, d, are computed, $1 \leq e \leq \Phi(n)$, $\gcd(e, \Phi(n)) = 1$ and $e*d \bmod \Phi(n) = 1$. After that, if the sender wants to sent message to the receiver, he or she has to get the public key of the receiver to encrypt the message from the equation: $c = m^e \bmod n$, m is a message, also known as plaintext, and c is a ciphertext that sent to the receiver. When the receiver gets c from the sender, he or she will recover m from the equation: $m = c^d \bmod n$.

3 Related Work

This section explains some integer factorization algorithms used for breaking RSA. If n is factored, then p and q are known. So, d can be recovered and RSA is broken. However, in this research, we will discuss some algorithms based on size of two large prime factors of n as follows:

3.1 Fermat's Factorization and Modified Fermat Factorization

FFA was discovered by Pierre de Fermat in 1600 [5]. This method looks for two integers, x and y, such that $x^2 - y^2 = n$, where $x = (p + q)/2$ and $y = (p - q)/2$. We know that p and q are odd prime numbers. Therefore, both of x and y are always integers.

Later, MFF was proposed to decrease time to factor n. It is a modified integer factorization modified from FFA. For the difference of squares equation, the equation of this method is rewritten as: $y = sqrt(x^2 - n)$, the value of x is started by the minimum integer which is more than or equal to the square root of n. After that, the value of y is computed. If y is not an integer, we have to compute: $x = x + 1$ and back to compute the new value of y again until the integer of y is found. Otherwise, we can conclude that $p = x - y$ and $q = x + y$ are two large prime factors of n.

3.2 Modified Fermat Factorization Version 2

Recently, we have proposed MFFV2 [3] to decrease time for factoring n in comparison to MFF. This method does not compute the square root of y^2 whenever the least significant digit of y^2 is equal to *2, 3, 7* or *8*. The least significant digit of the integer is a perfect square and must be equal to *0, 1, 4, 5, 6* or *9*. Thus, it is not time-consuming to compute the square root of y^2 in case of the least significant digit of y^2 is equal to *2, 3, 7* or *8*, because it is certainly not an integer.

3.3 VFactor

VFactor [7] is a new modified integer factorization algorithm proposed in 2012 which can finish all trivial and nontrivial values of n. The key of VFactor is finding two integer, x and y, where $y = floor(sqrt(n))$ and $x = y + 2$. The production of these integers, $m = x*y$ must be equal to n. However, if m is less than n, we have to continue computing $x = x+2$ and computing the new value of m until the value of m which is equal to n is found. On the other hand, if m is more than n, we have to continue computing $y = y-2$ and computing the new value of m until the value of m which is equal to n is found.

4 The Proposed Method

4.1 Theorem1

For any integer, k, the modulus, n, will be rewritten only as the form: $n = 6k - 1$ or $n = 6k + 1$

Proof: We know that, for any prime numbers which are more than or equal to *5* could be rewritten as $6k - 1$ or $6k + 1$ [4]. Therefore, assume *a* and *b* are any integers, there are four cases of *n* as follows:

Case 1: $p = 6a + 1$ and $q = 6b + 1$

$$n = p*q = 6(6ab + a + b) + 1 = 6k + 1, k = 6ab + a + b$$
$$x = (p + q)/2 = (6(a + b) + 2)/2 = 3(a + b) + 1 = 3m + 1, m = a + b$$
$$y = (p - q)/2 = 6(a - b)/2 = 3(a - b) = 3n, n = a - b$$

Case 2: $p = 6a + 1$ and $q = 6b - 1$

$$n = p*q = 6(6ab - a + b) - 1 = 6k - 1, k = 6ab - a + b$$
$$x = (p + q)/2 = 6(a + b)/2 = 3(a + b) = 3m, m = a + b$$
$$y = (p - q)/2 = (6(a - b) + 2)/2 = 3(a - b) + 1 = 3n + 1, n = a - b$$

Case 3: $p = 6a - 1$ and $q = 6b + 1$

$$n = p*q = 6(6ab + a - b) - 1 = 6k - 1, k = 6ab + a - b$$
$$x = (p + q)/2 = 6(a + b)/2 = 3(a + b) = 3m, m = a + b$$
$$y = (p - q)/2 = (6(a - b) - 2)/2 = 3(a - b) - 1 = 3n - 1, n = a - b$$

Case 4: $p = 6a - 1$ and $q = 6b - 1$

$$n = p*q = 6(6ab - a - b) + 1 = 6k + 1, k = 6ab - a - b$$
$$x = (p + q)/2 = (6(a + b) - 2)/2 = 3(a + b) - 1 = 3m - 1, m = a + b$$
$$y = (p - q)/2 = 6(a - b)/2 = 3(a - b) = 3n, n = a - b$$

From the four cases of two prime numbers, *p* and *q*, we can conclude that *n* is distinguished in two forms: $n = 6k - 1$ and $n = 6k + 1$. Moreover, if *n* has the form: $n = 6k - 1$, *x* could be expressed only the form: $x = 3m$, *m* is any integer. Nevertheless, if *n* has the form $n = 6k + 1$, *x* could be expressed as the form: $x = 3m - 1$ or $3m + 1$.

Because, for one iteration of the computation, a value of *x* in the form: $n = 6k - 1$ is increased more than a value of *x* in the other form; therefore, the performance of the form: $6k - 1$ is better than the other form when the size of prime factors of each algorithm is close.

4.2 Possible Prime Modified Fermat Factorization

P^2MFF is a new modified integer factorization algorithm modified from MFF and MFFV2 [3]. For P^2MFF, we do not choose all possible value of *x* which is in range *sqrt(n)* to *n* to find the integer *y*, because some value of *x* is not in the form: $n = 6k - 1$ or $n = 6k + 1$. For this reason, P^2MFF can reduce computation time to factor *n* when compare with MFF. Moreover, P^2MFF can increase computation speed by applying two concepts: One is reducing number of times to compute the square root of integer when its least significant digit is *2, 3, 7* or *8*, the method is used in MFFV2 [3]. The other is from the number theory that if an integer is a perfect square, the result of this integer modulo *4* must be only *0* or *1* [9]. The algorithm of P^2MFF is distinguished in two cases, $n = 6k - 1$ and $n = 6k + 1$, as follows:

Algorithm1:
Input: The modulus $n = 6k - 1$

```
1:    x = ceil(sqrt(n));
2:    While (x%3 != 0)       //Make x to be the form x = 3m
3:        x = x+1;
4:    EndWhile
5:    y = sqrt(x^2 - n);
6:    While (y is not an integer)
7:        x = x+3;
8:        y^2 = x^2 - n;
9:        a = y^2 % 10;
10:       If(a is equal 0, 1, 4, 5, 6 or 9)
11:               a2 = y^2 % 4;
12:              If(a2 is equal to 0 or 1)
13:                   y = sqrt(y^2);
14:              EndIf
15:       EndIf
16:   EndWhile
17:   p = x - y;
18:   q = x + y;
```

Outputs: p, q (two prime factors of n)

Algorithm 2:
Input: The modulus $n = 6k + 1$

```
1:    x = ceil(sqrt(n));
2:    While (x%3 == 0)       /*Make x to be the form x = 3m - 1 or
x = 3m + 1*/
3:        x = x+1;
4:    EndWhile
5:    y = sqrt(x^2 - n);
6:    xtmp = 1;         //xtmp is a result of x modulo 3
7:    If(x % 3 == 1)
8:        xtmp = 1;
9:    Else
10:       xtmp = 2;
11:   EndIf
12:   While (y is not an integer)
```

```
13:        If(xtmp == 2)
14:            x = x + 1;
15:            xtmp = 0;
16:        Else
17:            x = x + 1;
18:            y^2 = x^2 - n;
19:            a = y^2 % 10;
20:            If(a is equal 0, 1, 4, 5, 6 or 9)
21:                 a2 = y^2 % 4;
22:                  If(a2 is equal to 0 or 1)
23:                       y = sqrt(y^2);
24:                  EndIf
25:            EndIf
26:        EndIf
27:    EndWhile
28:    p = x - y;
29:    q = x + y;
```

Outputs: p, q (two prime factors of n)

Example 1: $n = 453899 = 541 * 839$
$n = 6 * 75650 - 1$, Form: $n = 6k - 1$
$x = ceil(sqrt(453899)) = 674$
$674 \% 3 = 2$, thus, increase x by 1 ($x = 674 + 1 = 675 = 3*225 = 3m$)
Iteration 1: $y^2 = 675^2 - 453899 = 1726$, because $1726\%4 = 2$, we do not compute $sqrt(y^2)$ and continue computing $= x = x+3 = 678$.
Iteration 2: $y^2 = 678^2 - 453899 = 5785$, $y = sqrt(5785) = 76.06$ that is not an integer. Thus, we continue computing $x = x+3 = 681$.
Iteration 3: $y^2 = 681^2 - 453899 = 9862$, because the least significant digit of y^2 is equal to 2, we do not compute $sqrt(y^2)$ and continue computing $x = x+3 = 684$.
Iteration 4: $y^2 = 684^2 - 453899 = 13957$, because the least significant digit of y^2 is equal to 7, we do not compute $sqrt(y^2)$ and continue computing $x = x+3 = 687$.
Iteration 5: $y^2 = 687^2 - 453899 = 18070$, because $18070\%4 = 2$, we do not compute $sqrt(y^2)$ and continue computing $= x = x+3 = 690$.
Iteration 6: $y^2 = 690^2 - 453899 = 22201$, $y = sqrt(22201) = 149$ here is an integer. Thus, $p = (690 - 149) = 541$ and $q = (690 + 159) = 839$ are two prime factors of n.

For Example 1, there are only 6 iterations to find the integer of y. However, if we use MFF, there are ((690 – 674) + 1) 17 iterations to find the integer of y. In the computations, we start with $x = 674$ and end with $x = 690$. Moreover, y^2 is computed its square root only in the 2nd and 6th iteration. Therefore, the computation time to find two large prime factors of n is decreased.

Example 2: $n = 441289 = 607 * 727$
$n = 6 * 73548 + 1$, Form: $n = 6k + 1$
$x = ceil(sqrt(453899)) = 665$
$665 \% 3 = 2$, thus, we can start with $x = 665$ ($x = 665 = 3*222 - 1 = 3m - 1$)
Iteration 1: $y^2 = 665^2 - 441289 = 936$, $y = sqrt(936) = 30.59$ that is not an integer. Thus, we continue computing $x = x+2 = 667$ ($665 \% 3 = 2$).
Iteration 2: $y^2 = 667^2 - 441289 = 3600$, $y = sqrt(3600) = 60$, here is an integer. Thus, $p = (667 - 60) = 607$ and $q = (667 + 60) = 727$ are two prime factors of n.

For Example 2, there are only two iterations to find the integer of y. However, if we use MFF, there are ((667 – 665) + 1) 3 iterations to find the integer of y, we start with $x = 665$ and end with $x = 667$. Therefore, the computation time to find two large prime factors of n is decreased.

4.3 Comparison of P²MFF, MFF and MFFV2

In P²MFF, for the form: $n = 6k - 1$, we can see that the step of x is up by 3. The other form: $n = 6k + 1$, the step of x is up by 1 but if x has the form: $x = 3m$, y is not computed and we continue incrementing x by 1. Therefore, we can conclude that the computation time of P²MFF is decreased when compared with MFF and MFFV2 which the step of x is always up by only 1.

5 Experimental Results

This section is for describing about experimental results of factoring n by using P²MFF. This method will be compared about computation time with MFF and MFFV2. The experiments were divided into two cases. One is for the same size of p and q. The other is for the difference size of p and q. In this research, the experiments were run on a 2.53 GHz an Intel® Core i3 with 4 GB of memory. All experiments in this study were implemented by using BigInteger Class.

Table 1. Comparison of the computation times of three integer factorization algorithms for the same size of the modulus which has the form in 6k – 1

n	Size p, q (Bits)	Computation Time (Sec.)		
		MFF	MFFV2	P²MFF
611184411990263 = 19554967 * 31254689	25	1.95	1.03	**0.28**
43170764358014549 = 178859767 * 241366547	28	6.37	4.67	**1.13**
819724297698120371 = 787898597 * 1040393143	30	24.34	17.04	**4.17**
8269813160041838417 = 2303447563 * 3590189459	32	210.17	107.56	**25.53**
39456286068624113159 = 5359683107 * 7361682637	33	234.38	183.34	**39.88**

Table 2. Comparison of the computation times of three integer factorization algorithms for the same size of the modulus which has the form in 6k + 1

n	Size p, q (Bits)	Computation Time (Sec.)		
		MFF	MFFV2	P^2MFF
687584152433803 = 20754931 * 33128713	25	2.11	1.22	**0.57**
46380698455907323 = 173928673 * 266665051	28	13.81	7.32	**3.67**
606043545959590699 = 648993353 * 933820883	30	35.12	25.52	**12.55**
8269813160041838417 = 2674520143 * 4140349063	32	227.64	164.28	**74.53**
47260507180106575051 = 5661639091 * 8347495561	33	395.27	284.04	**135.62**

Table 3. Comparison of the computation times of three integer factorization algorithms for the difference size of the modulus which has the form in 6k – 1

n	Size (Bits)		Computation Time (Sec.)		
	p	*q*	MFF	MFFV2	P^2MFF
84050314308563 = 2525293 * 33283391	22	25	24.17	12.71	**3.21**
43452790497629 = 525433 * 82699013	20	27	95.44	68.44	**15.82**
180848815076590481 = 192067027 * 941592203	28	30	398.01	292.74	**66.33**
223407579692036591 = 124401157 * 1795864163	27	31	1394.35	997.95	**231.28**
7401587937219993089 = 1843942433 * 4014001633	31	32	610.62	424.42	**97.26**
5361436430587906421 = 679570687 * 7889446283	30	33	6362.63	4399.65	**932.39**

From Table 1 – 4, the experimental results show that P^2MFF can factor the modulus faster than MFF and MFFV2 for all value of n. In addition, Table 1 and Table 3, the form of $n = 6k - 1$, show that the computation speed of P^2MFF is increased from 82 to 88 percents and from 73 to 79 percents in comparison to MFF and MFFV2, respectively. The other tables, Table 2 and Table 4 that are the form of $n = 6k + 1$, show that the computation speed of P^2MFF is increased from 64 to 74 percents and from 50 to 56 percents in comparison to MFF and MFFV2, respectively. So, from Table 1 and Table 3, the computation time of P^2MFF is very small in comparison to the information in Table 2 and Table 4. The reason is that the form of n in Table 1 and Table 3 is 6k – 1. Thus, x is increased by 3. While in Table 2 and Table 4, it is increased by 1 (when x%3 = 0) or 2 (when x%3 == 2), whenever the integer y is not found.

Table 4. Comparison of the computation times of three integer factorization algorithms for the difference size of the modulus which has the form in 6k + 1

n	Size (Bits)		Computation Time(Sec.)		
	p	q	MFF	MFFV2	P^2MFF
39670146539281 = 2106337 * 18833713	22	25	11.64	8.04	**3.80**
56899183657009 = 675083 * 84284723	20	27	94.22	68.77	**32.36**
134809991236867861 = 142625627 * 945201743	28	30	505.80	357.93	**172.11**
151826379800519143 = 73518587 * 2065142789	27	31	1923.01	1061.69	**508.98**
5496158094627417961 = 1819651331 * 3020445731	31	32	218.34	159.28	**72.69**
4639414977187852189 = 573626447 * 8087867987	30	33	6892.92	4906.96	**2168.66**

6 Conclusion

P^2MFF is a new modified integer factorization algorithm that improved from MFF and MFFV2 for breaking RSA. The aim of this algorithm is to decrease the iterations of computation. For P^2MFF, the value of x is increased more than *1*, x is increased by *3* for the form: $n = 6k - 1$ and x is increased by *2* (if x modulo *3* is equal to 2) or x is increased by *1* (if x modulo *3* is equal to *1*) for the form: $n = 6k + 1$. This algorithm is also the integrating between two techniques in order to reduce the number of times to compute the square root of an integer, the integer is not computed its square root when we can strongly confirm that it is not a perfect squares. The first technique is from the idea that an integer is certainly not a perfect square whenever the least significant digit of this integer is equal to *2, 3, 7* or *8*, this technique was implemented in MFFV2. The other technique is that, if the result of an integer divided by *4* yields a remainder of *2* or *3*, this integer is certainly not a perfect square. The experimental results show that the computation time of P^2MFF is decreased in comparison to MFF and MFFV2.

References

1. Rivest, R.L., Shamir, A., Adleman, L.: A method for obtaining digital signatures and public key cryptosystems. Communications of ACM 21, 120–126 (1978)
2. Bishop, D.: Introduction to Cryptography with java Applets. Jones and Bartlett Publisher (2003)
3. Somsuk, K., Kasemvilas, S.: MFFV2 and MNQSV2: Improved factorization Algorithms. In: International Conference on Information Science and Application, pp. 327–329 (2013)
4. Huang, Q., Li, Z.T., Zhang, Y., Lu, C.: A Modified Non-Sieving Quadratic Sieve For Factoring Simple Blur Integers. In: International Conference on Multimedia and Ubiquitous Engineering, pp. 729–732 (2007)
5. Ambedkar, B.R., Gupta, A., Gautam, P., Bedi, S.S.: An Efficient Method to Factorize the RSA Public Key Encryption. In: International Conference on Communication Systems and Network Technologies, pp. 108–111 (2011)
6. Pollard, J.: Monte Carlo methods for index computation (mod p). Math. Comp. 32, 918–924 (1978)
7. Sharma, P., Gupta, A.K., Vijay, A.: Modified Integer Factorization Algorithm using V-Factor Method. In: Advanced Computing & Communication Technologies, pp. 423–425 (2012)
8. Somsuk, K., Kasemvilas, S.: MVFactor: A Method to Decrease Processing Time for Factorization Algorithm. In: 17th International Conference on International Computer Science and Engineering Conference, pp. 339–342 (2013)
9. Silverman, J.H.: A Friendly introduction to number theory, Pearson International Edition (2005)

N-Gram Signature for Video Copy Detection

Paween Khoenkaw and Punpiti Piamsa-nga

Department of Computer Engineering, Faculty of Engineering, Kasetsart University,
Jatujak, Bangkok, 10900, Thailand
{g4885027,pp}@ku.ac.th

Abstract. Typically, video copy detection can be done by comparing signatures of new content with of known contents in database. However, this method requires high computation for both database generation and signature detection. In this paper, we proposed an efficient and fast video signature for video copy protection. The video features of a scene are extracted and then transformed to be a signature as a bit-wise string. All string signatures then are stored and manipulated by n-gram based text retrieval algorithm, which is proposed as a replacement with computation-intensive content similarity detection algorithm. The evaluation on the CC_WEB_VIDEO dataset shows that its accuracy is 85% where our baseline algorithms achieved only 75%; however, our algorithm is around 20 times as fast as the baseline.

Keywords: Similarity Measure, Video Matching, Video Search, Video Database.

1 Introduction

Exponential growth of video contents on the internet makes cyber life uncomfortable for users to select the most appropriate contents; however, redundant uploads/posts of the same contents make internet much less enjoyable. Redundant post on the internet is usually related to violation to copyright laws and website policies as well. To prevent redundant uploads, video similarity detection algorithms have been proposed [1]. There are two approaches to tackle this problem: frame-by-frame comparison method and signature based method. Although frame-by-frame comparison achieves high accuracy, it requires exhaustive computation; therefore, it is impractical for long video. The signature based method is about to represent video contents by a smaller vector and then that vector is used to be compared instead of the original content; therefore, it is faster and more practical even though it may lose accuracy.

Early signature-based technique was about to digest video contents using strong hash algorithm, such as SHA [2]. This method is efficient for comparing videos that contained exactly the same contents; however, it is not practical for real application since there are many existing formats of video and the video files usually contains other additional information [3]. Many other techniques have been proposed as following literatures.

S. Boonkrong et al. (eds.), *Recent Advances in Information and Communication Technology*, Advances in Intelligent Systems and Computing 265,
DOI: 10.1007/978-3-319-06538-0_33, © Springer International Publishing Switzerland 2014

There are many video features that can be used as video signature. Parts of video stream such as DCT coefficients extracted from MPEG streams can be used for internet video similarity detection [4]; however, this method is limited to MPEG encoded videos. Weighted color component [5], color histogram [6] and color with applying coherence vector [7] can be used as signature to detect the television commercials. Color deterioration in motion pictures usually degrades signature quality. Therefore, there are many proposed techniques to improve robustness of color. For example, Xian-Sheng et. al. proposed to down-sample video frame to create the temporal shape of relative intensity [8, 9] and Lifeng et. al. proposed to use local average of gray-level values for signature creation [10]. However, in many applications, color deterioration is a main trouble to video signature detection [8].

Other proposed techniques are based on high-level feature extraction, such as motion [3], shot durations, event length [11], camera behaviors [12-15], and object recognition [16, 17]. These types of signatures can represent its source videos; however, it suffers many problems. Color histogram is simple to extract but it is not robust to color deterioration. Motion is robust to color deteriorated but it required complex calculation to extract motion signatures, such as motion estimation and template matching. Combined features to create signature can gain more accuracy than the previous proposed ones but it required more computation power to prepare.

The previous research projects on signature matching were based on similarity measurement method between pair of signatures. Usually a signature is represented by a multidimensional vector of real numbers, which requires very high computation power in similarity measurement process.

In temporal dimension, it is very usual that the same video content may have different frame rates. Therefore, video similarity detection algorithm must be tolerant to small frame rate change. Typically, most algorithms use average values within sliding windows as signature representation or use dynamic programming in signature matching process [18]. However, this method requires high computation.

However, the longest processing time in signature detection is about similarity ranking. All methods in literatures need to compute similarity score between query and all records in database. The results are the highest ranked results. Therefore, the processing time is much worse if the database is larger.

In this paper, we propose an n-gram signature for video copy detection. All video sequences are transformed into a string representation, where each character represents features extracted from video. Then, all comparisons take advantages of text comparison, which is much faster than using multidimensional vectors of real numbers. N-gram comparison is alignment free, which is robust to a small number of insertions or deletions of a few characters in the string; therefore, it should be robust to small frame rate changes. We tested our proposed method on the a video database corpus (CC_WEB_VIDEO) [19]. We found that its accuracy is 85% where our benchmark is 75%. However, run time using by our algorithm is only 5% of the benchmark.

In following sections, we give brief introduction of N-gram in section 2. The proposed algorithm is described in section 3. Section 4 has details of experiments and their results. Finally, we conclude our wok in section 5.

2 N-Gram

N-Gram is a contiguous sequence of n items from a given sequence of data stream, such as text, speech and video. It is widely used in sequence matching algorithm on natural language processing [20, 21] and biological sequence analysis [22-24]. N-Gram is generated by cutting given sequence into small pieces with size varied from 1 to N, called gram; every possible n-gram in a sequence are stored in a table along with its frequency. The similarity between two sequences is about the similarity of frequencies of common n-grams. N-gram signature is robust to insertion, deletion and noise [24]. It is fast because it does not need to compare every possible n-gram and only high-frequency n-grams are required in matching process. Therefore, it requires less storage for n-gram signature.

3 Proposed Model

Out system is depicted in Fig. 1. First, features of interest are extracted from video stream. Second, those features are "texualized", which is to transform features in multidimensional vector to string representation. Third, those strings are extracted as "grams" and then stored in a database as indices. The lower part of Fig. 1 is about video querying. The query video has to be texualized and then used it to generate n-grams. The query n-grams are then used as video representatives to match in the video archive database.

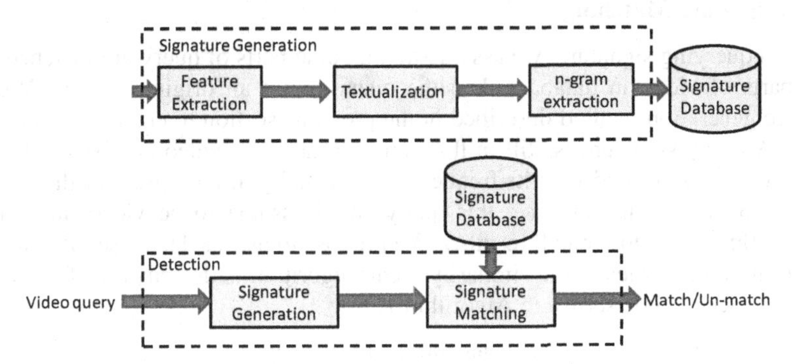

Fig. 1. Proposed model of video copy detection

3.1 Signature Generation

In order to generate a good signature, features must be carefully selected. Many previous research projects used temporal differentials of feature vectors to create signatures. This kind of features can amplify "landmark event" in video such as motion of object or scene cut detection in video sequence. However, landmark event cannot be a signature but a composition of small details around landmark can. In order to avoid problems of color deterioration problem and sampling-rate heterogeneity, we propose to use "luminance velocity" as feature. The signature generation is explained as follows. First, extract average luminance intensity of each video frame by converting image frame to gray scale and then compute average intensity of each frame to create a sequence of average luminance (***I***). Second, determine the "luminance velocity" by calculating the differences between adjacent elements of ***I*** using equation (1) where $\boldsymbol{I_v}$ is defined as "luminance velocity" and then its values are normalized to be in between [0..1] by equation(2) where ***D*** is normalized luminance velocity vector.

$$\mathrm{I_v} = \Delta I \tag{1}$$

$$D = \left(\frac{\mathrm{i_v} - I_{v_{min}}}{I_{v_{\max}} - I_{v_{min}}}\right) \tag{2}$$

Third, quantize the vector ***D*** into finite *l* levels, where $0 < l < 256$ (ASCII range), to create symbolic sequences that suitable to text processing algorithm. Practically, ASCII characters below 32 are the control code that can affect the text processing library in order to avoid that we suggest to shift the symbol to the range of the normal alphabet, where the first character is arbitrary *c* as described in equation (3) where sequence ***Q*** is a quantized vector.

$$Q = \lfloor D \times l \rceil + c \tag{3}$$

Finally, a signature string ***Q*** represents a video clip. Then Q is chopped and each piece is called candidate gram. In our experiments only gram that has frequency greater than 2 can be stored in database as signature (***S***).

3.2 Signature Matching

To match querying signature, we assume that if some parts of query are matched with some parts of video in database then the query is not an original video. We use signature generation method described in the previous section to create a signature of query. A query signature is still in the same format as of video archive, which is a set of n-grams associated with its frequencies. Actually, if an n-gram of the query is matched to the one in database, this query could assume to be video duplication; however, this brings low-recall results. We propose to use a selected set of *k* highest-frequent n-grams to detect the similarity. Our algorithm is described as follows and written in pseudo code shown in Algorithm 1.

Algorithm 1 Video query

Input: *k*, query signature($\boldsymbol{S_t}$), signature database(***S***)
Output: similar video($\boldsymbol{S_d}$)
1: $\boldsymbol{S_r}$=Select top *k* ***grams*** from $\boldsymbol{S_t}$ order by frequency, length(**gram**)
2: FOR each *gram* of $\boldsymbol{S_r}$
3: $\boldsymbol{S_d}$=$\boldsymbol{S_d}$+ Select videos from S where ***grams***=*gram*
4: ENDFOR

First, generate n-grams of query video (S_t) by process in previous section. Second, select the *k* most-frequent n-gram where length of n-gram is priority if frequency is equal. These *k* n-grams are denoted as $\boldsymbol{S_r}$. Third, every element in $\boldsymbol{S_r}$ is used to search in database ***S*** using binary search. If a single n-gram in S_r is found in S, the result is a set of matched videos.

4 Experiments and Results

The experiments are divided into two parts: to determine the most appropriate number of quantization levels (l) and number of grams in video query (l); and compare our method with a baseline algorithm [16, 17]. We assume that our method should maintain accuracy where using less computation power. We used the well-known CC_WEB_VIDEO, which is a near-duplicate web video dataset. This dataset contains a collection of viral videos, which is searched by 24 popular queries from YouTube, Google Video and Yahoo. All 12,790 videos are varied in encoding format, frame rate, bit rate, frame resolution, color/lighting change, overlay with logo, text and content modification. All queries in dataset are shown in Table 1 [17, 19, 25].

Table 1. Queries to build video collections

ID	Query	Number of Videos		Processing Time (Second)
		Total	Near-Duplicate	
1	The lion sleeps tonight	792	334	1.88
2	Evolution of dance	483	122	2.85
3	Fold shirt	436	183	0.39
4	Cat massage	344	161	0.21
5	Ok go here it goes again	396	89	1.87
6	Urban ninja	771	45	3.39
7	Real life Simpsons	365	154	1.07
8	Free hugs	539	37	2.89
9	Where the hell is Matt	235	23	0.84
10	U2 and green day	297	52	1.30
11	Little superstar	377	59	1.49
12	Napoleon dynamite dance	881	146	2.94
13	I will survive Jesus	416	384	0.38
14	Ronaldinho ping pong	107	72	0.65
15	White and Nerdy	1771	696	7.52
16	Korean karaoke	205	20	1.11
17	Panic at the disco I write sins not tragedies	647	201	2.71
18	Bus uncle (巴士阿叔)	488	80	1.70
19	Sony Bravia	566	202	1.42
20	Changes Tupac	194	72	1.15
21	Afternoon delight	449	54	1.45
22	Numa Gary	422	32	1.59
23	Shakira hips don't lie	1322	234	5.98
24	India driving	287	26	0.88
	Sum	**12790**	**3478**	**47.66**

To determine a number of quantization level l, we measure the performance (accuracy, precision, recall, and F-measures) of different l values, ranged from 0 to 200. the details of this effect are shown in Fig. 2. We found that effect of l can be divided into three regions. The first region is where $l < 80$. In this region, the result has high recall but low precision and accuracy. The second region is where between $80 \leq l \leq 160$. In this region, the accuracy is stable but the precision and recall are highly varied. Value of l in this region can be selected depending on type of an application whether it required either precision or recall. The third region is where $l >$ 160. In this region, the accuracy and precision is decreased where l is increased.

From the experiment, we consider the best value of parameter $\mathbf{l=150}$ because it produced the highest accuracy and the highest precision. Then, we will use this parameter for the rest of experiments.

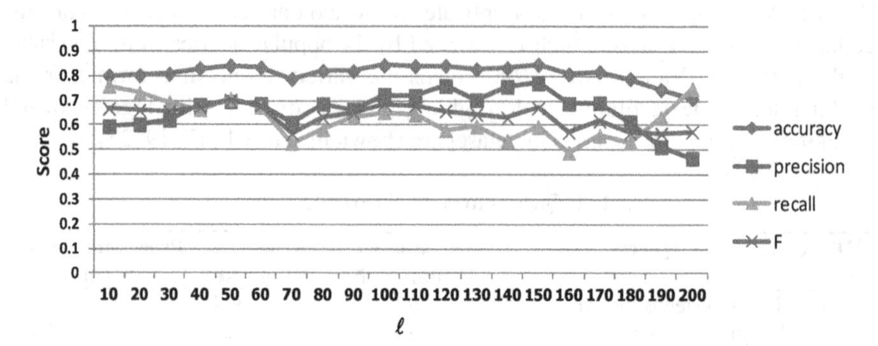

Fig. 2. Accuracy, precision, recall and F with different number of quantization levels (l)

To determine the most appropriate number of n-grams (k), we varied k from 2 to 70. The results are shown in Fig. 3. The result can be divided into three regions. The first region is where $k < 7$. We found that accuracy and precision is high but recall is low. Therefore, value of k in this region is appropriate to applications that require high precision. The second region is where $7 \leq k \leq 17$. In this region, accuracy is stable but precision and recall are converged. Therefore, this could be an optimal zone depending on types of applications. The third region is where $k > 17$. In this region, accuracy, precision and recall are stable. Therefore, $k > 17$ should not be used because the result will not get better and computation time varies linearly to k, shown in Fig. 4. From this experiment, we consider the best value of parameter $k = 8$ because it produces the highest accuracy while maintaining minimum computation and database load.

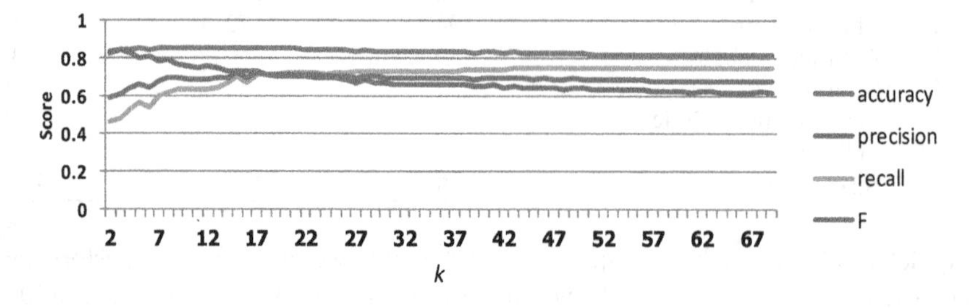

Fig. 3. Accuracy, precision, recall and F at different number of n-gram in query (k)

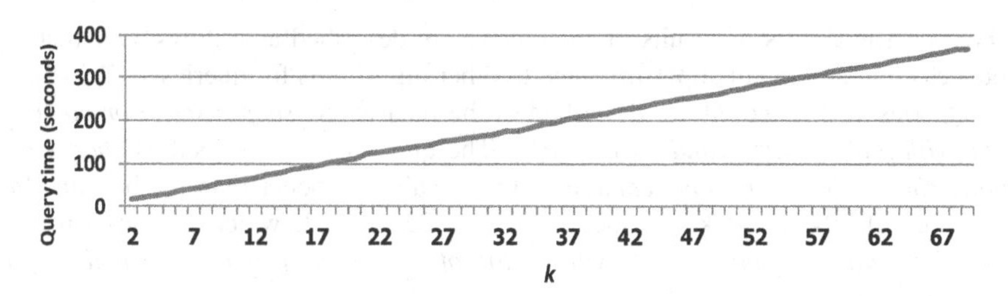

Fig. 4. Query times at different k of n-gram for all 24 queries

After the best practical values of parameters k and l are determined, we used it to set our algorithm to compare performances (precision and recall) with a baseline algorithm LSH-E [16] and a benchmark VK [17]. Computation time for signature generation is also measured. Locally sensitive hashing embedding on color moment (LSH-E), which is histogram based signature detection, is our baseline. VK (visual keyword), which is a high-level object recognition, is our benchmark. VK usually requires very high computation power because it is object recognition algorithm.

Fig. 5 depicts benchmark and baseline comparison results and averages of results in Table 2. Fig. 5 shows that precision, recall, and F from our method are higher than LSH-E; however, it cannot compete with the VK method in average. The average result of all 24 queries is **0.85** for accuracy, **0.72** for precision and **0.58** for recall, the average of precision and recall shown that our method is **10%** better than baseline method.

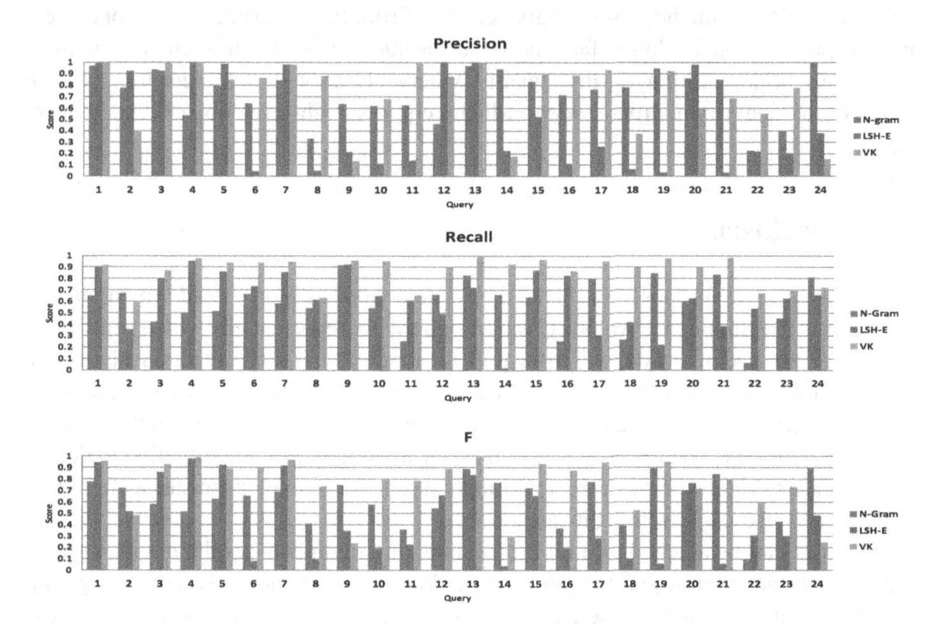

Fig. 5. Precision, recall and F-measure compared with LSH-E and VK

For specific details of results in each query are described as follows. Our method produces comparably equal performance to other algorithms for queries 1, 13, 18 and 24. Queries 1, 13, and 24 are described in the dataset as "*simple scene or complex scene with minor editing and variations*". The query 18 is described as "*bad video quality shot with a cell phone camera*". Our results are better than the baseline but worse than the benchmark for queries 10, 15, 22 and 23, which are described as "*extensive editing, modification with a lot of unrelated frames attached to the beginning and the end*".

Table 2. Average results

Algorithm	Precision	Recall	F-measure
N gram	0.726	0.580	0.624
LSH-E	0.474	0.622	0.450
VK	0.732	0.868	0.757

The implementation was run on a Windows-2008 server of Intel Xeon E5620 machine with 96-GB memory. Signature database is implemented on MySQL server version 5.1.50-community. The signature data is stored in memory-type table. Both baseline and benchmark algorithms are run on the same environment. Our method is almost four times as faster as VK method for signature generation process and 16 times as faster as query process. For LSH-E, the signature generation time is pretty much the same with our method. However, LSH-E requires ranking with all items in database while our method uses binary search. From the experiments we concluded that our proposed algorithm is fast and good for query with simple scene or complex scene with minor editing or low quality videos. However, our algorithm is still intolerant to heavily-modified video, which requires high-level object detection to analyze.

5 Conclusion

In this paper, we proposed a video copy detection method based on signature approach. The luminance velocity was proposed to represent the video signature. The query method was adapted from text document similarity detection algorithm using N-Gram algorithm. The evaluation was done with an internet video dataset that contains 12,790 near-duplicate videos which are results of 24 queries onto popular internet video sites. The result indicates that our method is fast and effective to detect near-duplicate videos, especially for simple scene or complex scene with minor editing and low quality or low bit-rate videos.

Our work can be adapted as a tool for eliminating redundant results in video query system and video piracy detection system. We are also currently advancing our algorithm for signature generation to gain more recall in complex scenes. Furthermore, we will enhance our method to target the online and real-time applications in the future.

References

1. http://www.youtube.com
2. Mogul, J.C., Chan, Y.M., Kelly, T.: Design, implementation, and evaluation of duplicate transfer detection in HTTP. In: Proceedings of the 1st Conference on Symposium on Networked Systems Design and Implementation, vol. 1, p. 4. USENIX Association, San Francisco (2004)
3. Hampapur, A., Hyun, K.-H., Bolle, R.: Comparison of Sequence Matching Techniques for Video Copy Detection (2000)
4. Wu, V.K.Y., Polychronopoulos, C.: Efficient real-time similarity detection for video caching and streaming. In: 2012 19th IEEE International Conference on Image Processing (ICIP), pp. 2249–2252 (2012)
5. BaoFeng, L., HaiBin, C., Zheng, C.: An Efficient Method for Video Similarity Search with Video Signature. In: 2010 International Conference on Computational and Information Sciences (ICCIS), pp. 713–716 (2010)
6. Sánchez, J.M., Binefa, X., Vitriá, J., Radeva, P.: Local Color Analysis for Scene Break Detection Applied to TV Commercials Recognition. In: Huijsmans, D.P., Smeulders, A.W.M. (eds.) VISUAL 1999. LNCS, vol. 1614, pp. 237–244. Springer, Heidelberg (1999)
7. Lienhart, R., Kuhmunch, C., Effelsberg, W.: On the detection and recognition of television commercials. In: IEEE International Conference on Multimedia Computing and Systems 1997, pp. 509–516 (1997)
8. Xian-Sheng, H., Xian, C., Hong-Jiang, Z.: Robust video signature based on ordinal measure. In: 2004 International Conference on Image Processing, ICIP 2004, vol. 681, pp. 685–688 (2014)
9. Cao, Z., Zhu, M.: An efficient video similarity search strategy for video-on-demand systems. In: 2nd IEEE International Conference on Broadband Network & Multimedia Technology, IC-BNMT 2009, pp. 174–178 (2009)
10. Shang, L., Yang, L., Wang, F., Chan, K.-P., Hua, X.-S.: Real-time large scale near-duplicate web video retrieval. In: Proceedings of the International Conference on Multimedia, pp. 531–540. ACM, Firenze (2010)
11. Mohan, R.: Video sequence matching. In: Proceedings of the 1998 IEEE International Conference on Acoustics, Speech and Signal Processing, vol. 3696, pp. 3697–3700 (1998)
12. Shivakumar, N., Indyk, G.I.P.: Finding pirated video sequences on the internet (1999)
13. Xie, Q., Huang, Z., Shen, H.T., Zhou, X., Pang, C.: Quick identification of near-duplicate video sequences with cut signature. World Wide Web 15, 355–382 (2012)
14. Ardizzone, E., La Cascia, M., Molinelli, D.: Motion and color-based video indexing and retrieval. In: Proceedings of the 13th International Conference on Pattern Recognition, 1996, vol. 133, pp. 135–139 (1996)
15. Flickner, M., Sawhney, H., Niblack, W., Ashley, J., Qian, H., Dom, B., Gorkani, M., Hafner, J., Lee, D., Petkovic, D., Steele, D., Yanker, P.: Query by image and video content: the QBIC system. Computer 28, 23–32 (1995)
16. Dong, W., Wang, Z., Charikar, M., Li, K.: Efficiently matching sets of features with random histograms. In: Proceedings of the 16th ACM International Conference on Multimedia, pp. 179–188. ACM, Vancouver (2008)
17. Wan-Lei, Z., Xiao, W., Chong-Wah, N.: On the Annotation of Web Videos by Efficient Near-Duplicate Search. IEEE Transactions on Multimedia 12, 448–461 (2010)

18. Junfeng, J., Xiao-Ping, Z., Loui, A.C.: A new video similarity measure model based on video time density function and dynamic programming. In: 2011 IEEE International Conference on Acoustics, Speech and Signal Processing (ICASSP), pp. 1201–1204 (2011)
19. Wu, X., Hauptmann, A.G., Ngo, C.-W.: Practical elimination of near-duplicates from web video search. In: Proceedings of the 15th International Conference on Multimedia, pp. 218–227. ACM, Augsburg (2007)
20. Dunning, T.: Statistical Identification of Language. Computing Research Laboratory. New Mexico State University (1994)
21. Khoo, C.S.G., Loh, T.E.: Using statistical and contextual information to identify two-and three-character words in Chinese text. J. Am. Soc. Inf. Sci. Technol. 53, 365–377 (2002)
22. Tomović, A., Janičić, P., Kešelj, V.: n-Gram-based classification and unsupervised hierarchical clustering of genome sequences. Computer Methods and Programs in Biomedicine 81, 137–153 (2006)
23. Pavlović-Lažetić, G.M., Mitić, N.S., Beljanski, M.V.: n-Gram characterization of genomic islands in bacterial genomes. Computer Methods and Programs in Biomedicine 93, 241–256 (2009)
24. Radomski, J.P., Slonimski, P.P.: Primary sequences of proteins from complete genomes display a singular periodicity: Alignment-free N-gram analysis. Comptes Rendus Biologies 330, 33–48 (2007)
25. Xiao, W., Chong-Wah, N., Hauptmann, A.G., Hung-Khoon, T.: Real-Time Near-Duplicate Elimination for Web Video Search With Content and Context. IEEE Transactions on Multimedia 11, 196–207 (2009)

Author Index